ELEMENTS OF

POLYMER SCIENCE & TECHNOLOGY

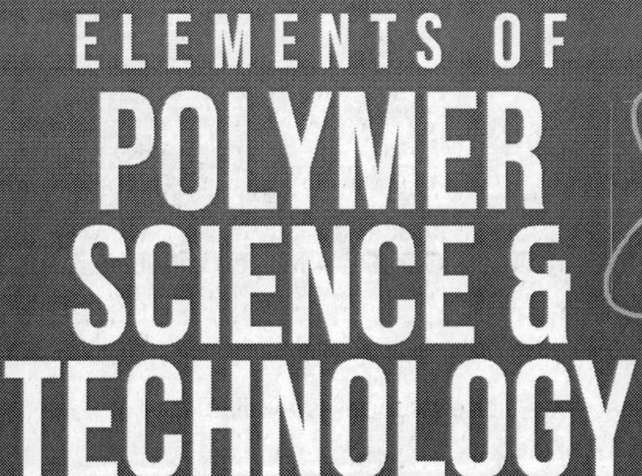

ELEMENTS OF

POLYMER SCIENCE & TECHNOLOGY

O.P. Gupta

KHANNA BOOK PUBLISHING CO. (P) LTD.

Publisher of Engineering and Computer Books

4C/4344, Ansari Road, Darya Ganj, New Delhi-110002
Phone : 011-23244447-48 **Mobile:** +91-9910909320
E-mail : contact@khannabooks.com
Website : www.khannabooks.com

Dear Readers,

To prevent the piracy, this book is secured with HIGH SECURITY HOLOGRAM on the front title cover. In case you don't find the hologram on the front cover title, please write us to at contact@khannabooks.com or whatsapp us at +91-99109 09320 and avail special gift voucher for yourself.

Specimen of Hologram on front Cover title:

Moreover, there is a SPECIAL DISCOUNT COUPON for you with EVERY HOLOGRAM.

How to avail this SPECIAL DISCOUNT:

Step 1: Scratch the hologram
Step 2: Under the scratch area, your "coupon code" is available
Step 3: Logon to www.khannabooks.com
Step 4: Use your "coupon code" in the shopping cart and get your copy at a special discount
Step 5: Enjoy your reading!

Thanks,
Khanna Publishing House

Price: ₹ 475.00

ISBN: 978-93-5538-456-0

Elements of
Polymer Science & Technology
by O.P. Gupta

First Edition: 2025

To view complete list of books, please scan the QR Code:

Published by:
Khanna Book Publishing Co. (P) Ltd.
CIN: U22110DL1998PTC095547

Visit us at: www.khannabooks.com
Write us at: contact@khannabooks.com

Printed in Bharat.

PREFACE

It gives me immense pleasure to present this book titled Elements of Polymer Science & Technology. This book is meant for the students, teachers, practicing engineers. To prepare a book on this subject in its true sense would be very exhaustive requiring many volumes to do real justice to the subject. However, attempt has been made to assimilate the essentials of the subject in lucid manner in a single volume.

This book covers the manufacture, properties & applications of important polymers-plastics, rubbers & synthetic fibres. Important informations about various raw materials used for polymer industries , various polymers processing technologies & status of Indian polymer industry are also included .It also contains short questions & answers and multiple choice questions & answers drawn from the examination papers of various engineering colleges for the benefits of the students.

This book is targeted to benefit the following:

- Diploma in engineering students
- Degree in engineering students
- AMIE, AMIIM, AMIICHE students.
- M.Sc. & B. Sc. students
- Faculty members and teaching staff
- Practising engineers/professionals

The scope and coverage of the subject matter presented can be gauged from glancing the contents portion of the book to judge its suitability and relevance for a particular target group of reader.

Extensive use of internet resources has been made in gathering the relevant informations and the subject matter has been presented in a simple, lucid and easy to understand manner for easy grasping of the subject.

Latest and updated informations/data/statistics pertaining to the subject matter has been included in this edition for the benefit of the readers.

Constructive criticisms and suggestion for further improvement of the book are solicited.

Ranchi, Jharkhand
15th August, 2023

Om Prakash Gupta

Contents

Appendices 207-326

CONTENTS

Syllabus for the B. Tech Course, Polymer Engineering and Technology

Historical developments in polymeric materials, Basic concepts & definitions: monomer & functionality, oligomer, polymer, repeating unites, degree of polymerization, molecular weight & molecular weight distribution.

Natural Polymers: Chemical & Physical structure, properties, source, important chemical modifications, applications of polymers such as cellulose, lignin, starch, rosin, shellac, latexes, vegetable oils and gums, proteins etc.

Raw material for synthetic polymers: Manufacturing of various fractions of crude petroleum important for polymer industry for (*a*) Raw Materials such as ethylene, propylene, butadiene, vinyl chloride, vinylidene dichloride, styrene, acrylic monomers like acrylic acid, acrylonitrile, methacrylic acid, methacrylates, acrylamide etc., (*b*) solvents such as alcohols, toluene, xylene, acetone, ketones, terpenes, chloromethanes etc. Evaluation of raw materials and reactants for synthesis & manufacturing of polymers. (*c*) Polyacids such as phthalic acid, terephthalic acid, isomers and anhydrides etc. (*d*) phenols, polyols and their modifications, (*e*) Isocyanates, (*f*) Amino Compounds, (*g*) Other petroleum based material

Textbooks/Sourcebooks:

1. Raw Materials for Industrial Polymers by H Ulrich, Hanser Publication 1989.
2. Principles of Polymer Science, by Bahadur and Sastry, Narosa Publishing House 2002.
3. Polymer Science by Gowarikar, Johan Wiley and Sons 1986.
4. Encyclopedia of Polymer Science and Technology, Johan Wiley and Sons, Inc 1965.
5. Encyclopedia of Polymer Science and Engineering, Johan Wiley and Sons, Inc 1988.
6. Petrochemicals The Rise of an Industry by Peter H. Spitz, Johan Wiley and sons 1988.
7. Polymer Chemistry by Malcolm P. Stevens, Oxford University Press, Inc, 1990.

Special 2 Polymer Science & Technology II

Classification of polymers thermoplastic/thermoset, addition/condensation, natural/synthetic, crystalline/amorphous, step growth/chain growth, commodity…specialty, homochain/heterochain, confirmation: homo & copolymers (detailed graft ,block alt, ladder etc. & nomenclature), configuration cis/trans; tacticity, branched/crosslinked, Classification of polymers based on end use etc. Molecular weight and its distribution determination (Mn to Mz & MWD), carothers equation, states of polymers, transition temperatures such as T_g, T_c, T_m, solubility parameter, solution properties, temperature, good/bad solvent, (12) Addition, condensation polymerization mechanism (7) Surface tension/energy & contact angle measurements of different polymeric systems & their wettability with other substances

Spl 4: Polymer Science & Technology III (Polymer)

(**Techniques of polymerization:** bulk, solution, suspension, emulsion, plasma etc. Different initiating systems such as free radicle polymerization, redox, cationic & anionic polymerization (different terms such as living polymers, inifers, telechelics). Their kinitics & control over structure of polymer. Condensation polymerization, different catalysts used case studies of condensation polymerization, Carothers equation, Comparison of these systems with advantages & disadvantages. (10) Copolymerization, reactivity ratios & kinetics of copolymerization (copolymer composition equation).

Rheological concepts of polymer solutions and melts, degradation plasticization Mixing operations: Typical agitation system, dissolution (10) Different advanced catalyst systems: Ziegler Natta catalyst & metallocene catalysts & their role in polyolefins, ATRP etc.

Technology of Elastomers I (Polymer) Definition of elastomers and requirements of polymer to be elastomer: Effect of molecular weight, tie points and glass transition temperature (T_g) characteristics. Different types of monomers used in synthesis of elastomers, classifications of elastomers, different processes used during life cycle of rubber like manufacture, storage, compounding, forming and vulcanization of rubbers, different ingredients used in it and functions of various compounding ingredient, various equipments used for compounding and their comparison Definitions of different terms like scorch, cure/over cure & study of curing. Different types of vulcanization systems used for compounding and fillers used in elastomers, measurement of mooney viscosity and state of cure for rubber compound. Synthesis of various rubbers natural rubber/synthetic

Polymer Science: Classification of polymers, definition of polymerization, chain polymerization (free radical, ionic and co-ordination polymerizations), step (condensation) polymerization, copolymerization methods of polymerization (bulk, solution, suspension, emulsion).

Polymerization Kinetics: Definition of reaction rate, order, molecularity, different theories of reaction rate, activation energy, kinetic expressions for simple first order & second order chemical reactions, kinetics of linear step reaction polymerization, kinetics of addition polymerization initiated by free radical initiator steady state assumption.

Processing Technology: Extrusion, injection moulding, blow moulding compression moulding, rotational moulding, thermoforming, calendaring .

Polymer Manufacturing: Industrial production methods of HDPE, LDPE, PP, PS, PVC, PMMA, Nylon 6 and Nylon 66.

GATE Syllabus for Polymer Science and Engineering is as follows:

Polymer Science and Engineering

Chemistry of high polymers: Monomers, functionality, degree of polymerizations, classification of polymers, glass transition, melting transition, criteria for rubberiness, polymerization methods: addition and condensation; their kinetics, metallocene polymers and other newer techniques of polymerization, copolymerization, monomer reactivity ratios and its significance, kinetics, different copolymers, random, alternating, azeotropic copolymerization, block and graft copolymers, techniques for copolymerization-bulk, solution, suspension, emulsion.

Polymer characterization: Solubility and swelling, concept of average molecular weight, determination of number average, weight average, viscosity average and Z-average molecular weights, polymer crystallinity, analysis of polymers using IR, XRD, thermal (DSC, DMTA, TGA), microscopic (optical and electronic) techniques.

Synthesis and properties: Commodity and general-purpose thermoplastics: PE, PP, PS, PVC, Polyesters, Acrylic, PU polymers.

Engineering plastics: Nylon, PC, PBT, PSU, PPO, ABS, Fluoropolymers

Thermosetting polymers: PF, MF, UF, Epoxy, Unsaturated polyester, Alkyds. Natural and synthetic rubbers: Recovery of NR hydrocarbon from latex, SBR, Nitrile, CR, CSM, EPDM, IIR, BR, Silicone, TPE.

Polymer blends and composites: Difference between blends and composites, their significance, choice of polymers for blending, blend miscibility-miscible and immiscible blends, thermodynamics, phase morphology, polymer alloys, polymer eutectics, plastic-plastic, rubber-plastic and rubber-rubber blends, FRP, particulate, long and short fibre reinforced composites.

Polymer technology: Polymer compounding-need and significance, different compounding ingredients for rubber and plastics, crosslinking and vulcanization, vulcanization kinetics.

Polymer rheology: Flow of Newtonian and non-Newtonian fluids, different flow equations, dependence of shear modulus on temperature, molecular/segmental deformations at different zones and transitions. Measurements of rheological parameters by capillary rotating, parallel plate, cone-plate rheometer. viscoelasticity-creep and stress relaxations, mechanical models, control of rheological characteristics through compounding, rubber curing in parallel plate viscometer, ODR and MDR. Polymer processing: Compression molding, transfer molding, injection molding, blow molding, reaction injection molding, extrusion, pultrusion, calendaring, rotational molding, thermoforming, rubber processing in two-roll mill, internal mixer.

Polymer testing: Mechanical-static and dynamic tensile, flexural, compressive, abrasion, endurance, fatigue, hardness, tear, resilience, impact, toughness. Conductivity-thermal and electrical, dielectric constant, dissipation factor, power factor, electric resistance, surface resistivity, volume resistivity, swelling, ageing resistance, environmental stress cracking resistance.

Fundamentals of Polymer Science & Technology

DEFINITION OF POLYMER

The word polymer has a Greek origin. which means many units (parts). Polymer is defined as a chemical substance of a high molecular weight (103 to 107) formed by the combination of a large number of simple molecules, called monomers. e.g.,

$$\underset{\text{Ethylene}}{n(CH_2 = CH_2)} \longrightarrow \underset{\text{Polyethylene}}{[-CH_2 - CH_2 -]_n}$$

A Polymer is a giant molecule formed by inter molecular linkage between same or different types of smaller molecules called **monomers**. Polymer is a high molecular mass molecule formed by linking up of two or more small molecules called monomers. Monomers are the small molecule which are capable of linking amongst themselves to form big molecules called polymers. In some polymers more than one type of monomers combine with each other to give the polymer. For example, a polymer may be obtained from two monomers (A) and (B) viz, $- A - B - A - B -$ or $-(-A - B -)_n-$

The process by which the monomers get combined and transformed into polymers. is known as polymerisation.

$$n[\text{Monomer}] \longrightarrow \text{Polymer}$$

Difference between Polymers and Macromolecules

Polymers are also called macromolecules due to their large size but converse is not always true. A macromolecule may or may not contain monomer units, e.g., chlorophyll ($C_{55}H_{72}O_5N_4Mg$) is a macromolecule but not a polymer since there are no monomer units present; so we can conclude that all polymers are macromolecules while all macromolecules may not be polymers in nature.

Types of Polymers

Depending upon the nature of the repeating structural units (monomers), polymers are divided into two broad categories viz., homopolymers and copolymers.

(*a*) **Homopolymer:** Polymer formed from only one kind of monomers is called homopolymer. Polyethene $-(-CH_2 - CH_2 -)_n-$ is an example of homopolymer.

(*b*) **Copolymer:** A polymer formed from more than one kind of monomer units is called copolymer or mixed polymer. For example, Buna-S rubber which is formed from 1, 3-butadiene

($CH_2 = CH - CH = CH_2$) and styrene ($C_6H_5CH == CH_2$) is an example of copolymer. If the polymer chain is formed from only one type of monomer units are called homopolymers. Alternatively, two of more different monomer units can combine to form a polymer chain is called copolymers. There are four types of copolymers known. First, alternate arrangement of monomer units results in alternating copolymer. Secondly, block arrangement of monomer units result in block copolymers. Thirdly, a random arrangement of monomer units results in random copolymers and finally, a graft copolymers results when one type of monomer units are arranged at the branches of the chain and the another type of monomer units still reside at the main chain.

```
An alternating copolymer   ABABABABABABABABABABABAB
A block copolymer          AAAABBBBAAAABBBBAAAABBBB
A random copolymer         AABABABBAABAAABBAABABBBA
A graft copolymer          AAAAAAAAAAAAAAAAAAAAAAAA
                              B         B         B
                              B         B         B
                              B         B         B
                              B         B         B
                              B         B         B
                              B         B         B
                              B         B         B
                              B         B         B
```

Types of Polymerization

The process by which the monomers get linked up is called polymerization. Depending upon the mode of reaction, polymerisation is classified as :

(*a*) Addition polymerization (*b*) Condensation polymerization.

Addition Polymerization

This process involves the addition of monomer units to themselves to form a growing chain by a chain reaction mechanism. It is for this reason that the process is also known as **chain growth polymerization.** Addition polymerization is achieved by adding a catalyst (known as initiator), which provides some reactive species like free radicals. For example, benzoyl peroxide provides free radical 'benzyl' to **initiate** the chain polymerization reaction. These free radicals then attack the unsaturated monomer and form a new free radical which goes on successively adding monomers and thus grows the chain, this is called **Chain propagation.** The final **termination** of the two growing chains leads to a polymer. This is called **chain termination.**

Monomer molecules are added individually to a growing active species in a chain reaction to form large polymeric molecules. The reactive species may be radical, anionic or cationic. Addition polymerisation is characterised by a gradual growth in an average molecular size (although each polymer molecule is produced very quickly).

There are no by-products but the starting materials (monomer and initiator) must be very pure and oxygen excluded from the reaction if high molecular weight polymer is required. Polymerisation ceases if monomer is removed from the system.

Examples of addition polymers are:

- Polystyrene
- Polyethylene
- Polypropylene
- Poly(vinyl chloride) (PVC)

Features of Chain Growth (addition polymerization) Reaction

- Reaction of unsaturated or cyclic monomers which link to form a continuously increasing chain. No. release of any co-products.
- Initiation is slow step. Propagation and termination steps are rapid.
- Monomer concentration decreases continuously during reaction.
- Increased reaction time has no effect on average molecular weight (M.W.) or chain length.
- Intermediates are transient and non-isolable.
- Examples are: polyethylene, polypropylene, poly (vinyl chloride), polystyrene.

Addition Polymerization Methods:

These methods are carried out with controlled thermal and catalyst conditions. They may be further classified as

(*i*) **Bulk Polymerization:** This mode of polymerization may be employed to obtain the purest form of polymer, and the greatest yield of polymer per unit volume may be obtained using this method. This method involves only the monomer molecule, an initiator and a chain transfer agent (if necessary). In a large-scale batch form, the process is to be run slowly or in continuous mode with large heat transfer area. Casting of objects of various shapes may be accomplished using the Batch Bulk Polymerization. Using continuous bulk polymerization processes, polystyrene and other thermoplastic compounds may be moulded.

(*ii*) **Solution Polymerization:** An inert solvent is added to the reacting components in the reaction vessel in this process. The solvent enhances the heat capacity, thereby reducing the viscosity and facilitating heat transfer. Some of the solvent may be refluxed to remove heat from the reaction vessel. But, much of the reactor space is taken up by the solvent. As such, compared to Bulk Polymerization, Solution Polymerization reduces both the reaction rate and the molecular weight of the compounds. Additional batch and continuous processes are used to separate the final polymer product and to recover and store the solvent used.

(*iii*) **Suspension Polymerization:** In order to control the enormous amount of heat release in Bulk Polymerization, Suspension Polymerization method was developed. The reaction mass is dispersed as minute droplets of size 0.01-1 mm in diameter in a continuous aqueous phase. Each of these droplets act as tiny bulk reactors. Heat transfer occurs from the droplets to the water having large heat capacity and low viscosity. Cooling jackets are used to facilitate heat removal. Agitators are used along with suspending agents in the aqueous phase in order to maintain a specific droplet size and dispersion.

The Suspension Polymerization process cannot be run in a continuous mode, since, such a system has stagnant corners where polymer accumulation may occur. On a commercial scale, Suspension Polymerization is carried out in jacketed, stainless steel or glass - lined stirred tanks, which may have a capacity of 75.5 m^3.

Suspension Polymerization produces small, uniform polymer spheres. These are used directly, or may be extruded and chopped to form larger, moulded pellets.

(*iv*) **Emulsion Polymerization:** This is the most widely used method of polymerization. This process overcomes the difficulty of heat control and low degree of polymerization. The organic reaction mass is emulsified with soap in a continuous aqueous phase. The dispersed particles are smaller in size than in Suspension Polymerization (< 0.1 μm). In addition, due to stabilizing action of soap, the emulsion obtained is stable and agitation may not be necessary. In some methods, a water - soluble initiator may be used. The main product of Emulsion Polymerization is latex, which forms the basis of the popular latex paints. By coagulating the latex with ionic salts and acids, solid rubber may also be obtained.

(*v*) **Homogeneous Polymerization:** In case of homogeneous bulk polymerization, the feed is a gas, liquid or solid monomer. No initiators or additives are used. For homogeneous Solution polymerization, the monomer is completely dissolved in a solvent.

(*vi*) **Heterogeneous Polymerization:** In heterogeneous Emulsion polymerization, the monomer molecules are emulsified in aqueous media in the form of micelles. For heterogeneous Suspension polymerization, the monomer is suspended in an aqueous or other type of media as large droplets.

Characteristics, Advantages and Disadvantages of Various Polymerisation Processes are Tabulated Below:

Table 1. Characteristics, advantages and disadvantages of various polymerization types

Polymerization Process	Characteristics	Advantages	Disadvantages
Bulk	Reaction mixture consists essential of monomer; and initiator in the case of chain reaction polymerization. Monomer acts as solvent for polymers	Products relatively pure due to minimum contamination. Enhanced yield per reactor volume	Exothermic nature of polymerization reactions (particularly chain reaction polymerization) makes temperature control of system difficult. Product has broad molecular weight distribution. Removals of tracers of unreacted monomer difficult.
Solution	Solvent miscible with monomer, dissolves polymer	Heat transfer efficiency greatly enhanced resulting in better process control Resulting polymer solution may be directly usable	Necessary to select an inert solvent to avoid possible transfer to solvent. Lower yield per rector volume. Reduction of reaction rate and average chain length. Not particularly suitable pure polymer due to difficulty of complete solvent removal.
Suspension	Monomer and polymer insoluble in water, initiator soluble in monomer	Heat removal and temperature control relatively easier. Polymer obtained in a form that is convenient and easily handled. Resulting polymer suspension or granules may be directly usable.	Need to main stability of droplets requires continuous and a minimum level of agitation. Possibility of polymer contamination by absorption of stabilizer on particle surface. Continuous operation of system difficult.

Emulsion	Monomer and polymer insoluble in water, initiator soluble in water. Emulsifier needs for stabilization of system component particularly at initial stages of polymerization.	Physical state of the system enhances heat transfer efficiency. Possible to obtain high rates of polymerization and high average chain lengths. Narrow molecular weight distribution. Latex (emulsion) often directly usable.	Difficult to get pure polymer due to contamination from other components of polymerization system. Difficult and expensive if solid polymer product is required. Presence of water lowers yield per reactor volume. Separation of product difficult and expensive. Catalyst systems are special and need careful preparation. Molecular weight distribution depends on type of catalyst
Interfacial	Polymerization occurs at interface of two immiscible solvents, usually water and an organic solvent	Polymerization is rapid and occurs at low temperatures. High conversions are not necessarily required to obtain high molecular weight. Unnecessary to start with stoichiometric quantifiers of reactants	Limited to highly reactive systems. Need appropriate choice of a solvent to dissolve reactants.

Condensation Polymerization

In this, the monomers combine with the elimination of a small molecule like H_2O, ROH or NH_3, etc. The reaction is called **(step growth)** condensation polymerization and the product formed is called condensation polymer. The process involves the elimination of by product molecules, therefore, the molecular mass of the polymer is not the integral multiple of the monomer units. For example, polyester or Terylene is a condensation polymer of ethylene glycol and terphthallic acid.

Monomer units condense to form large fragments which in turn condense to produce even larger fragments. Average molecular size increases exponentially. For commercial polymers with high molecular weight, it is essential to have pure starting materials, an exact balance of monomers and over 99% reaction.

In **elimination condensation** a small molecule (e.g., water) is produced at each reaction stage, causing problems in scale up processes. In Rearrangement Condensation, there is no by-product, which suits this polymerisation mechanism to in-situ moulding.

Examples of condensation polymers are:

Elimination condensation:

- phenol-formaldehyde
- urea-formaldehyde
- melamine formaldehyde
- polyesters
- polyamides (nylon)

Rearrangement condensation:

- epoxides
- polyurethanes

Features of Step Growth (Condensation Polymerization) Reaction

- Reaction of molecules containing specific functional groups that undergo condensation to a larger molecule, with release of H_2O, ROH, HCl] etc.
- Concentrations of monomer(s) decrease rapidly; concentration of polymer increases gradually.
- Increased reaction time results in increased average. M.W. and chain length.
- Isolable reaction intermediates.
- Examples are: polyesters, polyamides, polyurethanes, polycarbonates.

These two processes of making polymers have several characteristics which distinguish them from each other. These are tabulated below.

Table 2. Some differences between addition polymerization and condensation polymerization

S. No.	Addition Polymerization	Condensation Polymerization (Polycondensation)
1.	Involves unsaturated monomers like ethylene, vinyl chloride, styrene etc.	Involves substances with at least 2 functional groups like ethylene glycol (2-OH groups), adipic acid (2-COOH groups).
2.	Fast addition of monomers	Step-wise slow addition.
3.	At any instant, only monomer and polymers are present.	No monomer. Mixtures of dimers, trimers and tetramers etc. are present.
4.	Initiator is necessary to catalyse the polymerization.	Catalyst is not necessary.
5.	No small molecules are eliminated.	Small molecules like H_2O, HCl, CO_2, CH_3OH are often eliminated.
6.	Polymers made are, for example, polyethene, polypropylene, polybutadiene, polyvinylchloride.	Polymer made are, for example, terylene, nylon, Formaldehyde-resins, silicones.

Classification of Polymers

The polymers can be classified in a number of ways. Some of these are discussed below:

(*a*) **Classification of polymers on the basis of origin.** On the basis of origin, polymers are classified as : Natural polymers **and** Synthetic polymers

Natural polymers: The polymers obtained from nature (plants and animals) are called natural polymers. Starch, cellulose, natural rubber, proteins, etc. are some examples.

Synthetic polymers: The polymers which are prepared in the laboratories are called synthetic polymers. These are also called **man-made polymers.** Polyethene, PVC, nylon, Teflon, Bakelite, terylene, synthetic rubber, etc. are some examples.

Semisynthetic polymers: Polymers obtained by making some modification in natural polymers by artificial means, are known as semi synthetic polymers, e.g., cellulose acetate (rayon), vulcanised rubber etc.

(*b*) **Classification of polymers on the basis of structure:** On the basis of structure of polymers, these can be classified as: Linear polymers, Branched chain polymers and Cross-linked polymers.

Linear polymers: These are polymers in which **monomeric units** are linked together to form linear chains. These linear polymers are well packed and therefore, have high densities, high tensile (pulling) strength and high melting points. Polyethylene, nylons and polyesters are examples of linear polymers.

Branched chain polymers: These are polymers in which the monomers are joined to form long chain with side chains or branches of different lengths. These branched chains polymers are irregularly packed and therefore, they have lower tensile strength and melting points than linear polymers. For example, low density polyethene, glycogen, starch, etc.

Cross-linked polymers: These are polymers in which long polymer chains are cross-linked together to from a three-dimensional network. These polymers are hard, rigid and brittle because of the network structure. Bakelite, melamine and formaldehyde resin are some examples of this type.

Linear Polymers

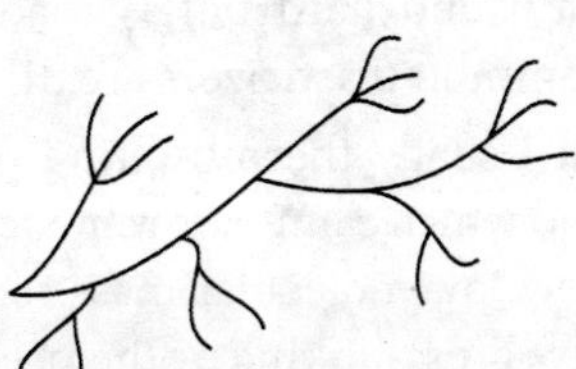

Branched chain polymers

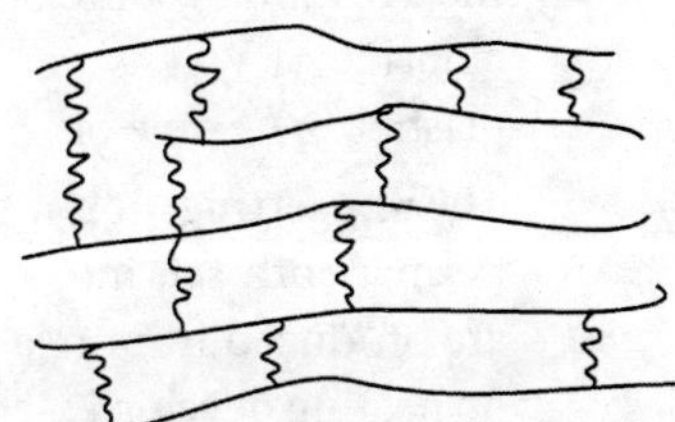

Cross linked polymers

(*c*) **Classification of polymers on the basis of method of polymerisation:** On the basis of method of polymerisation the polymers are classified as :

Addition polymer: A polymer formed by direct addition of repeated monomers without the elimination of any small molecule is called **addition polymer.** In this type, the monomers are unsaturated compounds and are generally derivatives of ethene. The addition polymers have the same empirical formula as their monomers. Examples are polyethene, polypropylene, polyvinyl chloride, etc.

Condensation polymer: A polymer formed by the condensation of two or more than two monomers with the elimination of simple molecules like water, ammonia, hydrogen chloride, alcohol, etc. is called **condensation polymer.** In this type, each monomer generally contains two functional groups. For example, nylon-66 is obtained by the condensation of two monomers; hexa methylene diamine and adipic acid with the loss of water molecules. Examples of condensation polymers are Nylon 66, terylene, bakelite, alkyl resins, etc.

(*d*) **Classification of polymers on the basis of molecular forces:** Depending upon the intermolecular forces between monomer molecules, the polymers have been classified into four types.

1. Elastomers
2. Fibers
3. Thermoplastics
4. Thermosetting

Elastomers: In case of elastomers the polymer chains are held together by weak **Van der Waals forces.** Due to weak forces, the polymers can be easily stretched on applying small stress and they regain their original shape when the stress is removed. This is due to the presence of few-'cross links' between the chains, which help the polymer to retract to its original position after the force is removed, as in vulcanized rubber. The most important example of elastomer is **natural rubber.**

Fibres: These are the polymers which have strong intermolecular forces between the chains. These forces are either hydrogen bonds or dipole-dipole interactions. Because of the strong forces, the chains are closely packed, giving them high tensile strength and less elasticity. These polymers can be drawn into long, thin and thread like fibres and therefore can be woven into fabrics. The common examples are nylon- 66, Dacron, silk, etc.

Thermoplastics: These are linear polymers with very few cross linkages or no cross linkages at all. The polymeric chains are held by weak VANDER WAAL forces and slide over one another. Due to lack of cross linkages these polymers soften on heating and harden or become rigid on cooling. Thus, they can be moulded to any shape. Polythene, PVC, polystyrene are addition type thermoplastics and Terylene, nylon are condensation type thermoplastics.

Plasticizers: Certain plastics do not soften much on heating. These can be easily softened by the addition of some organic compounds which are called **plasticizers.** For example, polyvinyl chloride (PVC) is very stiff and hard but is made soft by adding di-*n*-butyl phthalate (a plasticizer). Some other common plasticizers are dialkyl phthalates and cresyl phthalate.

Thermosetting polymers: Usually thermosetting polymer can be heated only once when it permanently sets into a solid which cannot be remelted and remoulded. Thermosetting polymers are produced from relatively low molecular mass semi fluid polymers (called polymers) which on heating develop extensive cross-linking by themselves or by adding some cross-linking agents and become infusible and insoluble hard mass. The cross links hold the molecules in place so that heating does not allow them to move freely. Therefore, a thermosetting plastic is cross linked and is permanently rigid. The common example are bakelite, melamine, formaldehyde resin, etc.

Some important differences in the properties of thermosetting and thermoplastic polymers are summarized in Table below.

Table 3. Difference between thermoplastics and thermosetting plastics

	Thermo plastics	Thermosetting plastics
1.	Thermoplastic pellets soften when heated and become more fluid as additional heat is applied. The curing process is completely reversible as no chemical bonding takes place. This characteristic allows thermoplastics to be remolded and recycled without negatively affecting the material's physical properties.	Thermoset plastics contain polymers that cross link together during the curing process to form an irreversible chemical bond. The cross linking process eliminates the risk of the product remelting when heat is applied, making thermosets ideal for high heat applications such as electronics and appliances.
2.	There are multiple thermoplastic resins that offer various performance benefits, but most materials commonly offer high strength, shrink-resistance and easy bendability. Depending on the resin, thermoplastics can serve low stress applications such as plastic bags or high stress mechanical parts	Thermoset plastics significantly improve the material's mechanical properties, providing enhanced chemical resistance, heat resistance and structural integrity. Thermoset plastics are often used for sealed products due to their resistance to deformation.

3.	**Advantages** • Highly recyclable • Aesthetically superior finish • High impact resistance • Remolding/Reshaping capabilities • Chemical resistance • Hard crystalline or rubbery surface options • Eco friendly manufacturing	**Advantages** • More resistance to high temperature than thermoplastics • Highly flexible design • Thick to thin wall capabilities • Excellent aesthetic appearance • High levels of dimensional stability • Cost effective
4.	**Disadvantages** • Generally, more expensive than thermoset • Can melt, if heated	**Disadvantages** • Cannot be recycled • More difficult to get a finished surface • Cannot be remoulded or reshaped

(*e*) **Polymer Classification According to Applications:** On the basis of applications, polymers can be further classified as:

(*i*) **Adhesives:** Some polymers can be used for bonding. They are usually of the resin type and require some water resistance. Some common adhesives are:

Cellulose adhesives; which consist of cellulose derivatives dissolved in a solvent. e.g., Duco cement.

Vinyls; these are rubber base type water-emulsified latex adhesives. Apart from these, some cheap natural products such as starch, dextrins, proteins and natural rubber may also be used for adhesive formulations.

(*ii*) **Coatings and films:** A large bulk of the polymers produced are used for manufacturing coatings and films. Free films of polyethylene and cellulosic types, protective and decorative coatings are the products of the polymer industry. Coatings can be manufactured by solvent evaporation followed by polymerization.

Emulsion and casting or extrusion of free films by mechanical methods can also produce coatings and films.

(*iii*) **Fibers:** These are formed by extrusion or spinning of linear monomer molecules into thin sections of diameter in the range of 10 - 50 microns.

Fibers have excellent tensile strength, creep and resilience. Nylon, Dacron, rayon are examples of synthetic fibre.

Fibers are extensively used in the textile industry. Cotton, wool etc. are the examples of some natural fibers.

Environmental Problems and Biodegradable Polymers

With the increasing use of polymers, the problem of disposal of waste of these products is posing alarming problems. Since most of the synthetic polymers are in the form of plastics, it is frequently used in abundance in the form of packing material and throw away bags. Since ordinary polymers do not degrade naturally by light, oxygen, water or micro- organisms, there is a serious problem of their disposal. The environmental problems caused by careless use of non-biodegradable polymers can be reduced by proper disposal of these materials and reusing them and remoulding them for other uses. Another way is to collect them and depolymerise them back to monomers. Though it has a limited application.

Another option is to produce biodegradable polymers which can be broken into small segments by enzyme-catalysed reactions. The required enzymes are produced by micro- organisms. It is a known fact that the carbon-carbon bonds of chain growth addition polymers are inert to enzyme catalysed reactions, and hence they are non-biodegradable. To make such polymers biodegradable we have to insert certain bonds in the chains so that these can be easily broken by the enzymes. When such polymers are buried as waste, microorganisms present in the soil can degrade the polymer, so that they do not cause any serious effects on the environment. One of the best method of making a polymer biodegradable is by inserting hydrolysable ester group into the polymer.

Some Biodegradable Polymers

A large number of bio-degradable polymers are now available and more are being added to the list. However, these are expensive, therefore, these find use in special situations where cost factor can be ignored. In future, as their cost reduces these will find greater use in daily life and will replace non-bio-degradable polymers.

Some important biodegradable polymers are PHBV, PGA, PLA and PCL.

(PHBV (Poly-Hydroxybutyrate): PHBV is a copolymer of 3-hydroxy butanoic acid, and (3-hydroxypentanoic acid), in which, the monomer units are connected by ester linkages. The properties of PHBV vary according to the ratio of both the acids. 3-Hydroxybutanoic acid provides stiffness and 3-hydroxypentanoic acid imparts flexibility to the copolymer.

PHBV is used in orthopedic devices and in controlled drug release. The drug put in PHBV capsule is released after this polymer is degraded by enzymatic action. It can also be degraded by bacterial action.

PGA(Polyglycolic acid): It is obtained by the chain polymerization of dimer of glycolic acid, HO — CH_2COOH.

$$\underset{\text{Glycolic acid}}{nHC-CHCO_2OH} \xrightarrow{\text{Heat}} \underset{\text{Polyglycolic acid (PGA)}}{\left(O-CH_2-\overset{\overset{O}{\|}}{C}-O \right)}$$

PLA (Polyactic acid): It is obtained by polymerization of the dimer of lactic acid (HO — $CH(CH_3)$ COOH) or by micro biological synthesis of lactic acid followed by the polycondensation and removal of water by evaporation.

$$\underset{\text{Lactic acid}}{HO-\underset{\underset{CH_3}{|}}{CH}-\overset{\overset{O}{\|}}{C}-OH} \xrightarrow{\text{Condensation}} \underset{\text{Polylactic acid (PLA)}}{\left[O-\overset{\overset{H}{|}}{\underset{\underset{CH_3}{|}}{C}}-\overset{\overset{O}{\|}}{C}-O \right]}$$

PCL [Poly (E–caprolactone)]: It is obtained by chain polymerization of the lactone of 6-hydroxy hexanoic acid.

$$\left(O-(CH_2)_5-\overset{\overset{O}{\|}}{C}-O \right)_n \xrightarrow{\text{Polymerisation}} PCL$$

- Most of the biodegradable polymers find use in stitching wounds and cuts.
- In medical goods such as surgical sutures.

- In agriculture materials such as films, seed coatings.
- In food wrappers, personal hygiene products, etc.

Some monomers and Corresponding polymers are tabulated below:

S. No.	Monomer	Polymer
1.	Ethylene $CH_2=CH_2$	Polyethylene $-(CH_2-CH_2)_n-$
2.	Propylene $CH_2=CH(CH_3)$	Polypropylene $-(CH_2-CH(CH_3))_n-$
3.	Styrene $CH_2=CH(C_6H_5)$	Polystyrene $-(CH_2-CH(C_6H_5))_n-$
4.	Vinyl chloride $CH_2=CH(Cl)$	Poly (vinyl chloride) $-(CH_2-CH(Cl))_n-$
5.	Vinylidene chloride $CH_2=C(Cl)_2$	Poly (vinylidene chloride) $-(CH_2-C(Cl)_2)_n-$
6.	Vinyl acetate $CH_2=CH(OCOCH_3)$	Poly (vinyl acetate) $-(CH_2-CH(OCOCH_3))_n-$
7.	Tetrafluoroethylene $CF_2=CF_2$	1 Poly (tetrafluoroethylene) $-(CF_2-CF_2)_n-$
8.	Methyl acrylate $CH_2=CH(COOCH_3)$	Poly (methyl acrylate) $-(CH_2-CH(COOCH_3))_n-$
9.	Methyl methacrylate $CH_2=C(CH_3)(COOCH_3)$	Poly (methyl methacrylate) $-(CH_2-C(CH_3)(COOCH_3))_n-$
10.	Acrylonitrile $CH_2=CH(CN)$	Polyacrylonitrile $-(CH_2-CH(CN))_n-$
11.	Formaldehyde $CH_2=O$	Polyformaldehyde or Poly (oxymethylene) $-(CH_2-O)_n-$

12.	Caprolactam $O=C-NH$ with $(CH_2)_5$ ring	Polycaprolactam (Nylon 6 polyamide) $\left[C(=O)-(CH_2)_5-N(H) \right]_n$
13.	Butadiene $CH_2=CH-CH=CH_2$	Polybutadiene $\left[CH_2-CH=CH-CH_2 \right]_n$
14.	Chloroprene $CH_2=C(Cl)-CH_2=CH_2$	Polychloroprene $\left[CH_2-C(Cl)=CH-CH_2 \right]_n$

Polymeric materials with specific properties and having different application areas are:

- **Plastics:** polymer characterized by rigidity and stiffness having low recoverable deformation
- **Rubber:** polymer characterized by flexibility and high amount of recoverable deformation.
- **Fiber:** polymer having high melting point with high amount of crystallinity when stretched.
- **Paints and adhesives:** polymer with tackiness enabling good bond strength.

Advantages of plastics are as follows:

- Properties can be altered to suit the needs
- Tough and durable
- Easy processability
- Fire retardant
- Resistant to microbes
- Good weatherability
- Reusable/Recyclable
- Good strength/weight ratio
- Not toxic
- Chemical/Moisture resistant
- Do not rust/corrode.
- Attractive colours
- Practically maintenance free
- Available at much cheaper price

Important and key polymers are:

- Poly ethylene
- Poly esters
- Poly styrene
- Poly propylene
- Polyvinyl chloride
- Poly carbonate

Important types of commodity plastics are tabulated below:

S. No.	Type of Commodity Plastics	Abbreviation	Major Uses
1.	Low-density polyethylene	LDPE	Packaging film, wire and cable insulation, toys, flexible bottles, housewares, coatings
2.	High-density polyethylene	HDPE	Bottles, drums, conduit, sheet, film, wire and cable insulation
3.	Polypropylene	PP	Automobile and appliance parts, furniture, cordage, webbing, carpeting, film packaging
4.	Poly (vinyl chloride)	PVC	Construction, rigid pipe flooring, wire and cable insulation, film and sheet
5.	Polystyrene	PS	Packaging (foam and film), foam insulation, appliances, Housewares, toys

Principal types of engineering plastics are listed below:

S. No.	Type of Engineering Plastics	Abbreviation
1.	Acetal[a]	POM
2.	Polymide[b]	—
3.	Polyamideimide	PAI
4.	Polyacrylate	—
5.	Polybenzimidazole	PBI
6.	Polycarbonate	PC
7.	Polyester[c]	—
8.	Polyether ether ok ketone	PEEK
9.	Polyetherimide	PEI
10.	Polyimide	PI
11.	Poly(phenylene oxide)	PPO
12.	Poly(phenylene sulfide)	PPS
13.	Polysulfone[d]	—

[a] Common name for polyform aldehyde, Abbreviation refers to" poly (oxyethylene).

[b] Principally nylons 6 and 66.

[c] Principally polyethylene terephthalate (PET) and poly (butylene terephthalate) (PBT).

[d] Several types marketed.

Principal types of thermosetting plastics are tabulated below:

S. No.	Type of Thermo-setting plastics	Abbreviation	Typical Uses
1.	Phenol–formaldehyde	PF	Electrical and electronic equipment, automobile parts, utensil handles, plywood adhesives Particle board binder
2.	Urea-formaldehyde	UF	Similar to PF polymers; also treatment of textiles, coatings
3.	Unsaturated polyester	UP	Construction, automobile parts, boat hulls, marine accessories, corrosion-resistant ducting, pipe, tanks, etc., business equipment
4.	Epoxy	—	Protective coatings, adhesives, electrical and electronics applications, industrial flooring, highway paving materials, composites
5.	Melamine-formaldehyde	MF	Similar to UF polymers; decorative panels, counter and table tops, dinnerware

Features of Some Common Thermoplastics

Polyethylene

Polyethylene (or polyethene, polythene, PE) is a family of similar materials categorized according to their density and molecular structure. For example, ultra-high molecular weight polyethylene (UHMWPE) is tough and resistant to chemicals, and it is used to manufacture moving machine parts, bearings, gears, artificial joints and some bulletproof vests. High-density polyethylene (HDPE), recyclable plastic no. 2, is commonly used as milk jugs, liquid laundry detergent bottles, outdoor furniture, margarine tubs, portable gasoline cans, water drainage pipes, and grocery bags. Medium-density polyethylene (MDPE) is used for packaging film, sacks and gas pipes and fittings. Low- density polyethylene (LDPE) is softer and flexible and is used in the manufacture of squeeze bottles, milk jug caps, retail store bags. and (LLDPE) as stretch wrap in transporting and handling boxes of durable goods, and as the common household food covering. XLPE or "PEX" (cross-linked polyethylene) is a semi-rigid/flexible material which has gained wide use in cold or hot water building heating/cooling applications (hydronic heating and cooling) due to its exceptional resistance to breakdown from wide temperature variations.

Polypropylene

Polypropylene (PP) is useful for such diverse products as reusable plastic food containers i.e., "microwave and dishwasher safe" plastic containers, ropes, carpets, plastic moldings, piping systems, car batteries, insulation for electrical cables and filters for gases and liquids. In medicine, it is used in hernia treatment and to make heat-resistant medical equipment. Polypropylene sheets are used for stationery folders and packaging and clear storage bins. Polypropylene is defined by the recyclable plastic number 5. Although relatively inert, it is vulnerable to ultraviolet radiation and can degrade considerably in direct sunlight. It may be worthy to note that (PP) is not as impact-resistant as the polyethylenes, **PP** is also somewhat permeable to highly volatile gases and liquids.

Polystyrene

Polystyrene is manufactured in various forms that have differing applications. Extruded polystyrene (PS) is used in the manufacture of disposable cutlery, CD and DVD cases, plastic models of cars and boats, and smoke detector housings. Expanded polystyrene foam (EPS) is used in making insulation and packaging materials, such as the "peanuts" and molded foam used to cushion fragile products. Extruded polystyrene foam (XPS), known by the trade name Styrofoam, is used to make architectural models and drinking cups for heated beverages. Polystyrene copolymers are used in the manufacture of toys and product casings.

Polyvinyl chloride

Polyvinyl chloride (PVC) is a tough, lightweight material that is resistant to acids and bases. Much of it is used by the construction industry, such as for vinyl siding, drainpipes, gutters and roofing sheets. It is also converted to flexible forms with the addition of plasticizers, thereby making it useful for items such as hoses, tubing, electrical insulation, coats, jackets and upholstery. Flexible PVC is also used in inflatable products, such as water beds and pool toys.

Teflon

Teflon is the brand name given by DuPont Corp. for a polymer called polytetrafluoroethylene (PTFE), which belongs to a class of thermoplastics known as fluoropolymers. It is famous as a coating for non-stick cookware. Being chemically inert, it is used in making containers and pipes that come in contact with reactive chemicals. It is also used as a lubricant to reduce wear from friction between sliding parts, such as gears, bearings and bushings.

Acrylic

Acrylic, a polymer called **poly(methyl methacrylate)** (PMMA), is also known by trade names such as Lucite, Perspex and Plexiglas. It serves as a sturdy substitute for glass for such items as aquariums, motorcycle helmet visors, aircraft windows, viewing ports of submersibles, and lenses of exterior lights of automobiles. It is extensively used to make signs, including lettering and logos. In medicine, it is used in bone cement and to replace eye lenses. Acrylic paint consists of PMMA particles suspended in water.

Nylon

Nylon, belonging to a class of polymers called polyamides, has served as a substitute for silk in products such as parachutes, flak vests and women's stockings. Its fibers are useful in making fabrics, rope, carpets, and musical strings. Whereas in bulk form, nylon is used for mechanical parts including machine screws, gear wheels and power tool casings. In addition, nylon is used in the manufacture of heat-resistant composite materials.

FEATURES OF SOME COMMON THERMOSETTING PLASTICS

Epoxy Resin

Used for casting and encapsulation, as an adhesive to bond other materials. Its a good electrical insulator. Very hard and brittle when not reinforced. Has the ability to resist chemicals. Has low shrinkage during curing. Provides better adhesion between fibre and matrix. The major disadvantages include the toxicity of the material with limited temperature application range upto 175°C. It absorbs moisture affecting its dimensional properties. It has high coefficient of thermal expansion.

Melamine Formaldehyde

These laminates are used for work surfaces, electrical insulation, table ware etc., It is stiff, hard, strong and resists chemicals.

Polyester Resin

Used for casting, encapsulation and bonding of materials. Its stiff, hard and brittle, good electrical insulator and resists chemicals.

Urea Formaldehyde

Stiff, hard, brittle and good electrical insulator

ENGINEERING PROPERTIES OF PLASTICS

Characteristic	Thermoplastic	Thermoset	Effect on Product
High temperature mechanical performance	Poor to good	Excellent	Increases survivability during high temperature
Coefficient of friction	Very low to moderate	Moderate to very high	Affects pulling tension and wire processing
Elongation	Poor to good	Good to excellent	Stretching property before breaking. Its encountered during installation, processing etc.
Tensile strength	Poor to excellent	Poor to excellent	Has impact on durability
Colourability	Good to excellent	Poor to excellent	Improved aesthetic look and protection of surface
Heat deformation	Poor to Good	Good to excellent	Becomes soft when it nears melting point
Cold temperature compatibility	Poor to excellent	Good to excellent	Affects, storage, handling, installation, and usage at extreme cold conditions

Principal types of synthetic fibres are tabulated below:

S. No.	Type of Synthetic Fibres	Description
	Cellulosic:	
1.	Acetate rayon	Cellulose acetate
2.	Viscose rayon	Regenerated cellulose
	Noncellulosic:	
3.	Polyester	Principally poly(ethylene terephthalate)
4.	Nylon	Includes nylon 66, nylon 6, and a variety of other aliphatic and aromatic polyamides
5.	Olefin	Includes polypropylene and copolymers of vinyl chloride, with lesser amounts of acrylonitrile, vinyl acetate, or vinylidene chloride (copolymers consisting of more than 85% vinyl chloride are called *vinyon* fibers)
6.	Acrylic	Contain at least 80% acrylonitrile; included are *modacrylic* fibers comprising acrylonitrile and about 20% vinyl chloride or vinylidene chloride

Principal types of synthetic rubbers are tabulated below:

S. No.	Type of rubber	Description
1.	Styrene-butadiene	Copolymer of the two monomers in various proportions depending on properties desired; called SBR forstyrene- butadiene rubber.
2.	Poly butadiene	Consists almost entirely of the cis-1,4 polymer
3.	Ethylene-propylene	Often abbreviated EPDM for ethylene-propylene-diene monomer; made up principally of ethylene and propylene units with small amounts of a diene to provide unsaturation.

4.	Polychloroprene	Principally the trans-1,4 polymer, but also some cis-1,4 and 1,2 polymers; also known as *neoprene* rubber.
5.	Polyisoprene	Mainly the cis-1,4 polymer; sometimes called "synthetic natural rubber".
6.	Nitrile	Copolymer of acrylonitrile and butadiene, mainly the latter.
7.	Butyl	Copolymer of isobutylens and isoprene, with only small amounts of the latter.
8.	Silicone	Contains inorganic backbone of alternating oxygen and methylated silicon atoms; also called polysiloxane
9.	Urethane	Elastomers prepared by linking polyethers through urethane groups.

Polymers Formulation

The properties of many plastics can be modified by varying their formulation. For example, one of the most versatile plastics is poly vinyl chloride (PVC). It can be made in either a flexible or a rigid form (and every combination between these extremes) by using various additives. Examples and functions of various additives used to enhance the properties of polymers is tabulated below.

Table 4. Additives used to enhance the properties of plastics

Additive	Examples	Function
Plasticiser	e.g., esters of benzene-1,2-dicarboxylic acid	Acts as a lubricant for polymer chains. Large amounts give a flexible product, low quantities produce a rigid one.
Stabiliser	e.g., lead carbonate (<1%), lead phosphate or, for non-toxic requirements, mixtures of metal octadecanoates and epoxidized oil	Prevents decomposition of polymer. Without a stabiliser, poly(chloroethene), for example, decomposes on heating to give a brittle product and hydrogen chloride. Some plastics become coloured (yellowing) when exposed to long periods of sunlight.
Extender	Chlorinated hydrocarbons	Extends the effect of the plasticiser, but generally cannot plasticise alone. They are cheaper than plasticisers, so help reduce costs.
Fillers	Chalk, glass fibre	Tailor the plastic for special requirements, or make it cheaper.
Miscellaneous Additives	Flame retardants, UV stabilisers, antistatics, processing aids, pigments	Impart specially required properties to the plastic for the manufacturing process or for end-use.

Processing of Polymers

Processing converts plastics into useful articles. Processing methods of plastics are tabulated below.

Table 5. Methods of processing of plastics

Process	Application
Compression moulding	Usually for thermosets - powder moulded under heat and pressure.
Injection moulding	Usually for thermoplastics - molten plastics injected into a mould under pressure. The mould surface detail can be accurately reproduced. Very widely used.

Rotational moulding	Usually for thermoplastics. The powder is heated in a closed mould which is rotated, fairly slowly, simultaneously about two axes. Surface detail is poor but this method can be used to make large hollow articles.
Reaction injection moulding	Usually for thermosets, polymerization takes place in the mould thereby producing the finished article directly from a resin.
Extrusion	Usually for thermoplastics - the molten plastics are fed by a screw through a die, which for sheet or film, for instance, is a slit. Various extensions to the process are possible - e.g. a tube may be inflated by air whilst still hot to produce a tubular film (for bags, etc.), or short lengths of hot extruded tube can be inflated in moulds to form bottles.
Calendering	Usually for thermoplastics - molten plastics squeezed between hot rollers to form foil and sheet.
Thermoforming	Heat-softened thermoplastic sheet is drawn into or over a mould. If a vacuum is used to 'suck' the sheet into a mould, the process is known as vacuum forming. This process is used for a variety of articles, ranging from chocolate box liners to acrylic baths.

Applications of Polymers

Major applications of some important polymers are tabulated below.

Table 6. Major applications of some important polymers

Polymers	Applications
Acrylics	Automotive parts, control knobs, dials and handles, pens and pencils, brush-backs, hospital equipment, display material, signs, lighting fittings, inspection panel covers, skylights
Cellulosic plastics:	
Cellulose acetate	Toys, beads, cutlery handles, electrical parts, knobs, steering wheels, shoe heels, packaging, sheeting, toothbrushes, cosmetics
Cellulose acetate butyrate	Moulded or extruded parts for metallization (reflectors, etc.), outdoor signs, automobile tail light covers, tool handles, tooth-brushes, pipe inspection traps, piping
Cellulose propionate	Automotive parts, pens and pencils, telephone housings, shoe heels, toys, radio/TV parts, tooth-brushes, sunglass frames
Fluorinated polymers:	
Fluorinated ethylene propylene (copolymer)	Coil formers, wire insulation, electronic components, encapsulations and fluidized bed coatings, non-stick valves
Polytetrafluorethylene (Teflon)	Gaskets, packings, valves, anti-adhesive coatings, rigid and flexible pipes, membranes, wire insulation, electronic engineering applications
Polytrifluorochlorethylene	Extruded sheet, profile and film, electronic parts, gaskets, pump sealants, dispersion coatings, liquid level indicators of particular use where resistance to aggressive chemicals is needed

Polymers	Applications
Melamine formaldehyde:	
Unfilled	Noted for durability, hardness and good electrical properties, suitable for appliance housings, dinnerware, closures, writing equipment, clock housings, knobs, handles, lighting fixtures, appliances, instruction panels
α-cellulose filled	As for unfilled material
Nylons:	
Nylon 6	Moulded mechanical parts, gear wheels, bushings, elevator gibs, sliding parts for storm windows, automobile and refrigeration door closures, mixer valves, switch housings, grommets, cable clamps, tubing, filaments, aerosol bottles
Nylon 66	Same as for Nylon 6
Phenol formaldehydes:	
Unfilled	Bottle caps, electrical parts, impact resistant, heat resistant and special purpose mouldings requiring rigid durable properties. Restricted colours, close tolerance in moulding articles
Wood flour filled	Same as for unfilled material
Polyacetals	Instrument panel mouldings, aerosol packs, moulded zip-fasteners, wheels for lawn mowers and push carts, impeller blades, cosmetic cases, gear wheels, conveyors
Polycarbonates	Business machine parts, electrical apparatus, sterilizable ware, draughtsman's instruments, lamp covers, safety helmets, tail lights, de- luxe housewares, engineering and industrial components
Polyethylenes:	
Low density	Housewares (bottles, bowls, buckets, containers) closures for squeeze tubes, spouts for detergent cans, carboys, shoe parts, toys, packaging film, garment bags, sheet, piping for domestic, industrial and agricultural use
Medium density	Same as for low density polyethylene
High density	As for low density polyethylene, materials have greater rigidity, specially suitable for large carrying cases, housings, closures, appliance parts, packaging film, sheet, piping, carboys, containers and bottles
Polypropylene	Domestic, hospital and laboratory ware, textile, motor, electrical and chemical industrial usages, containers and closures, shoe heels, combs and toys, blown containers, extruded sheet, pipe, film, filament and fibre
Polystyrenes	Conventional packaging (disposable and others), dishes and utensils, refrigerator parts, emblems, signs, displays, toys, novelties, combs, brush-backs, radio and TV cabinets, lighting fixtures, rigid containers, housewares, reels/spools for film/ tape, appliance panels, handles and switches

Polymers	Applications
Toughened	Wheels, helmets, valve parts, refrigerator parts, bobbins, radio cabinets, electric fan blades, toys, housewares, containers, battery cases, sheeting for refrigerator linings and trays
Styrene copolymers:	
Styrene/acrylonitrile (SAN copolymer)	Housewares (dishes, tumblers, cutlery handles), hair curlers, cosmetic packaging, fountain pen barrels, radio cabinets, battery cases, oil dispensing devices, laboratory utensils electric razor parts, vacuum cleaner parts and casings
Acrylonitrile/butadiene/styrene (ABS terpolymer)	Calendered sheet, electrical housings, office machinery components, industrial mouldings, luggage, automobile parts, domestic appliances
Urea formaldehyde (α-cellulose filled)	Same as for Melamine Formaldehyde. In laminate form surfacing for tables, etc.
Vinyl polymers:	
Rigid poly (vinyl chloride) PVC	Extrusion of piping, profiles and sheet in applications requiring chemical inertness and scuff-resistance combined with light weight
'(Rigid) vinyl chloride/vinyl acetate copolymer	Similar applications to those for PVC but widely used in the manufacture of calendered sheet used for toys, novelties, wall coverings, displays, templates etc.
Rubber-modified PVC	Same as for PVC
Vinylidene chloride polymers	Specialized applications including weather resistant screening: rigid pipes and fittings; in film form, uses include a variety of heat-sealing packaging applications

Raw Materials Used for Synthetic Polymers

Base petrochemicals derived from petroleum and natural gas as shown below in figure are the main raw materials for the production of various polymers.

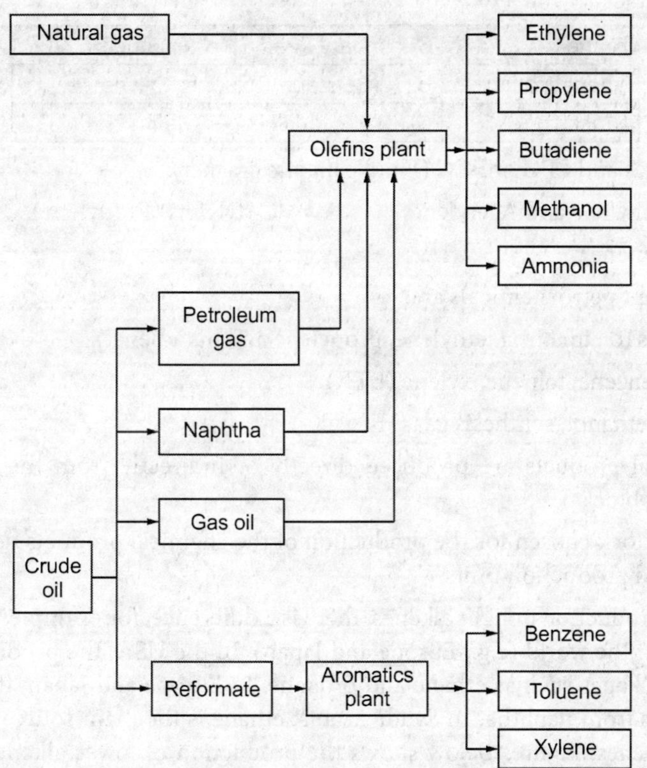

Fig. 2.1. Base petrochemicals derived from petroleum and natural gas

There are three basic group of raw materials that make up major part of Petrochemical industry:

- Synthesis gas from coal or natural gas hydrogen, ammonia, urea, methanol, GTL/CTL and derivatives
- **Aromatics:** Benzene, toluene and xylenes (BTX) and its derivatives
- **Olefins:** Ethylene, propylene and its derivatives

Basic feed stocks for petrochemicals are:

- Ethylene
- Propylene
- Para-xylene
- Butadiene

Basic Building Blocks for Petrochemicals are:

- Olefin
- Aromatics
- Synthesis Gas

Major Source of Olefins are:

- Naphtha Cracker
- Gas Cracker
- FCC (Fluid catalytic cracker)

Major sources of aromatics are:

- Naphtha catalytic reforming
- Pyrolysis gasoline from cracker

Table 2.1. Primary Feed Stock-naphtha

Route	Secondary Feed Stock/Products
Aromatization	BTX (Benzene, Toluene, Xylene)
Cracking	C_2, C_3, C_4 and Higher Olefins

Feed stock from gas and LPG of FCC (Fluid catalytic cracker):

- C_2 Olefins (Ethylene and Acetylene)
- C_3 Olefins (Propylene)
- C_4 Olefins (Butene)

The most important petrochemicals are:

- Lower alkenes (olefins): e.g. ethylene, propylene and butadiene.
- Aromatics: benzene, toluene, xylene (BTX)
- Ammonia, methanol, synthesis gas (H_2 + CO)
  - Most final products are produced directly or indirectly from the previous compounds (building blocks).
  - The feedstock chosen for the production of the chemical products depends on availability, prices and production units.
  - For the production of light alkenes, there is a difference, for example, between the USA and the rest of the world (e.g., Europe and Japan). In the USA, the production of light alkenes is from alkanes such as ethane and propane. In Europe and Japan, the production of light alkenes is from naphtha. In Saudi Arabia, ethane is the main source for the production of lower alkenes. Figures below shows the production of lower alkenes from crude oil and natural gas, respectively.

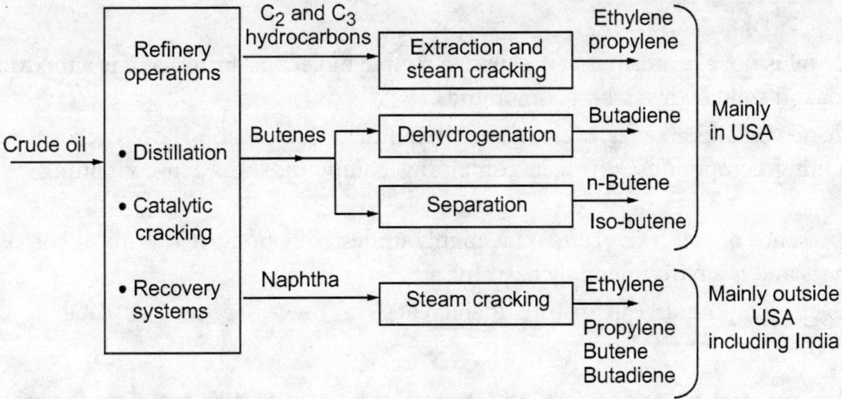

Fig. 2.2. Lower alkenes (olefins) production from oil.

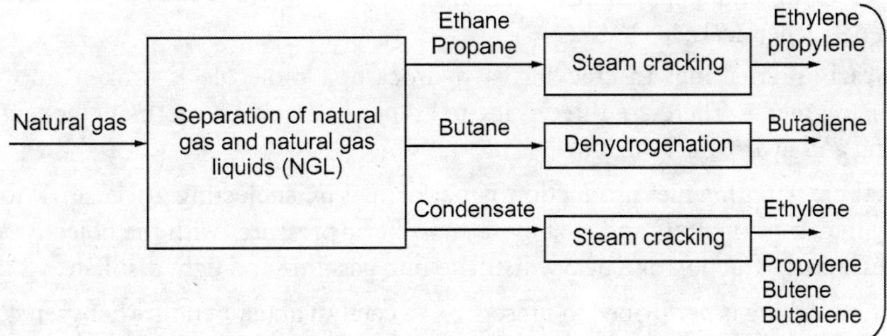

Fig. 2.3. Lower alkenes (olefins) from natural gas.

Natural Gas

- Natural gas is a mixture of hydrocarbons with methane as the main constituent (upto 90%). It can be found in porous reservoirs, either associated with crude oil (associated gas) or in reservoirs in which no oil is present (non-associated gas).
- Natural gas is important as a source of energy and also as a raw material for the petrochemical industry.
- Natural gas usually contains small amounts of non-hydrocarbon gases such as carbon dioxide, nitrogen, and hydrogen sulfide.
- Natural gas classifies as 'dry' or 'wet' natural gas. Dry natural gas contains only small amounts of condensable hydrocarbons (at ambient temperature.). Non-associated gas is usually dry.
- Natural gas contains substantial amounts of ethane, propane, butane, and C_5 + hydrocarbons, which liquefy on compression at ambient temperature called natural gas liquids (NGL).
- The terms 'sweet' and 'sour' natural gas denote the absence or presence of H_2S and mercaptans (RSH).
- Condensable hydrocarbons are removed from a wet natural gas and part can be sold as liquefied petroleum gas (LPG).

Crude Oil

- Crude oil is not a uniform material with a simple molecular formula. It is a mixture of gaseous, liquid, and solid hydrocarbon compounds.
- Crude oil contains alkanes, cycloalkanes (naphthenes), aromatics, poly-cyclic aromatics, sulfur-containing compounds, nitrogen-containing compounds, oxygen-containing compounds, ... etc.
- The presence of sulfur in crude oil is highly undesirable because it leads to corrosion, poisons catalysts and is environmentally harmful.
- Nitrogen compounds can disturb major catalytic processes, such as catalytic cracking and hydrocracking.

UNIT OPERATIONS IN MANUFACTURING PROCESSES OF PETROCHEMICALS

Hydrocarbons cracking and naphtha reformation are two main unit operations involved in them manufacture of petrochemicals:

(*a*) **Hydrocarbon cracking:** In cracking, a hydrocarbon molecule is broken into two or more smaller fragments. There are three principal types of cracking: thermal cracking, catalytically cracking and hydro cracking.

Thermal cracking for fuel production is performed by subjecting a feedstock to temperature usually in excess of 455°C and at above atmospheric pressures with the objective of converting a residual crude fraction or a heavy distillate into gasoline and light distillate.

Catalytic cracking is performed in presence of a catalyst at temperature between 455-540°C and at above atmospheric pressure. The process 'converts a distillate feedstock into gasoline as the primary product with production of light hydrocarbons.

Hydro cracking process operates at elevated pressure in the presence of hydrogen and catalyst at temperature generally less than 430°C.

(*b*) **Naphtha reforming:** The purpose of reforming naphtha is to rearrange or reform the molecular structure of hydrocarbon, particularly with the objective of producing aromatics. The chemical processes involved in reformation are as follows:

- Dehydrogenation of naphthenes to aromatics.
- Dehydrocyclization of paraffins to form aromatics.
- Isomerisation of paraffins to more highly branched isomers

As the reaction proceeds, the reformed products will contain increasing concentration of aromatics and decreasing concentration of heavy paraffins. These reforming reactions are regulated by metal catalysts in an environment of hydrogen under moderate pressure.

Petrochemicals product categories is shown below:

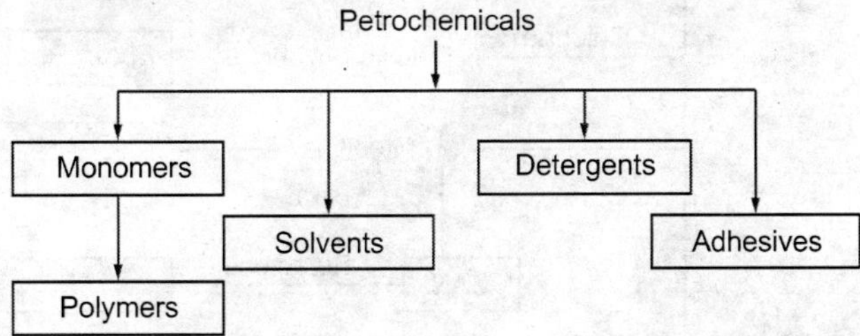

Types of polymers produced from petrochemicals is shown below:

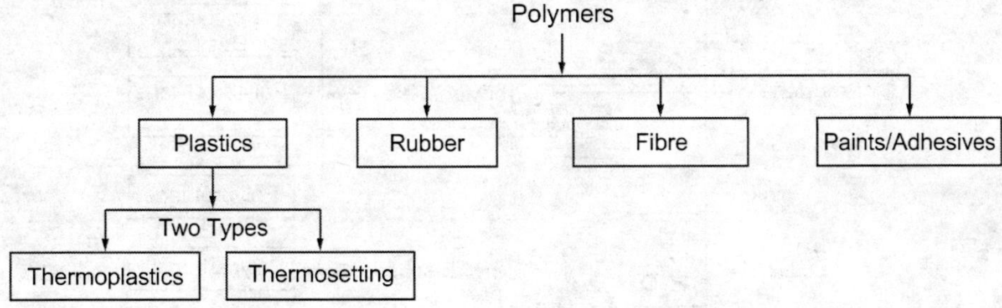

Each of these important petrochemicals are used for the production of different polymers as shown in figures below.

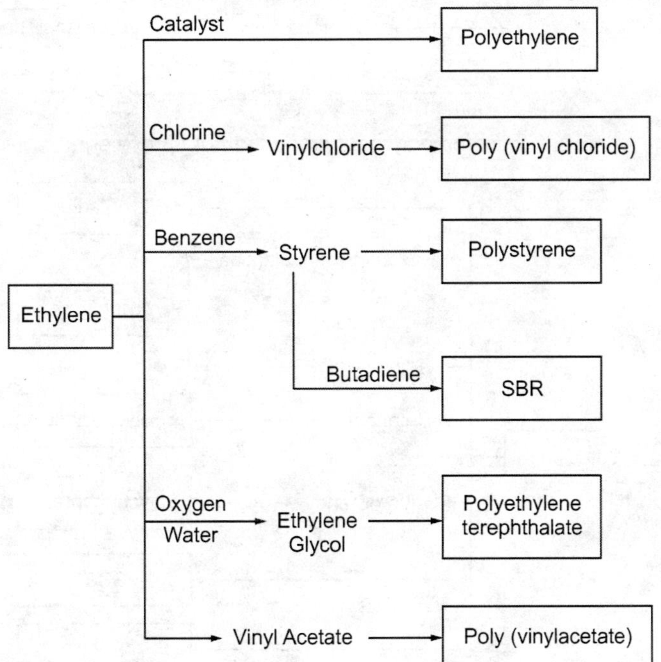

Fig. 2.4. Some polymers produced from ethylene.

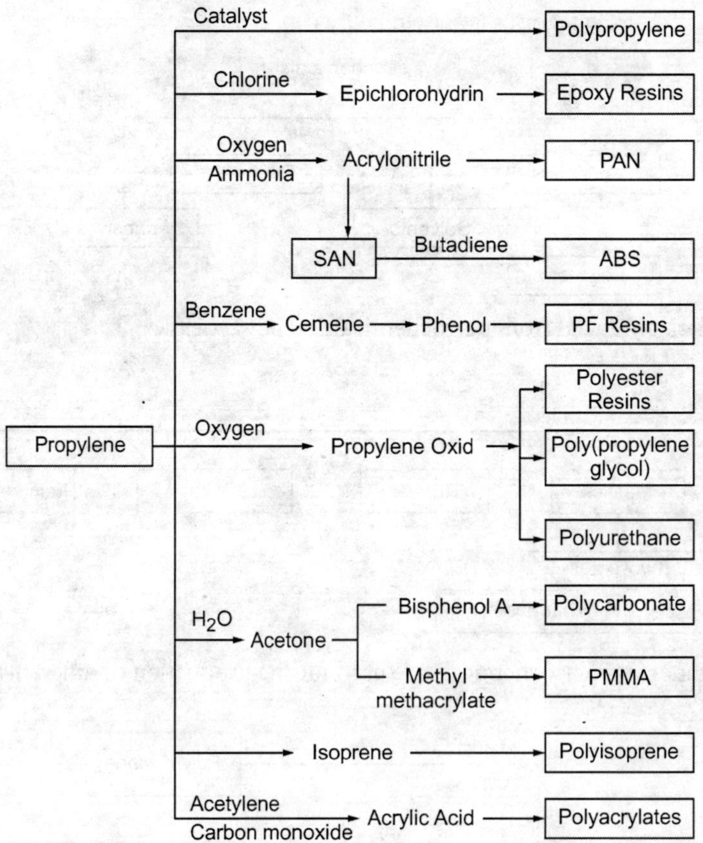

Fig. 2.5. Some polymers produced from propylene

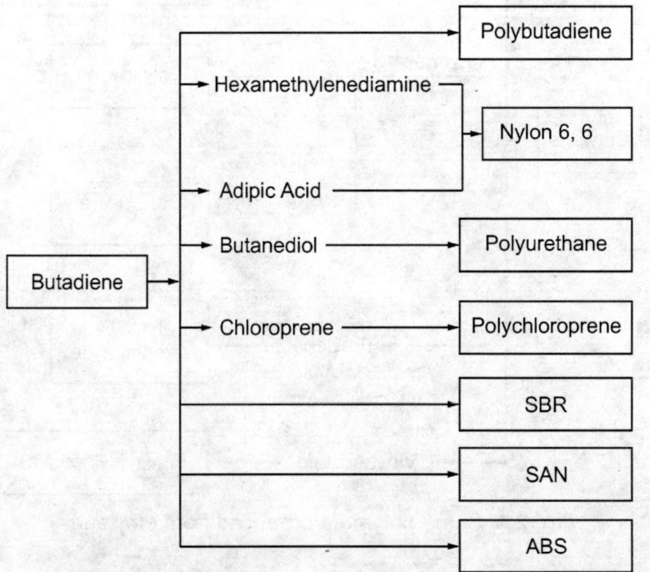

Fig. 2.6. Some polymers produced from butadiene

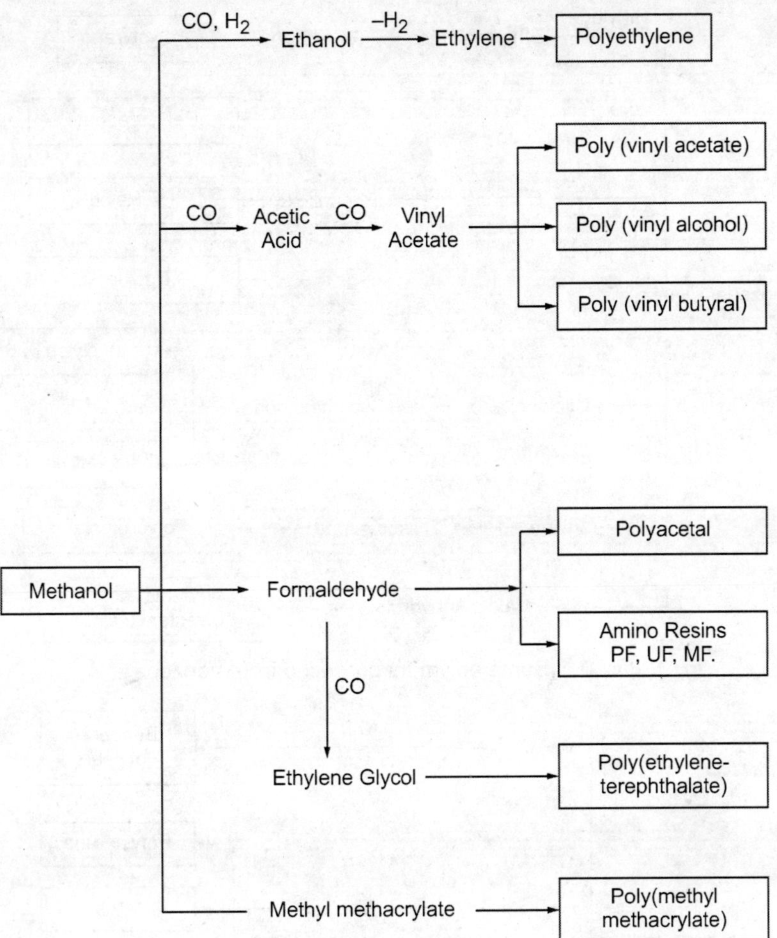

Fig. 2.7. Some polymers produced from methanol (methyl alcohol)

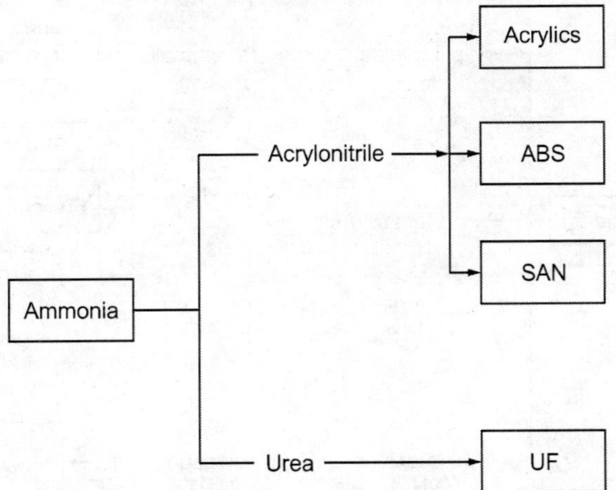

Fig. 2.8. Some polymers produced from ammonia

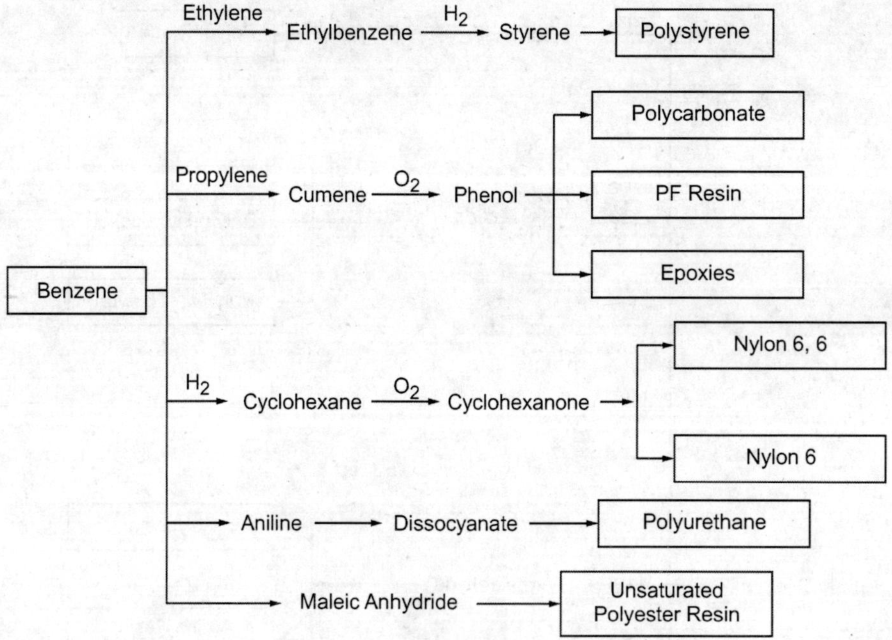

Fig. 2.9. Some polymers produced from benzene

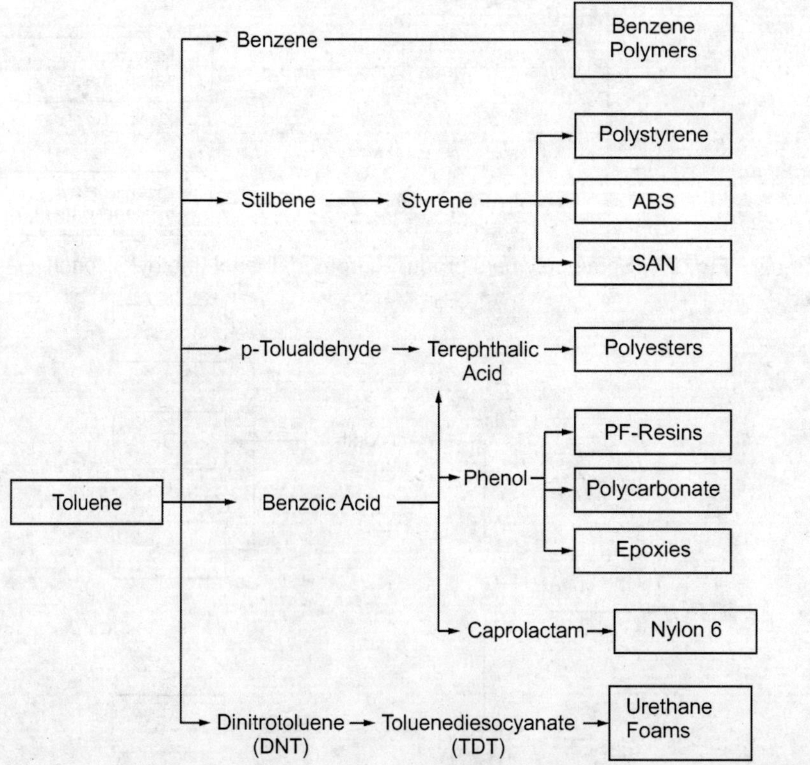

Fig. 2.10. Some polymers produced from toluene

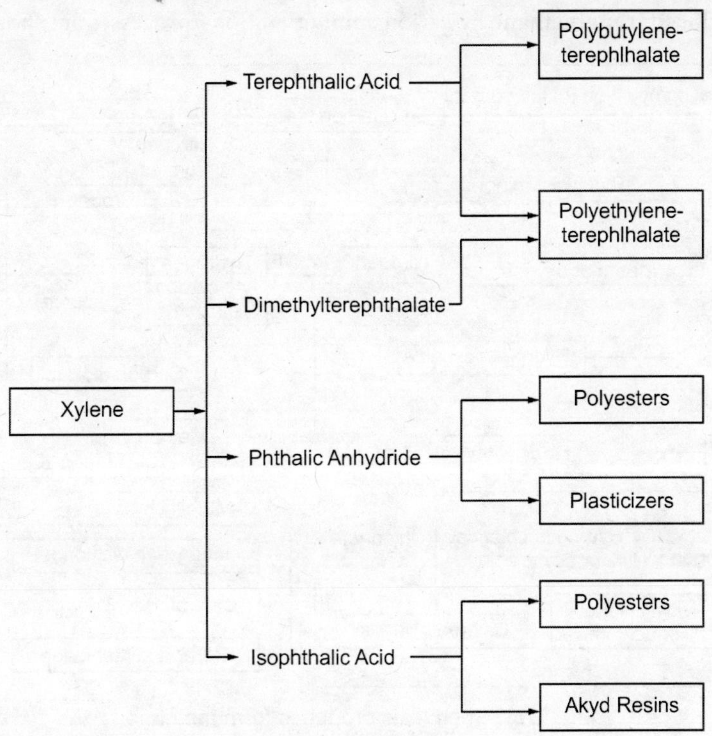

Fig. 2.11. Some polymers produced from xylene

SALIENT FEATURES OF VARIOUS RAW MATERIALS USED IN POLYMER INDUSTRY

Methane (Main Constituent of Natural Gas)

Methane Derivatives are shown below in figure.

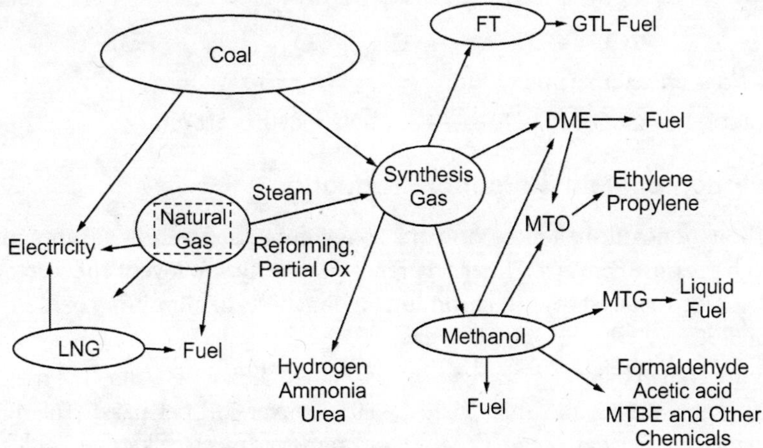

Fig. 2.12. Methane (Natural gas) derivatives

Chemicals produced from methane by various unit operations/processes are shown below in figure.

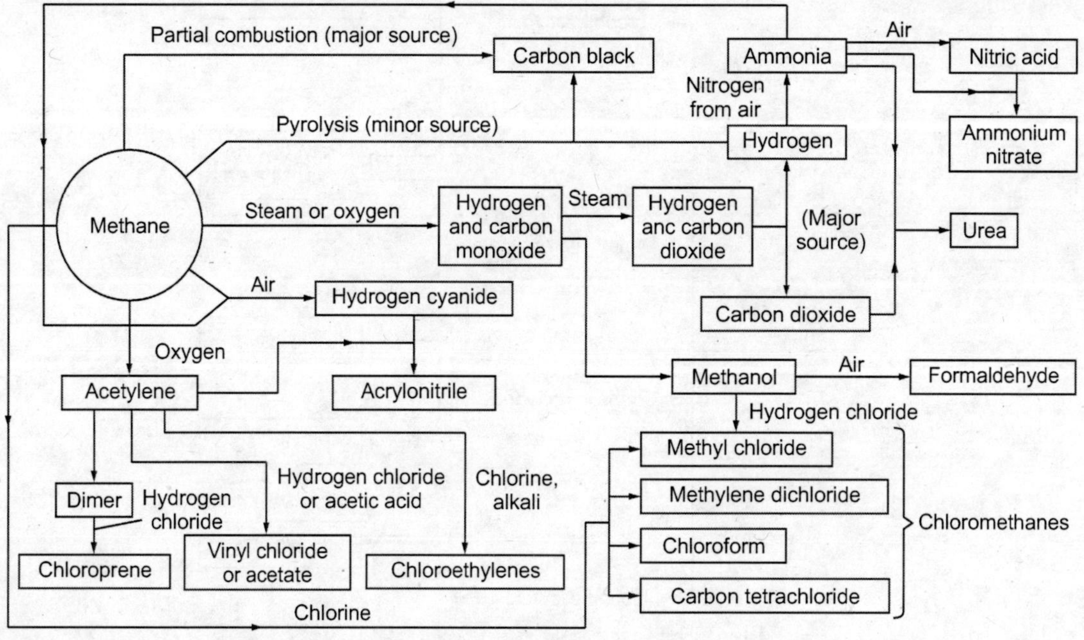

Fig. 2.13. Chemicals produced from methane

Methyl Alcohol (CH_3OH - Methanol)

When 'synthesis gas' ($H_2 + CO$) is subjected to high pressure and moderate temperature conditions, it converts to methanol.

Followed by this, the methanol is separated using a series of phase separators and distillation columns.

Chemical reactions involved in methanol production from synthesis gas ($CO + H_2$):

Desired/reaction: $CO + 2H_2 \longrightarrow CH_3OH$

Side reactions: $CO + 3H_2 \longrightarrow CH_4 + H_2O$

$2CO + 2H_2 \longrightarrow CH_4 + CO_2$

All above reactions are exothermic.

Undesired reaction: $zCO + aH_2 \longrightarrow \text{alcohols} + \text{hydrocarbons}$

Process Description of Methyl Alcohol Production

Hydrogen and carbon monoxide in a mole ratio of 2.25 is compressed to 200-350 atm mixed with recycle gas, and fed to a high pressure converter. Internal preheat is usually employed. The reactor is copper-lined steel and contains a mixed catalyst of zinc, chromium, manganese, or aluminum oxides. The temperature is maintained at 300-375°C.

The exit gases are cooled by heat exchange with reactants, then with water. The methanol condenses under full operating pressure to maximize yields (~50% conversion per pass). The liquid methanol is depressurized, purified by permanganate ($KMnO_4$) to remove traces of ketones, aldehydes, and other such impurities, sent to a stripper to remove light ends such as dimethyl ether, and to a fractionator to separate methanol from the higher mol. wt. compounds. ICI has developed a methanol process which operates at only 50 atmospheres pressure using a copper catalyst.

Properties of Methyl Alcohol (CH_3OH)

Mol. wt.: 32.04

M.P.: –97.8°C

B.P.: 64.7°C

Sp. gr.: 0.788 @ 20°C

Explosive limits: Lower = 6.0 vol. % in air

Upper = 36.5 vol. % in air

Toxicity limits: 200 ppm

Formaldehyde, the primary oxidation product of methyl alcohol, is the largest outlet for methyl alcohol. Formaldehyde is an important monomer in the plastics industry.

Approximately 50% of the demand of methyl alcohol comes from formaldehyde, about 25% from DMT(Dimethyl Terephthalate), and the balance from drugs, pesticides and chemicals such as acetic acid, methyl amines, esters, solvents, etc.

SYNTHESIS GAS (SYNGAS)

Synthesis gas (or syngas) is a general term used to describe mixtures of hydrogen (H_2) and carbon monoxide (CO) in various ratios.

Mixtures	Main Uses
H_2	Refinery hydrotreating and hydrocracking
$3H_2 : 1N_2$	Ammonia
$2H_2 : 1CO$	Methanol
$1H_2 : 1CO$	Aldehydes (Hydro formulation)
CO	Acids (formic and acidic)

Synthesis gas is used for the production of different materials and is also a source of pure hydrogen and pure carbon monoxide.

Syngas may be produced from a variety of raw materials ranging from natural gas to coal. The choice for a particular raw material depends on cost and availability of feedstock.

Syngas can be produced by any one of the following three processes:

- Steam reforming of natural gas or light hydrocarbons (e.g., naphtha) optionally in the presence of oxygen or carbon dioxide.
- Partial oxidation of (heavy) hydrocarbons with steam and oxygen.
- Partial oxidation of coal (gasification) with steam and oxygen.

Steam reforming is the reaction of hydrocarbons with steam in the presence of a catalyst.

e.g., $$CH_2 + H_2O \longrightarrow CO + 3H_2 \qquad (\Delta H^{\circ}_{298} = -206 \text{ KJ/mole})$$

Reforming (in gas industry) is commonly used for the conversion of a hydrocarbon by reacting it with O – containing molecules, usually H_2O, CO_2 and/or O_2.

Partial oxidation (also called steam/oxygen reforming) is a non-catalytic reaction of hydrocarbons with oxygen and usually also steam.

e.g., $CH_4 + 1/2O_2 \longrightarrow CO + 2H_2$ $(\Delta H^\circ_{298} = -206\ KJ/mole)$

Coal gasification is more common term to describe partial oxidation of coal. A combination of steam reforming (endothermic reaction) and partial oxidation (exothermic reaction) is referred to as (autothermic reforming).

A General Flowsheet for the manufacture of synthetic gas by different routes is given below:

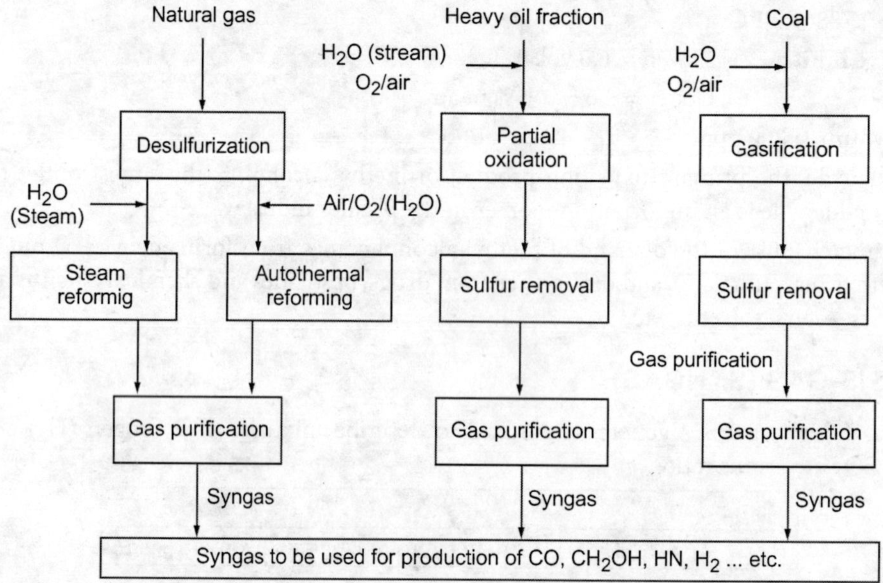

Fig. 2.14. General flow sheet for the production of syngas.

Synthesis Gas from Natural Gas

Natural gas consists mainly of methane and higher alkanes. For methane in presence of steam, the most important reactions are:

- The steam reforming reaction:

$$CH_4 + H_2O \longrightarrow CO + 3H_2$$

- The water-gas shift reaction:

$$CO + H_2O \longrightarrow CO + H_2$$

Steam Reforming Process

Steam reforming is carried out at high temperature (> 1000 K). A catalyst (supported nickel) is required to accelerate the reaction due to the very high stability of methane. The catalyst is contained in tubes, which are placed inside a furnace that is heated by combustion of fuel. The steam reforming consists of two sections: convection section and Radiation section. In the convection section, heat recovered from the hot flue gases is used for preheating of the natural feed process, the process steam and for generation of superheated steam. In the radiant section of the furnace, the reforming reactions take place.

FORMALDEHYDE (HCHO)

Formaldehyde is produced from methanol by catalytic oxidation-dehydrogenation of methyl alcohol. Chemical Reaction involved in production of formaldehyde (HCHO) from CH_3OH (methanol):

(*a*) **Oxidation:**

$$CH_3OH + 0.5\,O_2 \longrightarrow HCHO + H_2O$$

(*b*) **Pyrolysis:**

$$CH_3OH \longrightarrow HCHO + H_2$$

(*c*) **Undesired reaction:**

$$CH_3OH + 1.5\,O_2 \longrightarrow 2H_2O + CO_2$$

In the above reactions, the first and third are exothermic reactions but the second reaction is endothermic. The reactions are carried out in vapour phase.

Catalyst: Silver or zinc oxide catalysts on wire gauge are used.

Operating temperature and pressure: Near about atmospheric pressure and 500-600°C.

Process Description

- Air is sent for pre-heating using reactor outlet product and heat integration concept.
- Eventually heated air and methanol are fed to a methanol evaporator unit which enables the evaporation of methanol as well as mixing with air. The reactor inlet temperature is 54°C.
- The feed ratio is about 30-50 % for CH_3OH: O_2
- After reaction, the product is a vapour mixture with temperature 450-600°C.
- After reaction, the product gas is cooled with the heat integration concept and then eventually fed to the absorption tower.
- The absorbent in the absorption tower is water as well as formaldehyde rich water.
- Since formaldehyde rich water is produced in the absorption, a portion of the rich water absorbent solution from the absorber is partially recycled at a specific section of the absorber.
- From the absorber formaldehyde + methanol rich water stream is obtained as the bottom product.
- The stream is sent to a light end stripper eventually to remove any light end compounds that got absorbed in the stream. The vapors from the light end unit consisting of light end compounds can be fed at the absorption unit at specific location that matches with the composition of the vapors in the absorption column.
- Eventually, the light end stripper bottom product is fed to a distillation tower that produces methanol vapour as the top product and the bottom formaldehyde + water product (37 % formaldehyde concentration).

Flowsheet of formaldehyde production from methanol is shown below:

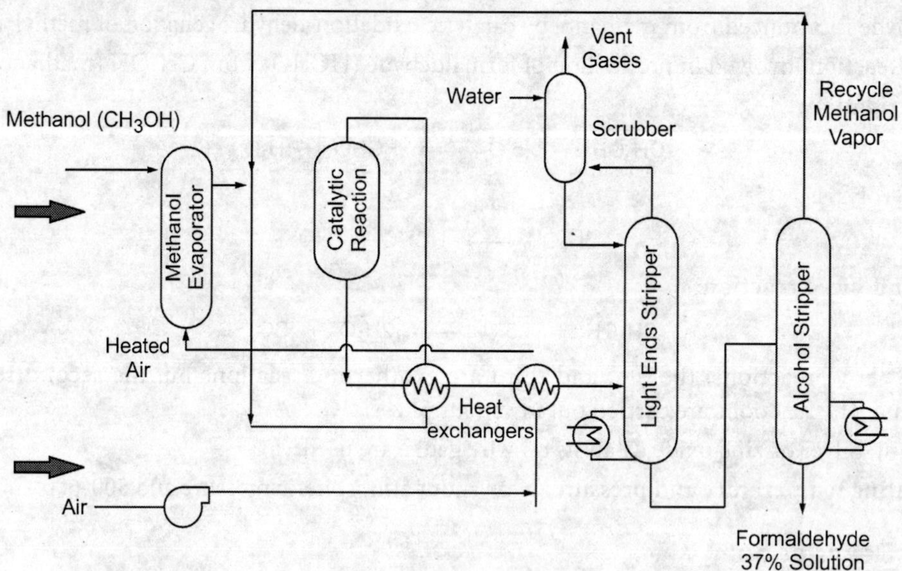

Fig. 2.15. Flow sheet of formaldehyde production from methanol

Properties of Formaldehyde (HCHO)

Mol. wt.: 50.49

M.P.: –118°C

B.P.: –19°C

Sp. gr.: 0.815(at 20°C)

Explosive limits: Lower = 7 vol. % in air
Upper = 73 vol. % in air

Toxicity limits: 10 ppm

Solubility: Soluble in water, alcohol and polar solvents, slightly soluble in hydrocarbons, chloroform, ether.

The principal use of formaldehyde is in the manufacture of phenolic, urea and melamine resins.

Raw material requirements:

Basis: 1 ton of formaldehyde: (90% yield)

Methyl alcohol (99 + % purity): 1.04 tons

Air: 1,760 Nm^3

ETHYLENE ($CH_2 = CH_2$)

Ethylene is known as the "king of petrochemicals" because more commercial chemicals are produced from ethylene than from any other intermediate. Ethylene is one of the most important fundamental chemicals in the petrochemical industry as it is the source material for a variety of products such as polyethylene resin, ethylene glycol, vinyl chloride resin, acetic acid, styrene, and alpha olefin which are produced by polymerization, oxidation, alkylation, hydration, or the addition of halogen.

Ethylene and its derivatives are shown below.

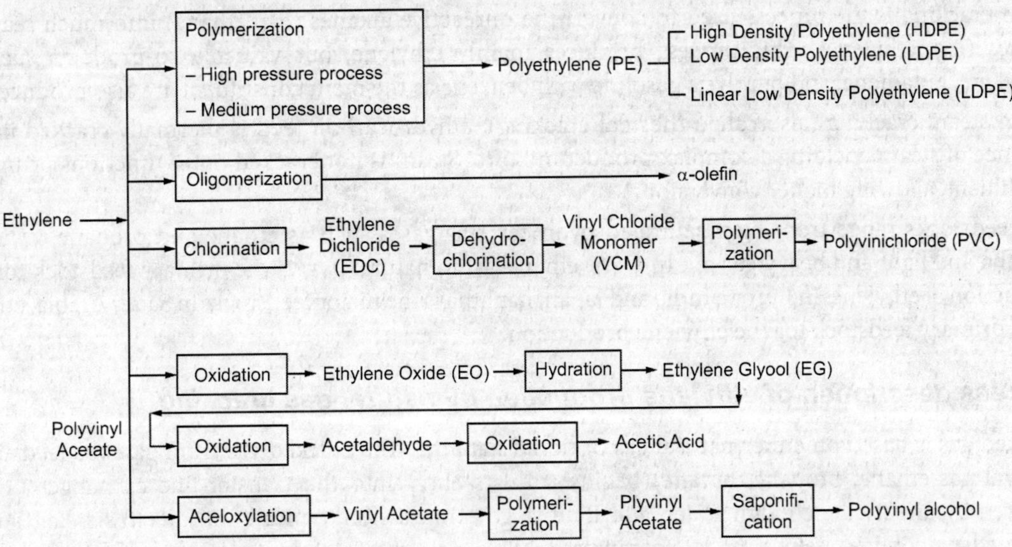

Fig. 2.16. Derivatives of ethylene

Properties of Ethylene:

Mol. wt.: 28.03

M.P.: –169.4°C

B.P.: –103.8°C

Density @ –103.8°C: 0.5699

Slightly soluble in water (25.6 cc per 100 gm H_2O at 0°C

Explosion limits: Lower = 3% in air, 2.9% in O_2

Upper = 29% in air, 79% in O_2

Demand of ethylene for various end uses via chemical intermediates is shown below in figure.

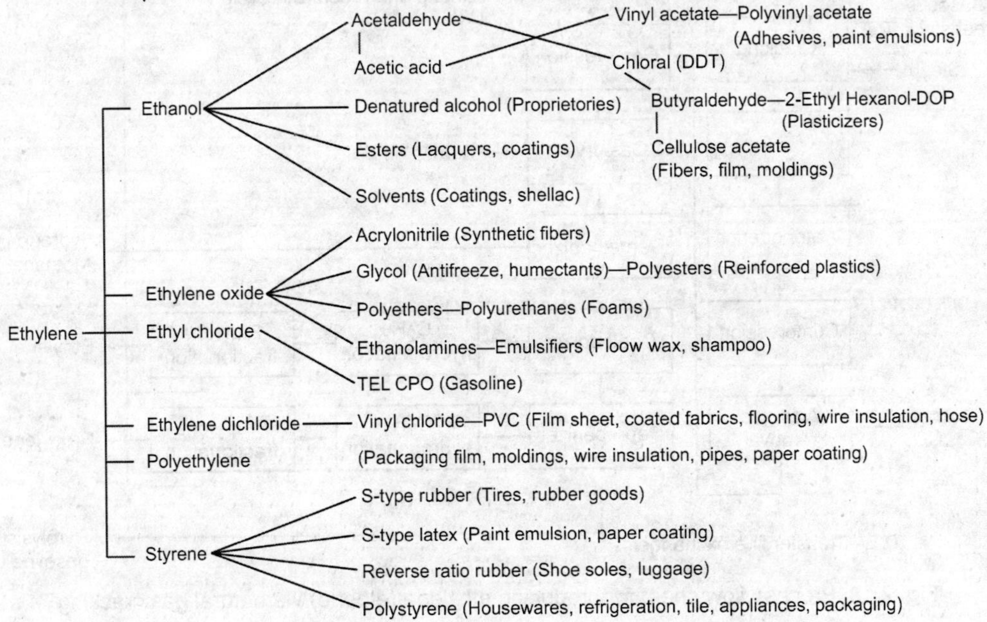

Fig. 2.17. Chemical intermediates and end products from ethylene

Production of ethylene by steam cracking of natural gas or liquid naphtha

Steam cracking is the process used to convert the unreactive alkanes (e.g. ethane) into much reactive alkenes (e.g. ethylene). This process produces mainly ethylene, but valuable co-products such as propylene, butadiene, and pyrolysis gasoline with benzene as the main constituent are also produced.

In steam cracking (also called thermal cracking), a hydrocarbon feed is thermally cracked in the presence of steam, yielding a complex product mixture. Steam is not cracked but it functions primarily as a diluent, allowing higher conversion.

Feedstocks range from light saturated hydrocarbons gases such as ethane and propane to liquid naphtha and light and heavy gas oil. In USA, ethane (from natural gas) is the primary feedstock for the production of ethylene. In Europe India and Japan, naphtha is the major feedstock. In Saudi Arabia, ethane is the primary feedstock for the ethylene production.

Process description of ethylene production by natural gas cracking

Cracked gas is based on either natural gas or liquid naphtha. The cracked gas from gaseous feedstocks (natural gas-ethane, propane, butane) begins with its entry into the transfer-line exchanger (TLE), followed by direct quench with water, and multistage compression, typically in four to six stages with intermediate cooling. Before the last compressor stage, acid gas (mainly hydrogen sulfide and carbon dioxide) is removed.

After the last compressor stage, water removal takes place by chilling and drying over zeolites. Subsequent fractionation of the cracked gas is based on cryogenic (temp. < 273 K) and conventional distillation under pressure (15-35 bar). Cryogenic distillation requires large amount of energy due to the need for a refrigeration system. In the purification section, separation of ethane/ethylene and propane/propylene is difficult.

Process flowsheet for ethylene production via natural gas cracking is shown below:

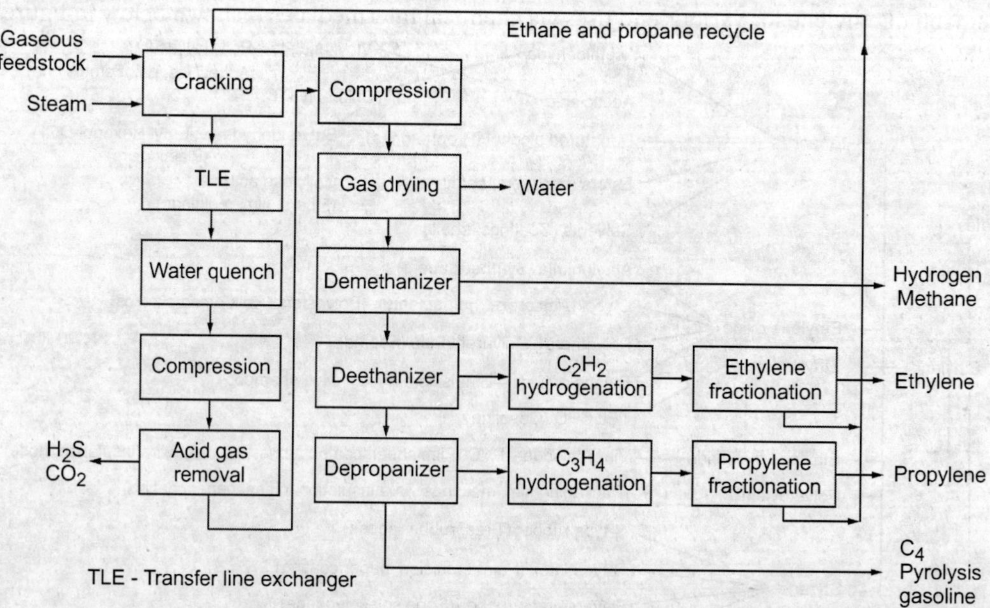

Fig. 2.18. Process flow sheet for producing ethylene (ethene) via natural gas cracking

Process description of ethylene production by liquid naphtha cracking

Naphtha is mixed with superheated steam and fed to a furnace fired with fuel gas + fuel oil as fuels to generate heat. The superheated steam is generated from the furnace itself using heat recovery boiler. The C_2-C_4 saturates are fed to a separate furnace fed with fuel gas + fuel oil as fuels to generate heat. In the furnace, apart from the steam cracking, steam is also generated. This is by using waste heat recovery concept from the combustion gases in the furnace.

After pyrolysis reaction, the products from the furnace are sent to another heat recovery steam boiler to cool the product streams (from 700-800 °C) and generate steam from water. After this operation, the product vapours enter a scrubber that is fed with gas oil as absorbent. The gas oil removes solids and heavy hydrocarbons. Separate set of waste heat recovery boiler and scrubbers are used for the LPG furnace and Naphtha steam cracking furnaces.

After scrubbing, both product gases from the scrubbers are mixed and fed to a compressor. The compressor increases the system pressure to 35 Kg/cm^2. The compressed vapour is fed to a phase separation that separates the feed into two stream namely the vapour phase stream and liquid phase stream. The vapour phase stream consists of H_2, CO, CO_2C_1-C_3 + components in excess. The liquid phase stream consists of C_3 and C_4 compounds in excess. Subsequently, the vapour phase and liquid phase streams are subjected to separate processing.

A flowsheet of ethylene (co-product-acetylene) production by steam creaking of naphtha is shown below:

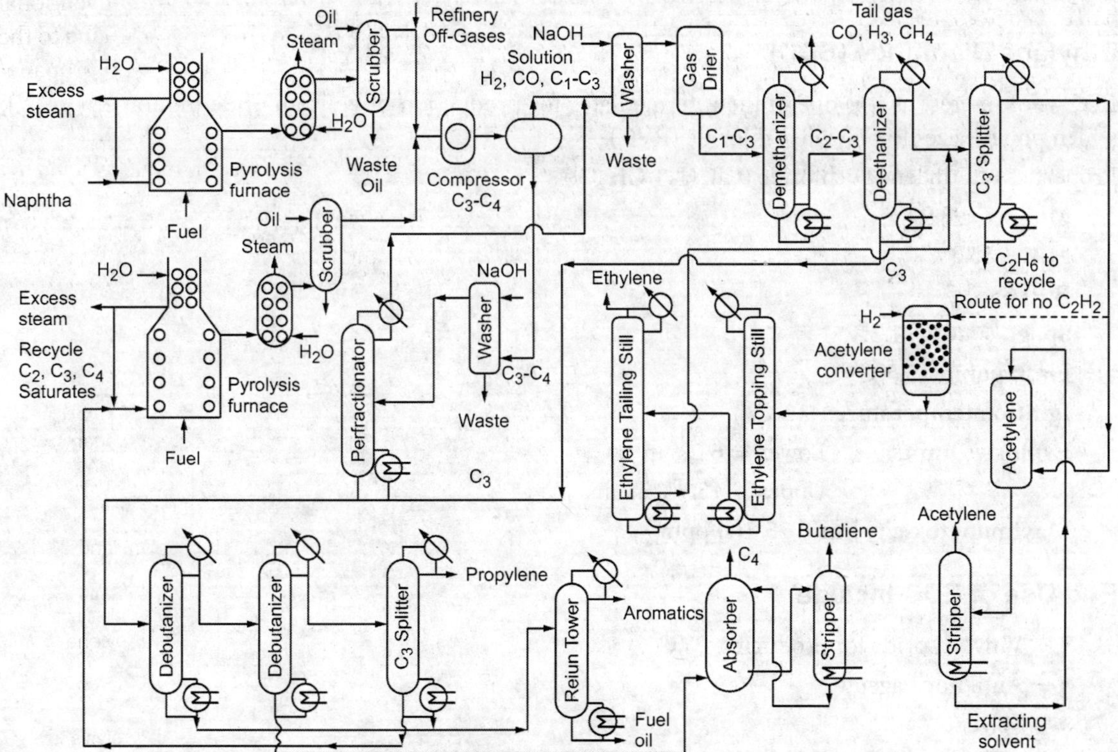

Fig. 2.19. Ethylene production by steam cracking of naphtha

Olefins derivatives are listed below.

Ethylene (C_2H_4)	Propylene (C_3H_6)	Butadiene (C_4H_6)
Polyethylene	Polypropylene	Polymers and copolymers with styrene and ACN
EB-Styrene	Acrylonitrile	ABS
Ethylene Oxide, Ethylene glycols	Cumene-Phenol	*n* Butene-MEK
EDC, VCM	Propylene Oxide	Isobutene-TBA
Vinyl acetate	Epichlorohydrin, glycerol	Disobutene triisobutene and polyisobutene
Ethyl alcohol acetaldehyde	Nonylphenol Tetramer	Isobutene-MTBE
		Motor Fuel Alkylate

ETHYLENE DICHLORIDE (EDC) AND VINYL CHLORIDE MONOMER (VCM)

Ethylene dichloride (EDC) is an intermediate product used primarily in the manufacture of Vinyl chloride monomer. The process consists of a direct chlorination operation, a cracking/pyrolysis operation (for hydro-de-chlorination of EDC to VCM), an oxy-chlorination step (for utilising HCl removed in hydro-de-chlorination stage), product finishing and waste treatment operations. The raw materials for the direct chlorination process are chlorine and ethylene. Oxy-chlorination involves the treatment of ethylene with oxygen and HCl. Oxygen for oxy- chlorination generally is added by feeding air or purified oxygen as feed material.

Ethylene Dichloride (EDC)

EDC is of interest as it is one of the intermediates for production of vinyl chloride monomer (VCM), which polymerizes to polyvinyl chloride (PVC).

Properties of Ethylene Dichloride ($ClCH_2 \bullet CH_2Cl$)

Mol. wt.: 98.97

M.P.: –35.3°C

B.P.: 83.7°C

Sp. gr. @ 20°C: 1.257

Flash point: 15.5°C

Ignition temperature: 412°C

Explosive limits: Lower = 6.2% in air
Upper = 15.9% in air

Maximum toxicity limit: 75-100 ppm

End Use of EDC Include

- Vinyl chloride for producing PVC
- Antiknock agent
- Solvent

Methods of Production of EDC

- Reaction of chlorine with ethylene in liquid or vapor phase
- By-product of direct chlorination of ethane to ethyl chloride
- By-product of chlorinated hydrocarbons

Ethylene + Chlorine Reaction:

$$CH_2 = CH_2 + Cl_2 \longrightarrow ClCH_2 \cdot CH_2Cl \text{ (EDC)}$$

Raw Material Requirements

Basis: 1 ton of $C_2H_4Cl_2$ (95% yield)

Ethylene: 0.30 ton

Chlorine: 0.75 ton

Ethylene dibromide catalyst: Trace

Cooling water: 48 tons

Co-products: HCl, propylene chloride, polychloroethanes

Process Description of EDC Production

Ethylene (with or without C_2H_6, CH_4 and H_2 diluent) is mixed with chlorine and bubbled through a liquid phase reactor. Ethylene dichloride product serves as the reacting medium. Heat of reaction is controlled by external heat exchange and recycle, or by coil or jacket heat transfer, to hold the reactor at 45-50°C with a pressure of 1.5-2 atms. Traces of ferric chloride or ethylene dibromide as catalyst gives 90-95% yield with little dimer formation.

The gaseous products are cooled to strip the acid gas of EDC. The liquid product is alkali washed and fractionated.

Vinyl Chloride Monomer (VCM)

Properties of VCM (CH_2Cl):

Mol. wt.: 62.50

M.P.: –153.8°C

B.P.: –13.81 °C

Sp. gr. @ 20°C: 0.983

Soluble in CCl_4, $(C_2H_5)_2O$, and alcohols; slightly soluble in water

Flash point: –77 °C

Explosive limits: Lower = 5% in air

Upper = 23% in air

Maximum toxicity limits: 500 ppm

Vinyl chloride monomer (VCM) is strictly an intermediate in polyvinyl chloride (PVC) production.

Methods of Production of VCM

- Ethylene dichloride thermal pyrolysis
- Ethylene dichloride-caustic reaction
- Acetylene-HCl reaction

The first two processes are described since they are often combined to keep a chlorine balance.

Ethylene Dichloride Thermal Pyrolysis Process of VCM Production:

Chemical reaction involved in VCM production:

$$CH_2Cl \cdot CH_2Cl \xrightarrow{500°C} CH_2{=}CHCl + HCl$$

Raw Materials Requirements

Basis: 1 ton of vinyl chloride (99.5% pure, 95% yield)

Ethylene dichloride: 1.65 tons

Cooling water: 30 tons

Fuel gas: 5.7 Nm^3

Heating steam: 2.0 tons

Electricity: 1.5 KWH

Co-product: 0.65 ton HCl

Process Description of VCM Production by Pyrolysis of EDC

Ethylene dichloride (EDC) vapor at 4 kg/cm^2 is dried by silica gel and sent to a stainless steel tubular cracking furnace. This is externally flue gas fired and controlled at 480-520°C. The contact surface catalyst within the tubes is pumice or charcoal. The conversion per pass is around 50% and the ultimate yield is 95%. Spray quenching with cold EDC prevents backward reaction. Uncondensed gases are sent to a surface heat exchanger to remove the balance of the EDC and vinyl chloride monomer (VCM).

The non-condensable containing HC1 are either sent to the acetylene-HCl process in an adjacent process area or water scrubbed to recover HCl.

The condensate is fractionated with the EDC bottoms returned for recycle and vinyl chloride monomer taken from overhead, stabilized and sent to storage.

Flowsheet of vinyl chloride monomer (VCM) production by pyrolysis of ethylene dichloride (EDC) is given below.

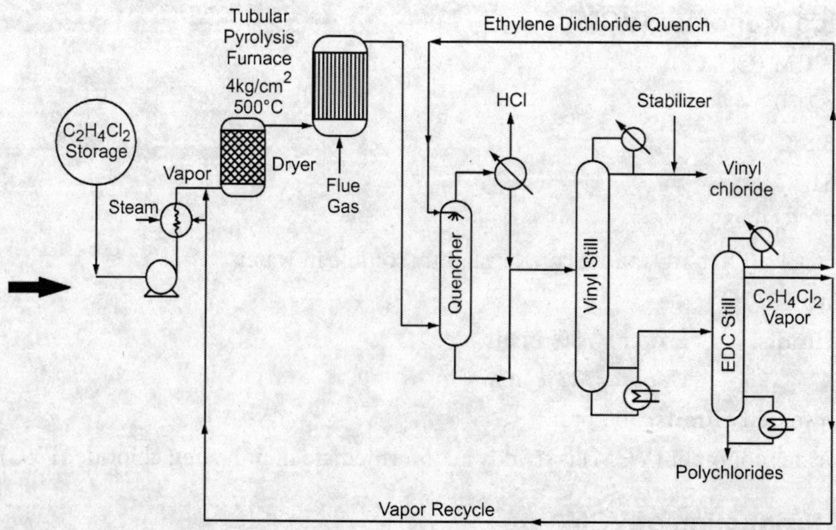

Fig. 2.20. Flowsheet of Vinyl chloride production via ethylene dichloride pyrolysis

VCM Production by Acetylene-HCl Reaction

Chemical reaction involved is as follows:

$$C_2H_2 + HCl \xrightarrow{1\,atm,\,160\text{-}200°C} CH_2 = CHCl$$

Raw Materials Requirements

Basis: 1 ton of vinyl chloride (99.5% pure, 97% yield)

Acetylene: 0.462 ton

Hydrogen chloride: 0.60 ton

Process Description of VCM Production by Acetylene-HCl Reaction

Acetylene and dry HCl in 5-10% molar excess are vapor blended by jet mixing in a pipe and passed through a tubular catalytic reactor containing carbon pellets impregnated with $HgCl_2$. The temperature is maintained at 160°C and is gradually raised to 200°C as the catalyst deteriorates. The effluent gases contain vinyl chloride which is separated from unreacted acetylene plus hydrogen chloride, these unreacted materials being recycled.

A flowsheet of vinyl chloride monomer (VCM) production by acetylene-HCl reaction is given below:

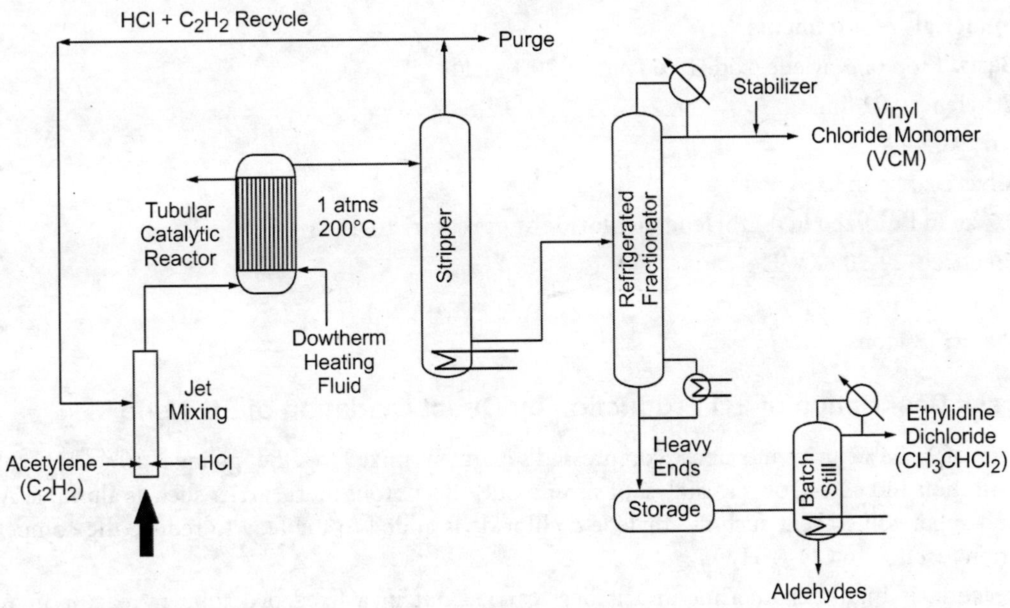

Fig. 2.21. Flowsheet of vinyl chloride production via acetylene-HCl reaction

ETHYLENE OXIDE (EO)

Properties of Ethylene Oxide (CH_2—CH_2 bridged by O)

Molecular weight: 44.05

M.P.: –111.7°C

B.P.: 10.7°C

Sp.gr. @ 0°C: 0.896

Flash point: –15°C

Ignition temperature: 430°C

Explosive limits: Lower = 3% in air

Upper = 80% in air

Maximum toxic limit: 25-100 ppm

Principal outlet in the U.S. is for production of ethylene glycol, a permanent antifreeze for automobile radiators. India has very little use for this derivative, but instead finds use in ethylene glycol polyesters, non-ionic detergents, and ethanolamines which are useful products derived from ethylene oxide.

Methods of production of Ethylene Oxide (EO)

- Direct oxidation of ethylene
- Chlorohydrination of ethylene

Direct Oxidation of Ethylene to Produce EO

Raw materials requirements

Basis: 1 ton of ethylene oxide (99% purity, 70% yield)

Ethylene: 0.92 ton

Air: 9.0 tons

Silver: 0.3 kg in fixed bed

0.7 kg in fluidized bed Ethylene dichloride suppressor: 10-15 kg

Electricity: 1,500 KWH

Steam: 0.1 ton

Water: 180 tons

Process Description of EO Production by Direct Oxidation of Ethylene

Ethylene of 95-98% purity and air are compressed separately, mixed together giving 3-10% C_2H_4 volume concentration and passed over a catalyst of silver oxide on a porous inert carrier such as alumina. A side reaction suppressing agent such as ethylene dichloride is added to the feed to reduce the competitive oxidation reaction to $CO_2 + H_2O$.

The reaction is highly exothermic and is best carried out in a fixed bed tubular reactor in which heat transfer salt or Dowtherm is pumped around the tube within the shell to maintain a 250-300°C temperature. Heat is recovered in a waste-heat steam boiler. A short residence time of 1 sec in plug flow is the design with an ultimate yield of 60-70%.

The effluent gases from the reactor are water-washed under pressure. The absorbed ethylene oxide is sent to a packed bed desorber-fractionator tower and taken overhead. It still contains a large amount of water vapor plus some impurities. This stream is compressed to 4-5 atms., and fractionated twice to remove light ends, water, and high boiling polymers.

A Flowsheet for production of EO by oxidation of ethylene is given below:

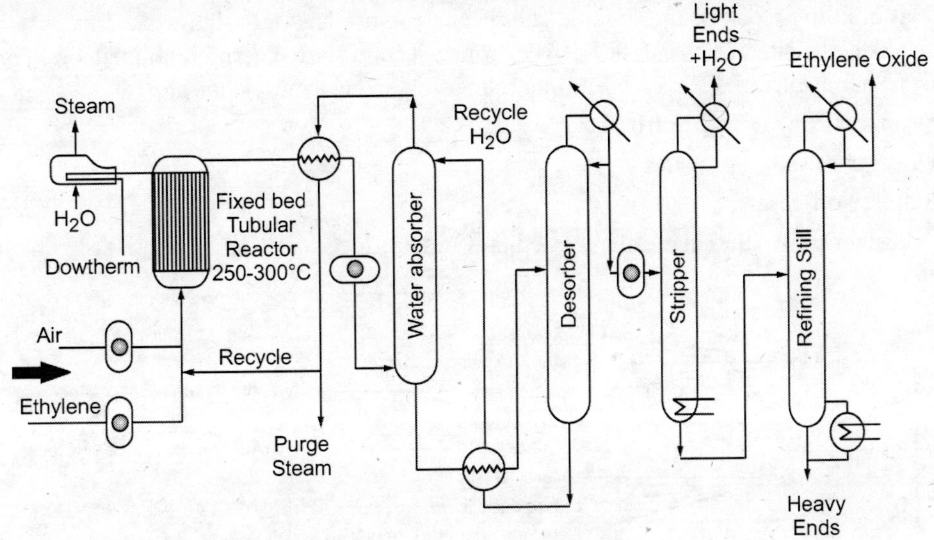

Fig. 2.22. Flowsheet of production of ethylene oxide (EO) by oxidation of ethylene

Ethylene Glycol (EG) manufacture from Ethylene oxide (EO)

The crude EO from re-absorber/purified EO product is hydrolyzed with water in a pipe reactor. The large amount of excess water is removed in a multiple effect evaporator. The water removed from the evaporator (stream 13) is used back in MEG reactor. The Ethylene glycol is then taken to a MEG Column where MEG is separated from higher glycols. The column bottoms (containing higher glycols) are further taken for separation of di-ethylene glycol and tri-ethylene glycol column residues are sold as heavier glycols/polyethylene glycols.

Flow diagram of ethylene glycol manufacture from ethylene oxide is given below in figure.

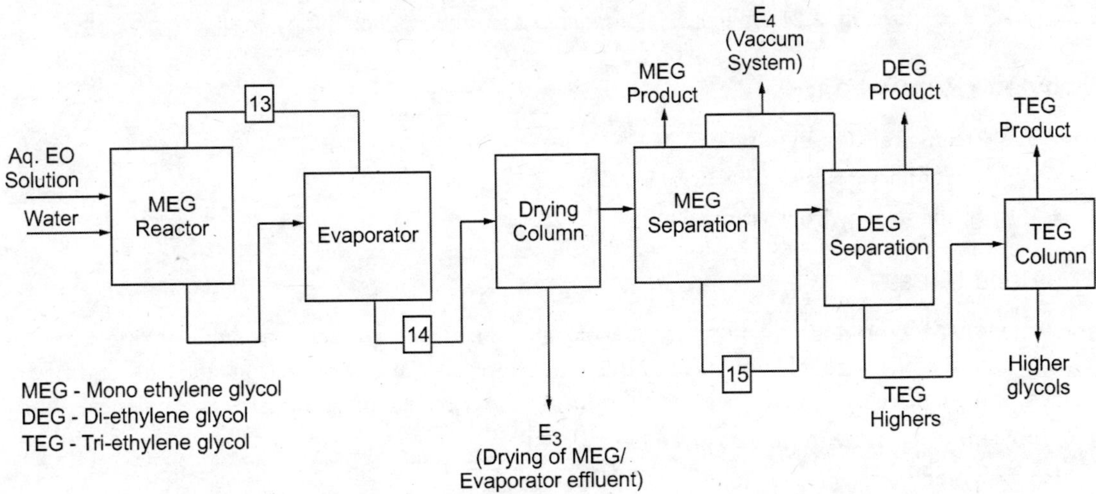

Fig. 2.23. Flow Diagram of ethylene Glycol Manufacture from EO

PROPYLENE (C_3H_6)

Propylene as building block in the organic chemical structure has only slightly less importance than ethylene. Unlike ethylene, which can be derived from a second source, fermentation alcohol, propylene must come from petroleum. Processes within a refinery which give propylene are:

- Steam cracking of hydrocarbons.
- Direct pyrolysis of propane.
- Refinery off-gases.

Following figure gives the chemical intermediates and products for various uses made from propylene.

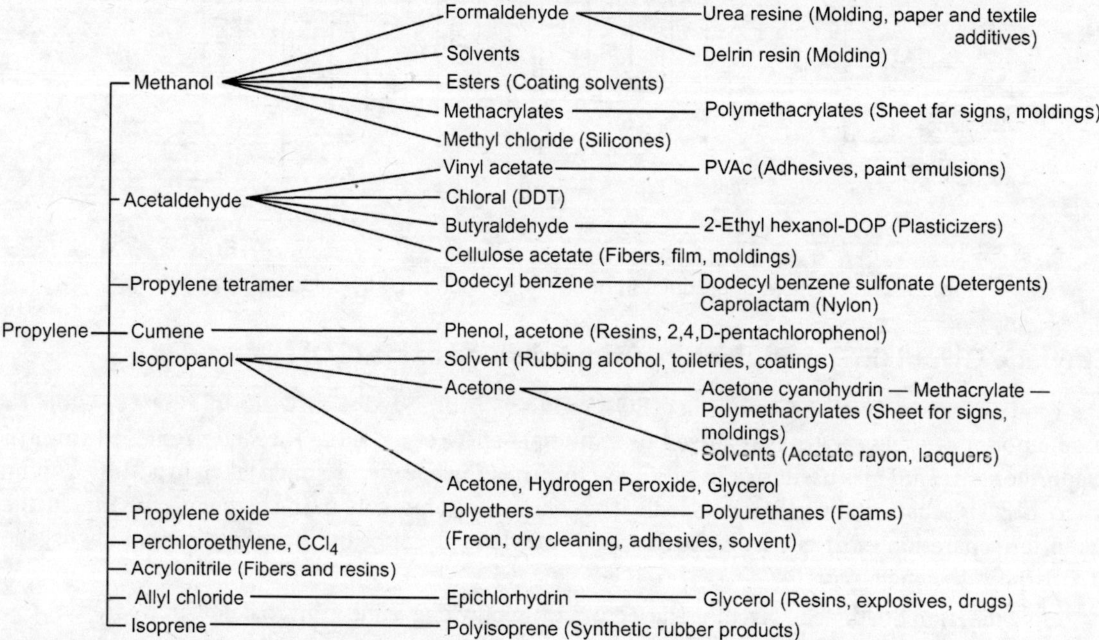

Fig. 2.24. Chemical intermediate and products from propylene

Propylene Production:

- By steam cracking/pyrolysis: 65%
- From refinery gases (FCC): 30%
- By propane dehydrogenation: 5%

Propylene Uses:

Propylene is used to produce polypropylene resin, acrylonitrile, acrylic acid, propylene oxide, isopropyl alcohol, and acetone through polymerization, oxidation, alkylation, hydration and the addition of halogen. Propylene is as important a basic chemical in the petrochemical industry as ethylene.

Propylene use picture in India is given below.

(*a*) **Polymer:**

- Polypropylene

(*b*) **Fibres/intermediates:**

- Acrylonitrile

(*c*) **Petrochemicals:**

- Phenol
- Propylene glycol
- Acrylates
- Propylene oxide
- Oxo alcohol

Propylene and Its Derivatives produced by various unit operations/processes are shown below.

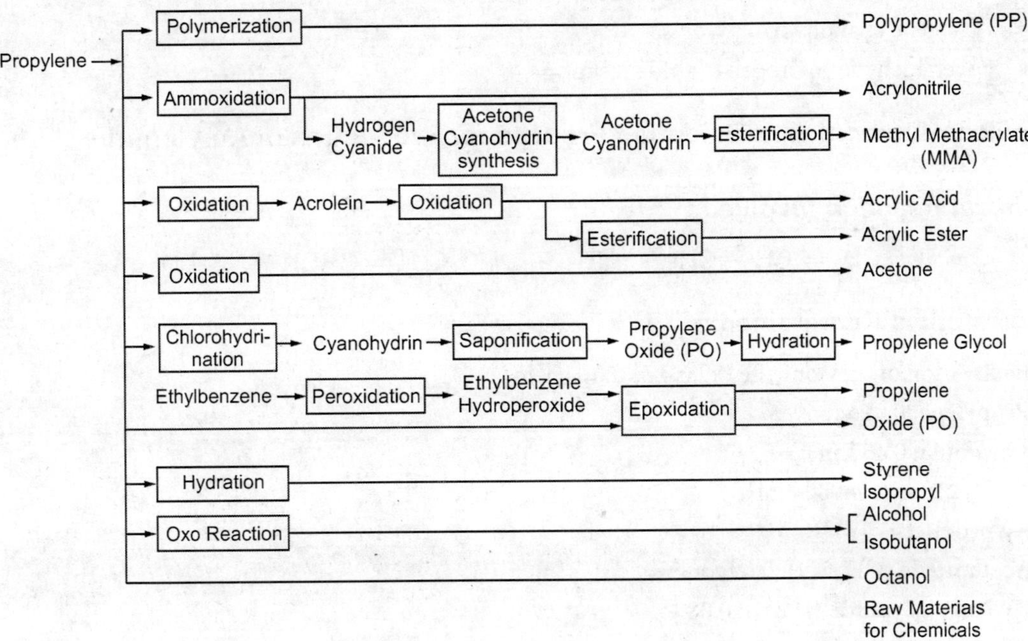

Fig. 2.25. Propylene and Its derivatives

ACRYLONITRILE ($CH_2 = CHCN$)

Properties of Acrylonitrile ($CH_2 = CHCN$)

Mol. wt.: 53.06

M.P.: –82°C

B.P.: 78.5°C

Sp. gr. @ 20°C: 0.807

Soluble in most organic solvents and water

Flash point: 0°C

Ignition temperature: 480°C

Explosive limits: Lower = 3.1% in air
Upper = 17.0% in air

Toxic limit: 15 ppm

Acrylonitrile is the cheapest of the acrylic monomer group and has good reactivity at C ≡ N bond and the double bond. Its end uses are largely for polymers.

The worldwide pattern of acrylonitrile use are for the production of acrylic fibres, ABS and SAN plastics, nitrile rubber, adiponitrile, acrylamide etc.

Methods of Production of Acrylonitrile

- Propylene-ammonia-air oxidation reaction.
- Acetylene-hydrogen cyanide reaction
- Ethylene cyanohydrin pyrolysis
- Acetaldehyde-hydrogen cyanide reaction

Propylene-Ammonia-Air Oxidation Process for Production of Acrylonitrile

Chemical Reaction Involved

$$CH_2 = CH - CH_3 \ + \ NH_3 \ + \ 3/2O_2CH_2 = CHCN \ + \ 3H_2O$$

Raw Materials Requirements

Basis: 1 ton of acrylonitrile (99.0 + %, 80% yield)

Propylene: 0.996 ton

Ammonia: 0.45 ton

Air: 5.2 tons or 4,000 Nm^3

By-products:

Acetonitrile (CH_3CN): 0.09 ton

Hydrogen cyanide (HCN): 0.08 ton

Process Description

A C_3 refinery stream is introduced along with ammonia and air via spargers into a fluidized bed catalytic reactor. Conditions within the reactor are 1.5-3 atms. pressure, a temperature of 400-500°C, and a contact time of several seconds.

The reactor effluent is scrubbed with water to remove the desired products in an aqueous solution which is further fractionated to give wet acrylonitrile and acetonitrile. Both are further purified by azeotropic and conventional distillation.

A flowsheet for production by acrylonitrile from propylene is shown below.

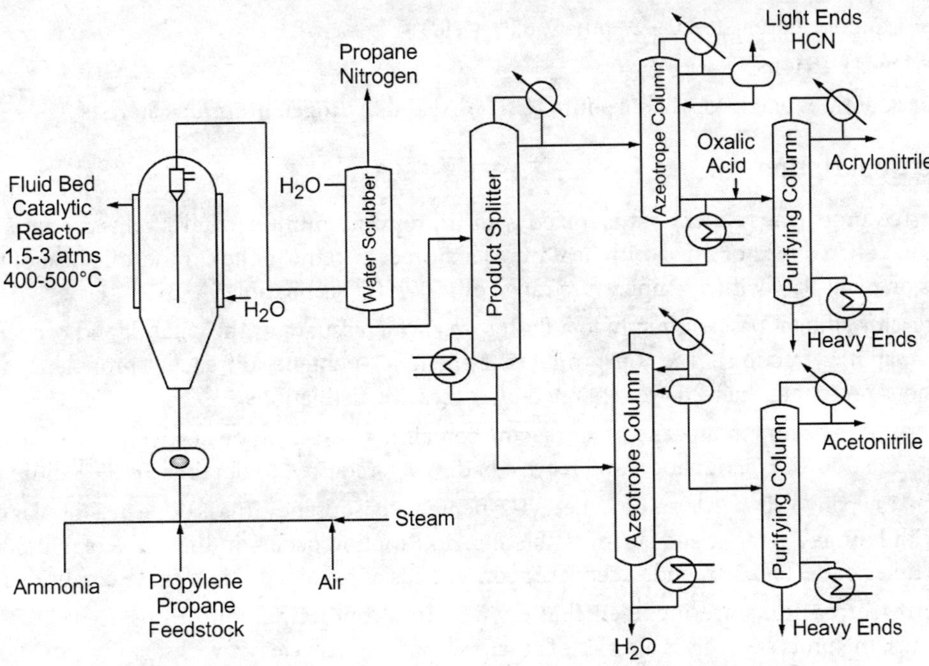

Fig. 2.26. Flowsheet of acrylonitrile production from propylene

ISOPRENE (C_5H_8)

Properties of Isoprene ($CH_2 = CH - C(CH_3) = CH_2$)

$$CH_2 = CH - \underset{\displaystyle CH_3}{\underset{|}{C}} = CH_2$$

Mol. wt.: 68.11

M.P.: –120°C

B.P.: 34°C

Sp. gr. @ 20°C: 0.681

Miscible with alcohols, ether.

Flash point: –53 °C

Ignition temperature: 220 °C

Toxic limit: 250 ppm

Polymerized isoprene (polyisoprene) closely resembles natural rubber chemically and adds flexibility to the synthetic rubber. This is the only market for the isoprene monomer.

Methods of Production of Isoprene (C_5H_8)

- Isomerization of propylene dimer
- Isobutylene-formaldehyde reaction
- Methyl ethyl ketone-formaldehyde reaction
- Acetone-acetylene reaction
- By product from isopentane/isopentene dehyodrogenation

Propylene dimer isomerisation process is the most economical process which produces high purity grade isoprene required for polymerisation.

Raw Material Requirements

Basis: 1 ton of isoprene (99.0 + % purity, 85% yield)

Propylene: 1.46 tons

Minor quantities of tripropyl aluminum $(C_3H_7)_3Al$ and hydrogen bromide catalysts.

Process Description

A compressed propylene refinery cut is mixed with tripropyl aluminum, catalyst plus carrier solvent at 200 atms. in a stirred reactor fitted with heat transfer surface to remove the exothermic heat of reaction. A conversion of 60-95% with a temperature range of 150-250°C is possible.

The reactor effluent passes directly to a flash tower which separates the catalyst and carrier solvent (saturated naphtha fraction). The top products, 2-methyl-1-pentene, unreacted propylene, and light hydrocarbons present in the feed are separated by successive distillations.

Isomerization occurs in a packed bed reactor containing an acid type catalyst. Process conditions range from 150-300°C. Fractionation and recycle of dimer is required to obtain a 98-99% ultimate yield.

The final step involves pyrolysis of 2-methyl-2-pentene to isoprene. This is done in a tubular cracking furnace with HBr as a catalyst and steam as the diluent. Reaction occurs in the range of 650-800°C and residence times of 0.3 to 0.5 second. Acetylenic compounds are avoided under these conditions.

Since three reactions are involved that have distinct operating conditions as well as catalyst requirements, the process involves a series of reactor-separator-recycle networks.

A flowsheet of isoprene production from propylene is shown below in figure.

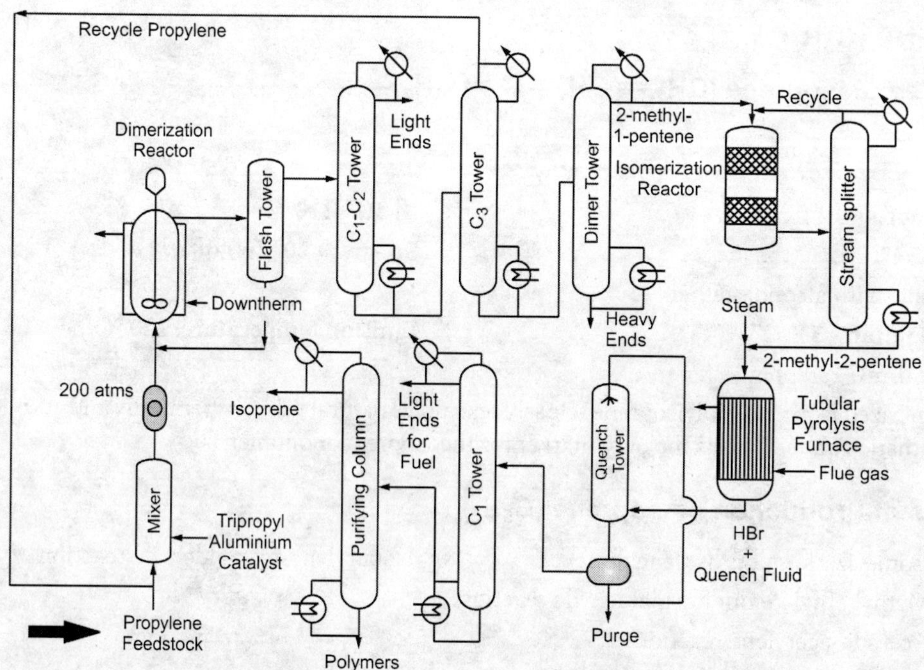

Fig. 2.27. Flowsheet of isoprene production from propylene

EPICHLOROHYDRIN AND PROPYLENE OXIDE

Epichlorohydrin

(1-chloro-2,3-epoxypropang) is another propylene derivative of industrial importance. Its manufacture involves the following stages:

- High temperature chlorination of propylene to Allyl chloride.
- Conversion of allyl chloride to dichlorohydrin with hypochlorous acid in an aqueous medium.
- Selective dehydrochlorination of dichlorohydrins to epichlorohydrin with an alkali.

Propylene Oxide

Properties of Propylene Oxide ($CH_3 — CH — CH_2$)

(with O bridging CH and CH_2)

Mol. wt.: 58.08

M.P.: –104.4°C

B.P.: 33.9°C

Sp. gr. @ 20°C: 0.831

Soluble in water, alcohols, ether

Flashpoint: –34°C

Explosive limits: Lower = 2.1% in air
Upper = 22% in air

Toxic limit: 250 ppm

Propylene oxide is an important chemical to study because of its similarities and differences as compared with ethylene oxide. For example, the toxic character of ethylene glycol, the hydrate of ethylene oxide, has created markets for propylene glycol in cellophane, food, and drug industries. The largest use for the glycol is in polyester resins.

Use of Propylene Oxide include the Following:

(*a*) Polypropylene glycol and propoxylated products for polyurethanes

(*b*) Propylene glycol

(*c*) Miscellaneous: glycol ethers, glycerin, isopropanol amine, etc.

Methods of Production of Propylene Oxide

Classification of processes:

- Chlorohydrin route
- Direct oxidation

CHLORHYDRIN PROCESS OF PROPYLENE OXIDE MANUFACTURE

Chemical Reactions

(*a*) **Chlorhydrination:**

$$CH_3CH{=}CH_2 + HOCl \longrightarrow \underset{\text{Chlorohydrin}}{CH_3{\cdot}CHCl{\cdot}CH_2OH}$$

(*b*) **Dehydrochlorination:**

$$CH_3{\cdot}CHCl{\cdot}OH + Ca(OH)_2 \longrightarrow 2CH_3CH{-}CH_2 \text{ (with O bridging, Propylene oxide)} + CaCl_2 + 2H_2O$$

Propylene oxide

Raw Material Requirements

Basis: 1 ton of propylene oxide (99 + % purity, 80% yield)

Propylene: 0.91 ton

Chlorine: 2.0 tons

Lime (100% CaO): 1.6 tons

Electricity: 150 KWH

Steam: 12 tons

Water: 240 tons

By-products: 0.97 ton of calcium chloride

0.10-0.15 ton of propylene dichloride

Process Description

Propylene, chlorine and water are introduced into the bottom of a packed tower where reaction forms the chlorhydrin. The reaction mechanism is formation of hypochlorous acid (HOCl or HClO) which reacts rapidly with propylene. The chlorine/propylene ratio of gaseous feed to the tower is chosen so that the liquid effluent contains 4-5% chlorhydrin. The reaction is exothermic and maximum tower temperature is held at 50°C. A conversion of 25-40% is attained, the ultimate yield of chlorhydrin is 87-90% with most of the by-product as propylene dichloride. Unreacted propylene is scrubbed with caustic to remove HCl, which also forms, and then recycled.

Manufacture of Propylene Oxide by Direct Oxidation of Propylene

Processes for the direct manufacture of propylene oxide by the single step oxidation of propylene have been the subject of intensive commercial development.

One possible route which is now being commercialized involves peroxidizing an organic compound, reacting the peroxide with propylene to give propylene oxide and then recovering a useful organic product from the peroxide.

The first commercial plant for direct manufacture of propylene oxide from propylene was built at Bayport, Texas.

A flowsheet of propylene oxide manufacture by chlorhydrin route is given below:

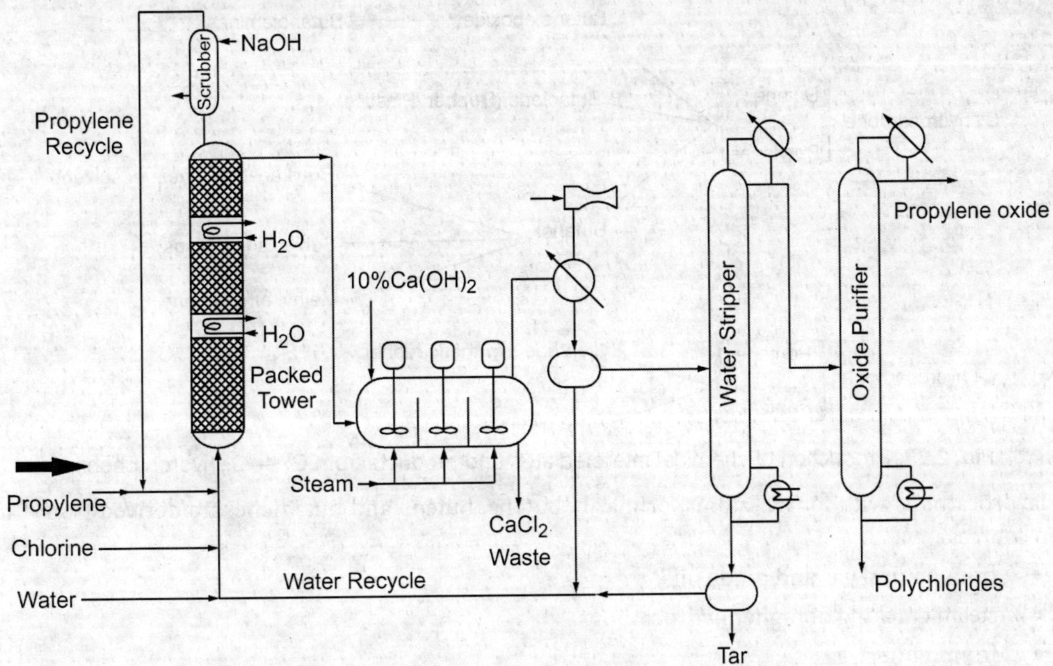

Fig. 2.28. Manufacture of Propylene oxide by chlorhydrin route

BUTADIENE (C_4H_6)

Properties of Butadiene ($CH_2 = CH \cdot CH = CH_2$)

Mol. wt.: 54.09

M.P.: –108.9°C

B.P.: –4.41 °C

Sp. gr. @ 20°C: 0.621

Soluble in alcohols, ether; insoluble in water

Explosive limits: Lower = 2.0% in air

Upper = 11.5% in air

Toxic limit: 5,000 ppm

The major use for butadiene is in the manufacture of synthetic rubber (styrene butadiene rubber)

Methods of Production

- Dehydrogenation of butane
- Dehydrogenation of butylene
- Dehydragenation–dehydration of ethanol
- Steam cracking of hydrocarbons

Chemical intermediates and products derived from C_4 and C_5 hydrocarbons are shown below:

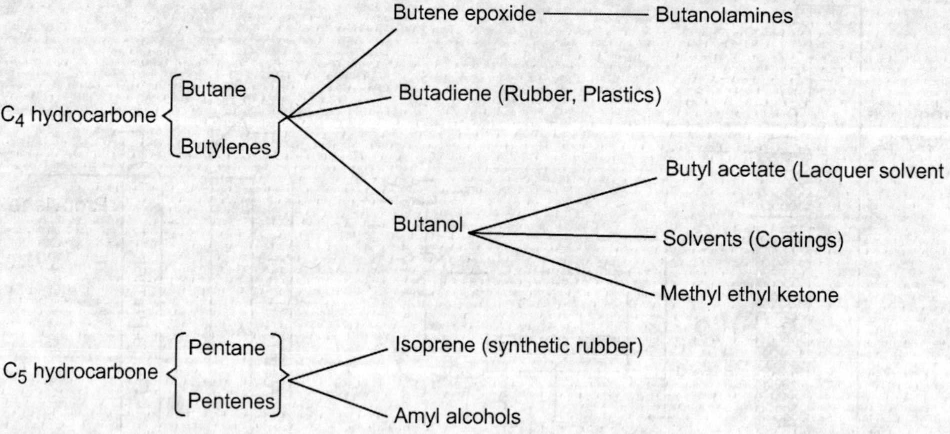

Fig. 2.29. Production of chemical intermediates and products from C_4 — C_5 hydrocarbons.

C_4–Hydrocarbons with four C-atoms, primarily butane, butene and butadiene, are derived from three main sources:

- from natural gas and crude oil.
- steam cracking of higher hydrocarbons.
- from refinery gases.

Schematic representation of production of C_4 hydrocarbons is shown below.

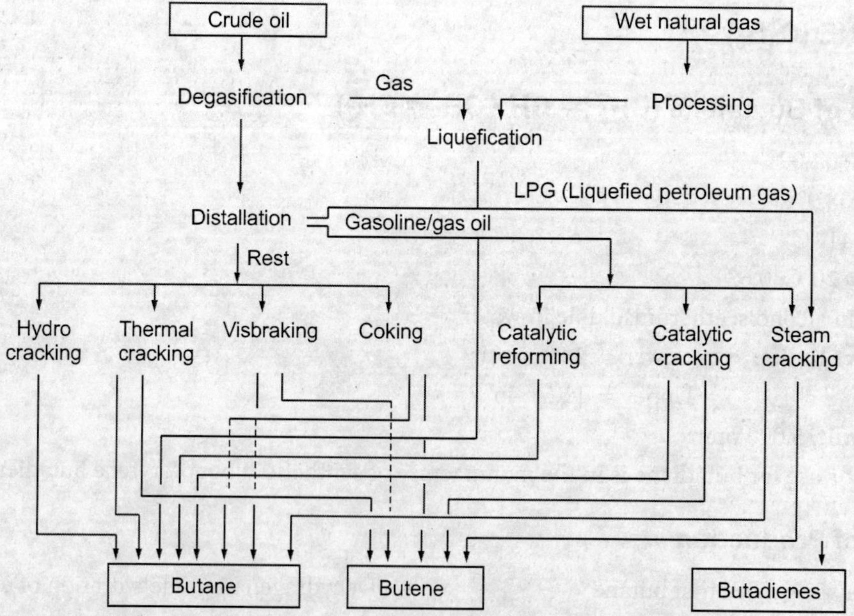

Fig. 2.30. Schematic representation of production of C_4 hydrocarbons

C_4/C_5 fractions and its derivatives are shown below:

Using naphtha as a raw material, an ethylene plant produces highly reactive materials in C_4 fractions or C_5 fractions as by-products.

Butadiene in the C_4 fractions and isoprene in the C_5 fractions in particular, are useful chemicals as they are used to produce synthetic rubber tires for cars.

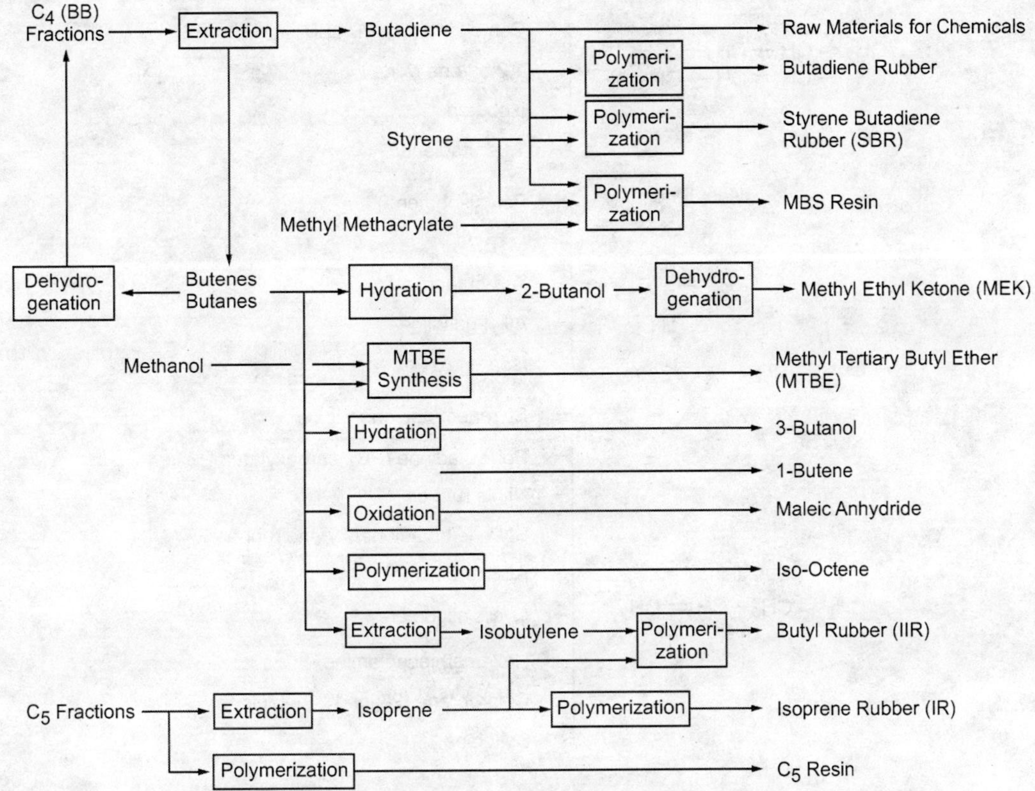

Fig. 2.31. C_4 and C_5 fractions and its derivatives

The most important products based on C_4-hydrocarbons are shown below.

Butadiene is mostly converted into Styrene Butadiene or Styrene Butadiene Rubber (SBR) more well known as synthetic rubber.

Polybutadiene is also used in tyres and can be used as an intermediate in the production of acrylonitrile-butadiene styrene (ABS).

IMPORTANT PRODUCTS BASED ON C_4 HYDROCARBONS

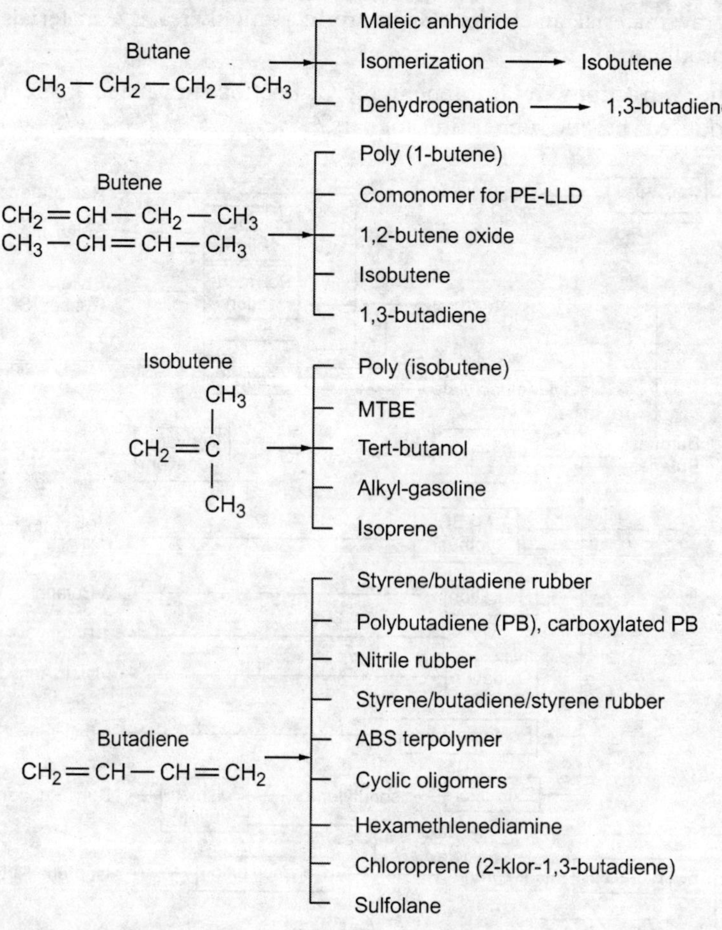

Butadiene Production by Dehydrogenation of Butane

Chemical reactions:

(*a*) **Main reaction:**

$$C_4H_{10} \longrightarrow \underset{\text{Butadiene}}{CH_2 = CH - CH = CH_2} + 2H_2$$

Side reaction:

$$C_4H_{10} \longrightarrow \underset{n\text{-butylenes}}{CH_4 + CH_2}$$

Raw Material Requirements

Basis: 1 ton of butadiene (98 + % and 60% yield)

Butane: 1.80 tons

By-products: 0.65 ton

Process Description

A refinery gas of C_4/C_5 cut containing predominantly n-butane with some isopentane is mixed with recycle gas and preheated to reaction temperature prior to contact with a catalyst in a fixed bed, regenerative-heating reactor system. A pair of reactors form an adiabatic cycle with the heat of reaction required during the 5-15 minutes "make" period equal to that supplied by the combustion of carbon deposit on the catalyst during the regenerative period. The temperature of reaction at the start of the "make" period is 650°C, dropping to 550°C at the end before switching to regeneration. The pressure is low, 120-150 mm Hg absolute, to force the reaction to the right.

The product gases are oil-quenched, compressed, cooled and separated from the light ends by absorption in naphtha followed by stripping. The overhead is fractionated to yield crude butadiene at the top which is purified by (1) absorption using cuprous ammonium acetate (CAA), (2) extractive distillation with furfural, or (3) azeotropic distillation with ammonia.

A flowsheet of butadiene manufacture by dehydrogenation of butane is given below:

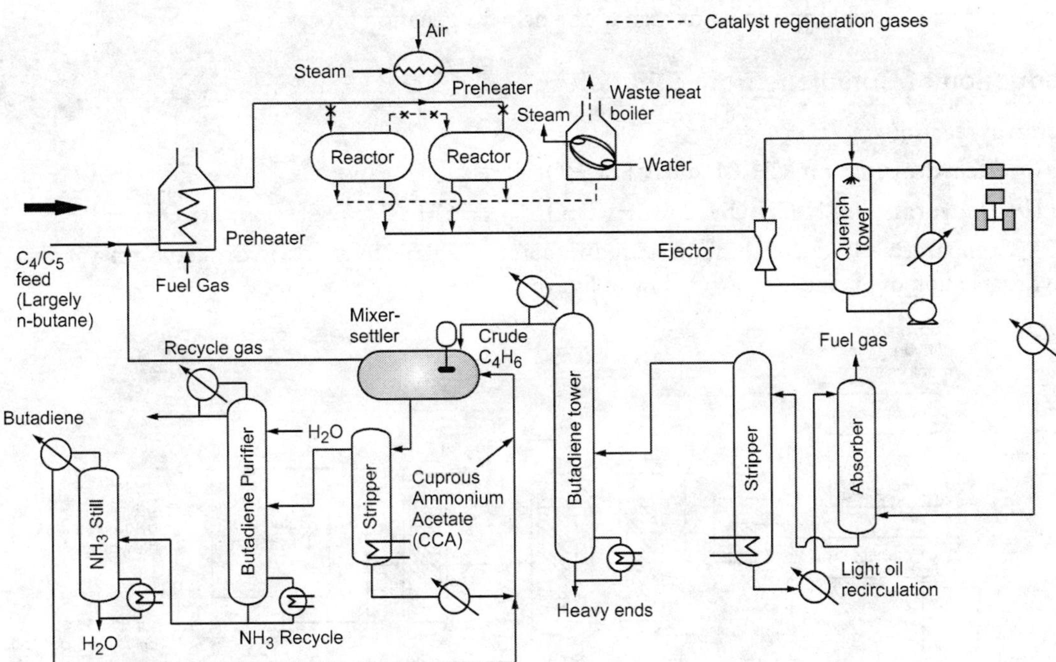

Fig. 2.32. Flowsheet of Butadiene manufacture by dehydrogenation of butane

The more common absorption process involves contact of the close-boiling butadiene- butene fraction with lean CAA solution which dissolves butadiene. A desorption step at higher temperature is followed by distillation, compression, and liquefaction of butadiene to give 98-99% product purity. Ammonia is recovered in the distillation tower by water addition; it is then separated as substantially anhydrous NH_3 by fractionation.

Production of Butadiene by Oxydehydrogenation of Butane (C_4H_{10})

Since the equilibrium in the reaction

$$C_4H_{10} \rightleftharpoons C_4H_6 + 2H_2$$

is reversible, it is obvious to add enough oxygen to the reacting mixture to pull the equilibrium far to the right and increase conversions from 20% or so (requiring much expensive recycle) to nearly 100%.

Chemical reaction:

$$C_4H_{10} + O_2 \longrightarrow C_4H_6 + 2H_2O$$

Process Description

Butanes or butenes are fed with a carefully controlled amount of oxygen, together with a small amount of a halogen (bromine or iodine) promoter into a reactor containing a heterogeneous catalyst. The higher yields simplify the subsequent product separation step greatly. Decrease in plant size for equivalent capacity is made possible by adding oxygen to the dehydrogenation process.

Production of Butadiene from Ethanol

Chemical reactions:

(*a*) **Dehydrogenation:** $CH_3CH_2OHCH_3CHO + H_2$

(*b*) **Dehydration:** $CH_3CH_2OH + CH_3CHOCH_2{=}CH{\bullet}CH{=}CH_2 + 2H_2O$

A Comparative layout of butadiene manufacturing plant by oxydehydrogenation vs. straight dehydrogenation of butane is shown below in figure.

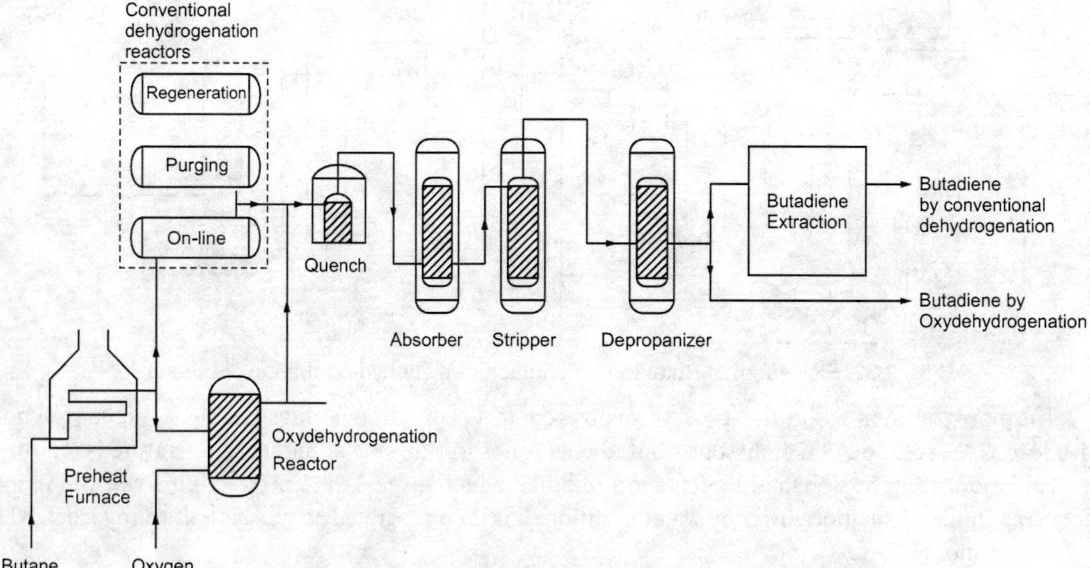

Fig. 2.33. Comparative plant layout for making butadiene from butane by oxydehydrogenation versus straight dehydrogenation

Raw Material Requirements

Basis: 1 ton of butadiene (98+ % purity and 70% yield)

Ethanol: 2.88 tons

By-products: Ethylene, butene, ethyl ether, ethyl acetate, butanol

Process Description

Ethanol (92-95 % grade) is vaporized and passed over a dehydrogenation catalyst at 260- 290°C to produce acetaldehyde: 90-94% yield on a 30-50% per pass basis.

Excess alcohol is mixed with acetaldehyde (3 : 1) and fed as a liquid to a fixed bed converter containing tantala-silica catalyst. Pressure is essentially atmospheric; the temperature is 325- 350°C. A conversion of 28-30 per cent based on acetaldehyde is obtained with an ultimate yield in this step of 75%.

The reaction gases contain numerous compounds in addition to unreacted ethanol- acetaldehyde. These are separated by a series of fractionating columns with the crude butadiene ultimately being purified.

Oxidation of ethanol to acetaldehyde using air is also a fully commercialized process. It has the disadvantage that H_2 cannot be recovered, but this is offset by not having to supply external heat for this exothermic reaction.

A one-step route is being sought to give butadiene directly from ethanol. So far, the correct dehydrating catalyst has not been found.

A flowsheet of butadiene production by dehydrogenation of ethanol (ethyl alcohol) is given below:

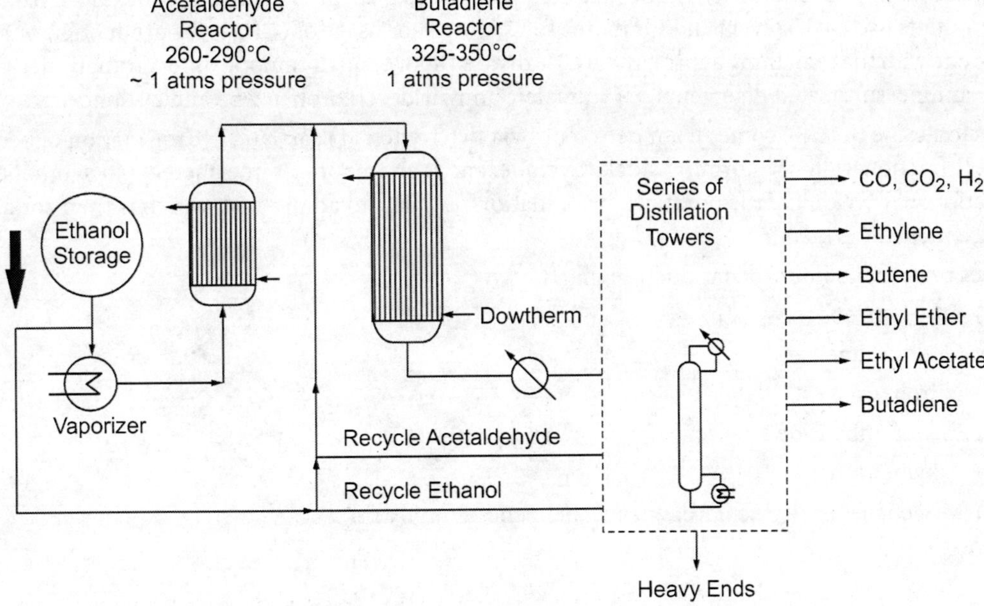

Fig. 2.34. Production of butadiene from ethanol dehydrogenation.

The future of the butadiene business depends on synthetic rubber production. The SBR plant at Bareilly, started in 1963, is based on butadiene from alcohol, but no further alcohol- based plants are contemplated. This plant will be uneconomic as soon as ethylene, butadiene, and benzene from petroleum raw materials become available at a central petrochemical supply base located closer to end user customers.

Utilisation of butadiene is mainly for SBR, polybutadiene rubber (BR), adiponitrile, chloroprene, ABS polymers, nitrile rubber (NBR) etc.

AROMATIC COMPOUND: BENZENE, TOLUENE AND XYLENE (BTX)

Important aromatics for chemical synthesis are Benzene (B) Toluene (T) and xylene (X) (Called BTX) and naphthalene (N).

Coking Coal carbonization is used for metallurgical coke production. The volatiles (coke oven gas) are stripped of tar by electrostatic precipitators, scrubbed with dilute H_2SO_4 to remove ammonia. The remaining crude benzol fraction is absorbed in a stable oil such as wash oil. This benzol has a proximate composition of 70% B, 15% T, 7% X, 4% N, solvent naphtha 4% and is fractionated to separate these components in benzol rectification plant.

The production of aromatic raw materials from petroleum is derived from (1) solvent extraction of distillate fractions, (2) the reforming process by fractionating the reformate, (3) naphtha cracking and (4) hydrodealkylation.

Benzene (C_6H_6)

Benzene ranks ahead of toluene, xylene, and naphthalene as seen in the above classification. Except for small amounts used as a solvent and for motor fuels, the major usage of C_6H_6 is for production of phenol and styrene with the remaining approximate 40% divided about equally among other aromatic derivatives such as aniline, sulfonated detergents, DDT, maleic anhydride, chlorobenzene and cyclohexane.

Cyclohexane (C_6H_{12}) comes from petroleum via two routes: (1) catalytic hydrogenation of benzene, (2) direct fractionation of reforming stock. Cyclohexane is an important intermediate in synthetic fiber preparation. 6,6-Nylon fiber is produced by oxidation of C_6H_{12} to adipic acid, which is then condensed with hexamethylene diamine.

Uses of benzene is for the production of:

(*a*) Ethylbenzene (styrene)

(*b*) Cyclohexane

(*c*) Nitrobenzene

(*d*) Maleic anhydride

(*e*) Alkylbenzenes

(*f*) Miscellaneous (*e.g.*, chlorobenzene, benzenesulfonic acid)

End uses of various aromatic conversion products are given below in figure:

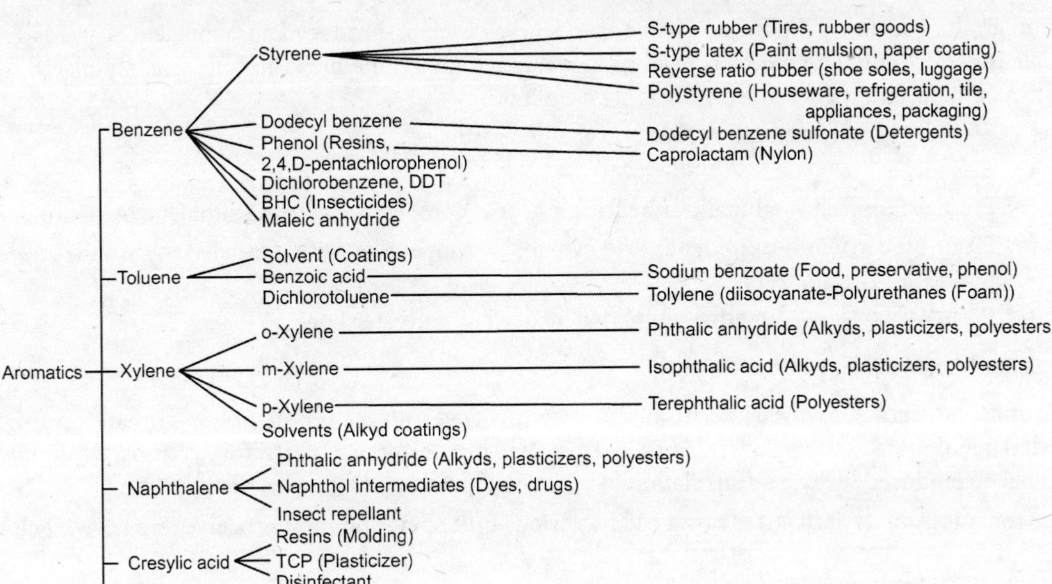

Fig. 2.35. End uses of chemicals and intermediates produced from aromatics(BTX).

Toluene ($C_6H_5 \cdot CH_3$)

Toluene is produced in yields about equal to that of benzene and of xylene in petroleum reforming operations. The demand for toluene is not as much as for benzene.

The major end uses of toluene are:

- Solvents—paints, rubber and plastic cements.
- Chemicals—benzyl chloride and its peroxide for polymerization catalysts; benzoic acid; as plasticizer; phenol from benzoic acid; toluene disocyanate in urethane foams.
- Detergents— toluene sulfonates.

These uses cannot absorb all of the toluene production; so other uses plans for using up the excess toluene are:

- Blending in motor petrol to upgrade octane number; this produces low income yield as fuel value.
- Production of phenol
- Dealkylation with H_2 to yield benzene.

The principal source of toluene is catalytic reforming of refinery streams. This source accounts for 87% of the total toluene produced. An additional 9% is separated from pyrolysis of gasoline produced in steam crackers during the manufacture of ethylene and propylene. Other sources are an additional 2% recovered as a by-product from styrene manufacture and 2% entering the market from separation from coal tar.

Xylene [$C_6H_4 \cdot (CH_3)_2$]

A similar story to that of toluene exists for xylene as each is produced in about equal yields from petroleum reforming. Some 60% is used as low-value, high-octane blending fuel (mainly in U.S.) the balance of important high-value products are as follows:

- Solvents—high boiling type, largely for alkyd resins.
- Chemicals

(*a*) From orthoxylene—phthalic anhydride and acid competitive with naphthalene oxidation.

(*b*) From metaxylene—isophthalic acid which is competitive with phthalic acid for reinforced plastics and plasticizers.

(*c*) From paraxylene—dimethyl terephthalate used in polyester fibre

Naphthalene ($C_{10}H_8$)

Bulk of naphthalene supply derived from coke oven gas goes into the production of phthalic anhydride which is mainly used in the production of alkyd resin and plasticizers and partly for dyestuffs. Naphthalene can also be produced by hydrodealkylation of petroleum products (e.g., alkyl naphthalene).

Aromatics and its derivatives produced by various unit operations and processes are shown below:

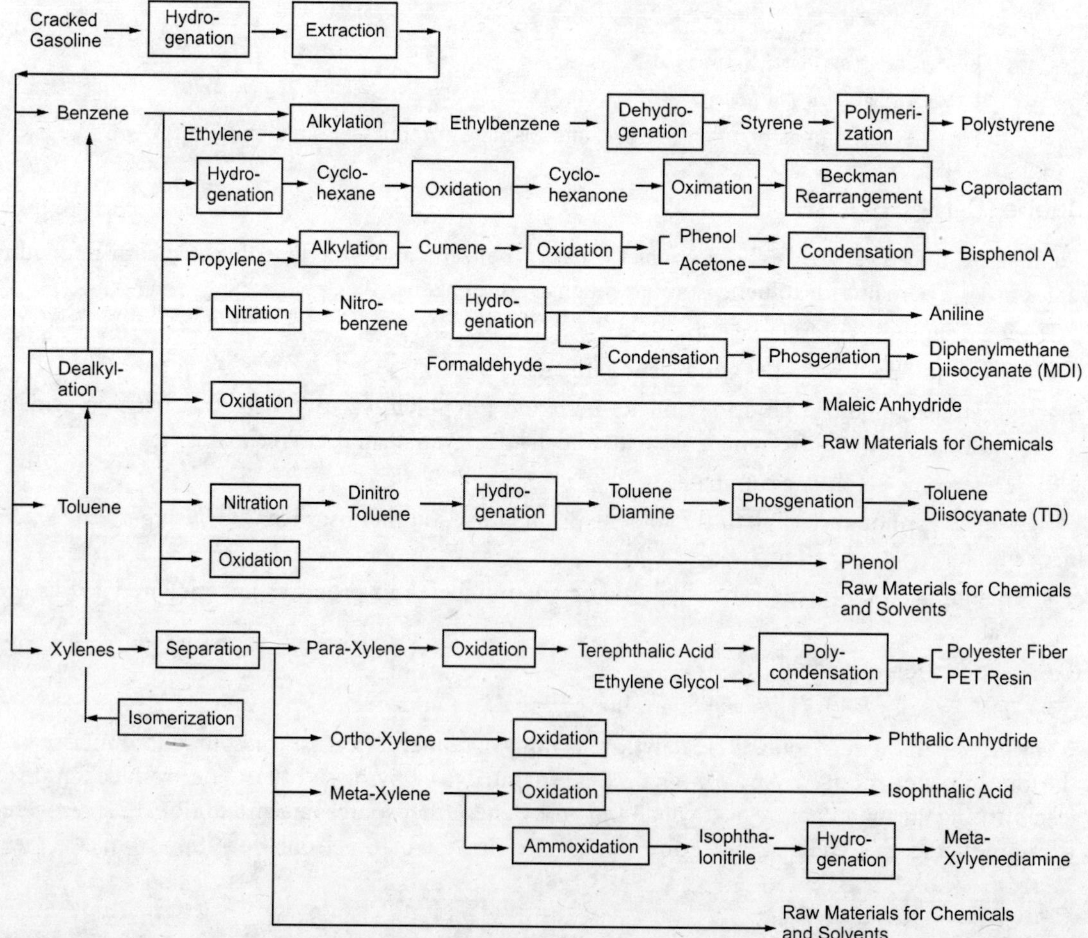

Fig. 2.36. Aromatics and Its Derivatives

Using naphtha as a raw material, an ethylene plant produces aromatic derivatives (BTX fractions) such as benzene (B), toluene (T), and xylenes (X) as cracked gasolines.

Since cracked gasolines produced from ethylene plants only are not enough to meet the demand for these products, supply from reformulated gasolines in the oil refining industry is increasing.

Of the BTX fractions, benzene and xylenes are particularly used in large quantities in general-purpose resins and fibers after being processed with polystyrene, caprolactam, and terephthalic acid.

Important Benzene, toluene and Xylene (BTX) Derivatives are listed below:

Benzene	Toluene	Xylene
EB-Styrene	Toluene diisocynate	pX-PTA-Polyester
Cumene-Phenol-phenolic resins	Motor Fuel-Octane Enhancement	oX-Phathalic anhydride
LAB	Solvent	mX-isophthalic acid
Cyclohexane	Conversion to Bz-pX	
Cyclohexanol		oX/mX Conversion to pX
Caprolactam-Nylon		
Adipic acid-Nylon		

BENZENE AND TOLUENE

Benzene and Toluene are Produced By Variety of Processes viz

- By extraction/fractionation from Pyrolysis Gasoline(PyGas).
- Reforming of Naphtha (in Petroleum Refineries).
- Distillation of coke oven light oil (in coke oven plants).

Benzene can also be produced by toluene de-alkylation and trans alkylation/ disproportionation.

Manufacture of Benzene and Toluene from Pyrolysis Gasoline

The pyrolysis gasoline (pygas) is a stream produced by the high temperature, short residence time cracking of naphtha in petrochemical plants. According to cracking severity, the pygas contains typically 20-40% benzene, 16-20% toluene, 10-13% higher aromatics (e.g. ethyl benzene, xylenes, styrenes etc.) with a balance of non-aromatic hydrocarbons (mainly olefins and diolefms).

The benzene extraction from pyrolysis gasoline is carried out in the following steps:

- Hydro-treating
- Secondary Hydrogenation
- Solvent Extraction
- Fractionation

Simplified flow diagram is presented in figure below and each of the process steps are described below:

(*a*) **Hydro-treating:** Hydro-treating is the process by which the quality of pyrolysis gasoline stream is improved by subjecting them to mild or severe conditions of hydrogen pressure in the presence of a catalyst. The process flow diagram is depicted in figure below.

The pyrolysis Gasoline feed is heated in a furnace and combined with recycled hydrogen gas. The combined feed is passed through a reactor containing a catalyst bed where the hydrogenation reaction takes place. Upon leaving the reactor, the stream is cooled and moved to a separator vessel where recycle or net hydrogen is removed. The liquid then moves to a stabilizer or stripper, which removes, hydrogen, hydrogen sulfide, ammonia, water and light hydrocarbons, dissolved in the separator liquid.

(*b*) **Secondary Hydrogenation:** The pyrolysis gasoline normally undergoes second stage hydrogenation (generally carried out at high temperature of 300°C) in which olefins are saturated, organic sulfur forms hydrogen sulfide, combined nitrogen is converted to ammonia, and oxygenated compounds are reduced to hydrocarbons and water. After these parallel reactions have been completed, the gases and liquid are separated. The liquid is then stripped of gaseous impurities, such as hydrogen sulfide and remaining light hydrocarbons.

(*c*) **Separation of Benzene and Toluene from Hydrogenated Pyrolysis Gasoline (pygas):** The hydro-treated pyrolysis gasoline is rich in aromatics such as benzene (typically about 40%) toluene (typically 20%) and has small amount of xylene as well as non-aromatics of similar boiling points. Therefore, separation of these aromatics from other process streams is achieved by extractive distillation and liquid-liquid extraction. Liquid-liquid extraction is the most commonly used extraction method. If only one aromatic (e.g., benzene, toluene or xylene) is to be recovered in pure form, extractive distillation is preferred as the capital and operating expenses are less than those associated with liquid-liquid extraction.

Liquid-Liquid Extraction

The solvents commonly used in liquid-liquid extraction are sulfolane and tetra ethylene glycol. It is adopted by three units in India (Reliance-Hazira, IPCL-Vadodara and NOCIL-TTC). Processes using other solvents such as tetra ethylene glycol, di-ethylene glycol, dimethyl sulfoxide (DMSO) are not adopted in India. The hydrogenated pyrolysis gasoline stream is fed in the extractor. Where it is contacted with the solvent in the extractor, the non-aromatics (the raffinate phases) leaves through the top while a mixture of aromatics, solvent and light non-aromatics leaves through the bottom. The raffinate is taken to a raffinate wash column where it is washed with water. Raffinate obtained is sent to storage, the bottom is sent to a water stripper. The aromatics/solvent and light non-aromatic stream is transferred from the extractor directly to a stripper. The stripper removes solvent, leaving behind the desired aromatic, which is then sent to a solvent recovery column for recovering the solvent if any. The aromatics phase is passed through a clay tower (to remove olefins) and then taken for fractionation to produce pure benzene and toluene.

Extractive Distillation

Extractive distillation methods are preferred when only one aromatic (e.g., benzene, toluene, or xylene) is to be recovered in pure form. Extractive distillation methods separate components by introducing a solvent to a mixture. The solvent suppresses the relative vapour pressure of some components compared to other components in the mixture allowing the desired component to be extracted. Hydro-treated PGH (pyro gasoline hydrocarbons) is processed to remove lighters (less than C_5) and heavier (greater than *o*-xylene) hydrocarbons. The crude BTX stream and the extractive solvent are fed to an extractive distillation column where non-aromatics (the raffinate) are separated. The aromatic/solvent stream is transferred to a stripper, which removes pure Benzene at the top leaving behind solvent, which is reused. Some of the solvents commonly, used in extractive distillation units are di-methyl-formamide, *n*-formyl-morpholine, *n*-methyl-pyrrolidone and sulfolane. HPL, Haldia uses *n*-methyl pyrrolidone to extract benzene from pyrolysis gasoline. The non-aromatics (raffinate) maybe further distilled in a small column to recover any solvent. Thus, the extractive distillation is simpler in design comprising primarily of the two columns (one for extractive distillation/one for stripping). The pressure in the coloums is close to atmospheric or vacuum and the working temperature is 175°C.

Flowsheet of extractive distillation process is given below in figure.

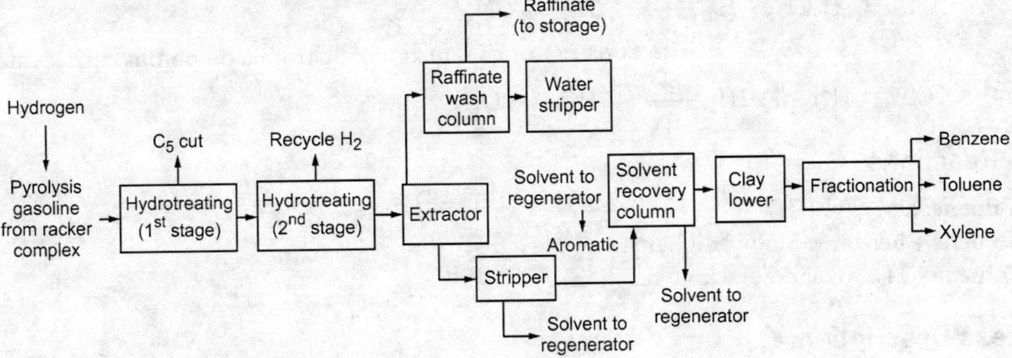

Fig. 2.37. Benzene/toluene liquid extraction from pyrolysis gasoline

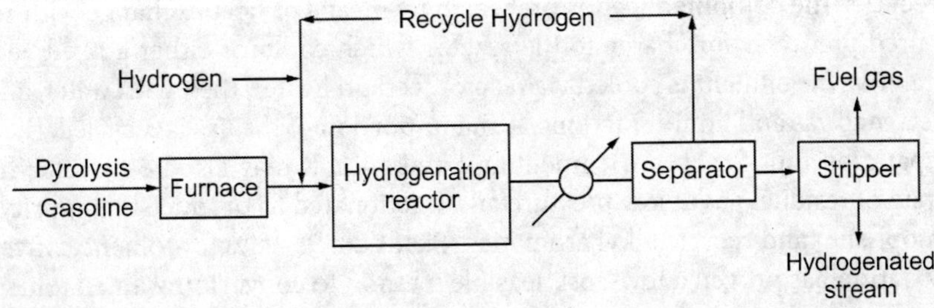

Fig. 2.38. Process flowsheet for hydro-treating to improve the quality of pyrolysis gasoline

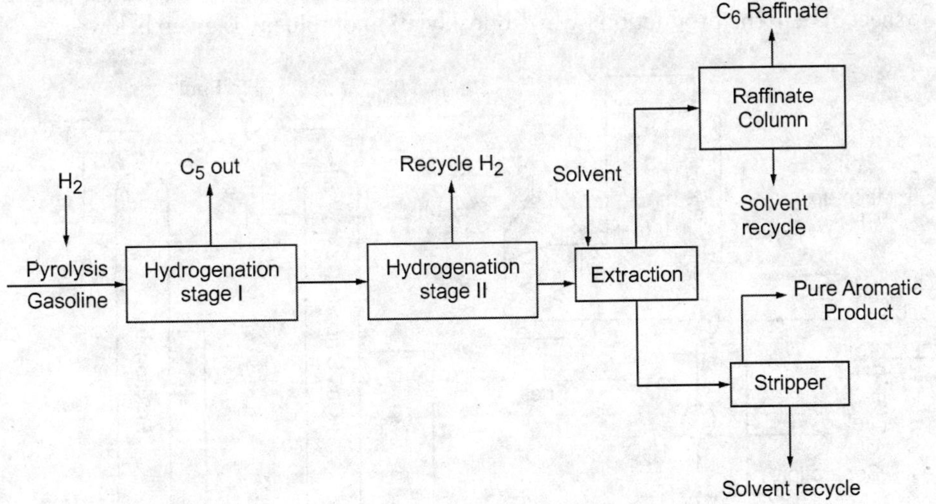

Fig. 2.39. Benzene/toluene extraction from pygas using extractive distillation

PRODUCTION OF BENZENE BY HYDROEALKYLATION OF TOLUENE

Hydrodealkylation is a method of converting toluene and dialkylbenzenes to benzene. Toluene is reacted with H_2 to produce benzene and methane as toluene does not have much market value.

Main Chemical Reactions Involved

$$C_6H_5CH_3 + H_2 \longrightarrow C_6H_6 + CH_4$$

(600-650°C and 35-40 kg/cm^2, chromia on porous carrier catalyst)

$$C_6H_4(CH_3)_2 + 2H_2 \longrightarrow C_6H_6 + 2CH_4$$

Side Reactions

Toluene: diphenyl + H_2
Toluene + benzene: Methyl diphenyl
Toluene + H_2: Alkanes (cracking reaction)

Process Description

Hydrogen-rich make-up gas and recycle gas are compressed and mixed with the liquid alkyl aromatic feedstock plus recycle. The combined feed is preheated by means of heat exchange with reactor effluent and finally in a fired preheater for charge to the reactor which contains either a fixed bed catalyst or is tubular, non-catalytic. The effluent is cooled via reactor feed preheater, then with water, and separated in a high pressure stage to gas and liquid fractions. A major portion of the gas is recycled, the balance going to a methane separation unit for H_2 enrichment and make-up. It may also be used for fuel. The liquid fraction is stripped of residual gas at low pressure and fractionated to produce high purity benzene (less than 0.5 ppm thiophene) and recycle alkyl aromatics. Yields are 95-98% for toluene conversion.

Hydrodealkylation has proven a low-cost, feasible method to convert unwanted toluene to benzene, thereby avoiding downgrading of toluene to fuel values. The process is readily handled within the refinery operations so that high grade benzene is available to consumer plants in the immediate vicinity.

A flowsheet of benzene production by hydrodealkylation of toluene is given below:

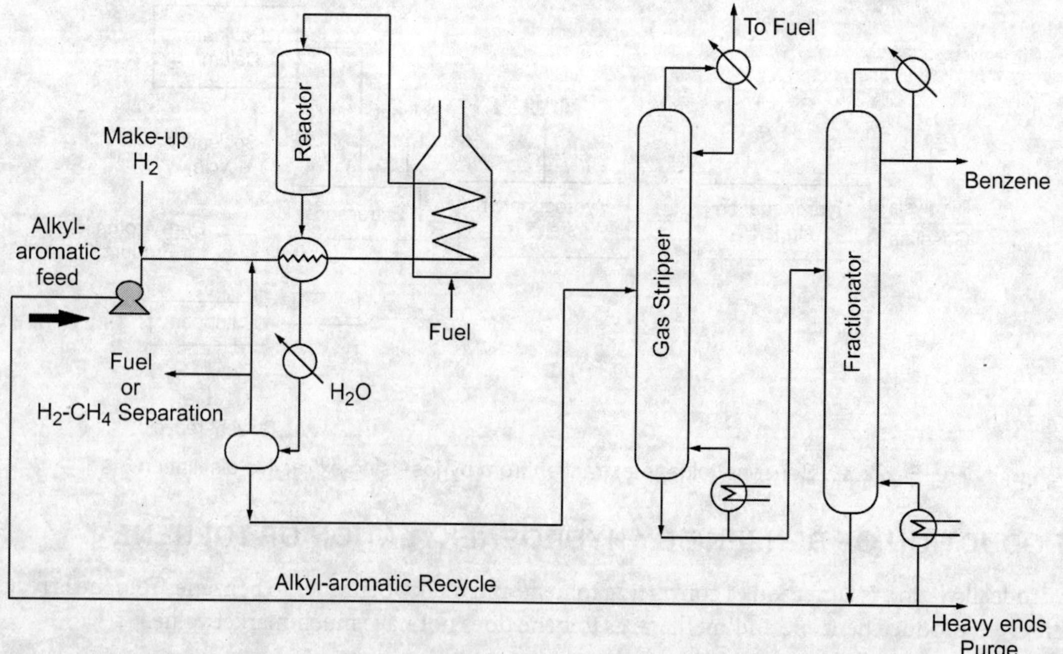

Fig. 2.40. Flowsheet of benzene production by hydrodealkylation of toluene

Note: For better heat economy in thermal process, reactor effluent can be used for stripper and fractionator reboiler heat supply.

XYLENE

Xylenes are produced in petrochemical complexes by the following processes:

- Reforming of Naphtha
- By extraction/fractionation from pyrolysis gasoline (pygas)
- Toluene disproportionation/toluene trans alkylation process:

The xylene produced are separated in individual components by separation processes.

Also, isomerisation processes are used to increase the yield of particular xylene isomer.

Xylene Manufacture by Naphtha Reforming

Two types of processes are observed.

- Optimized to produce both ortho-xylene and para-Xylene (IPCL, Vadodara and RIL, Jamnagar)
- Optimised to produce para-Xylene only (RIL-Patalganga)

The process adopted for production of both ortho and para-Xylene is described below:

Feed for the processes are petroleum hydrocarbons with boiling range of 105-170°C (straight run light naphtha) or hydro cracked naphtha having similar boiling range.

This is carried out in the following reaction steps:

- Hydro-treating of naphtha to improve its quality
- Reforming of naphtha to produce BTX rich stream
- Separation of mixed Xylene from aromatic streams
- Separation and purification of desired xylene Isomer (ortho, meta and para-xylene)

Flow diagram is presented below in figure and each of the process steps are briefly described below:

(*a*) **Hydro-treating of naphtha:** It Hydro-treating, schematically illustrated in Fig below is the process by which the quality of naphtha is improved by subjecting it to mild or severe conditions of hydrogen pressure in the presence of a catalyst.

The naphtha feed is preheated, heated in a furnace, and combined with recycled hydrogen gas. The combined feed is passed through a reactor containing a catalyst bed where the hydrogenation reaction takes place. Upon leaving the reactor, the **stream** is cooled and moved to a separator vessel where recycle or net hydrogen is removed. The liquid then moves to a stabilizer or stripper, which removes, hydrogen, hydrogen sulphide, ammonia, water and light hydrocarbons, dissolved in the separator liquid.

(*b*) **Reforming of naphtha to produce BTX rich stream:** Reforming is a platinum catalysed high temperature vapour phase process which converts a relatively non-aromatic C_6-C_{12} hydrocarbon mixture (naphtha) (typically containing less than 1% aromatics) to an aromatic product called reformate.

The main process units are depicted in Fig and described below:

- Reactors, which contain fixed bed catalysts.

- Heaters to bring the naphtha and recycle gas to reaction temperature and to supply heats of reaction.
- A product cooling system and a gas-liquid separator.
- A hydrogen-gas recycle system.
- A stabilizer to separate light hydrocarbons dissolved in the receiver liquid.

The naphtha is combined with recycled hydrogen, preheated, heated to the reaction temperature in a fired heater and then transferred to a series of catalyst-containing reactors. Because the reaction is endothermic, a series of three or four reactors with inter-stage reheat furnaces are necessary to achieve the required conversion. The reactors normally contain increasing amounts of catalyst in each stage.

The effluent from the last reactor is cooled and transferred to a receiving unit (e.g., the flash drum) where the hydrogen is separated from the liquid reformate. The hydrogen gases are compressed and recycled to the reactors while the reformate is moved to a stabilizer fractionator. The fractionator removes lighter hydrocarbons (heptane) to produce a stabilized reformate. The stabilized reformate is used as a feedstock for further recovery of benzene, toluene and xylene.

(*c*) **Separation of mixed xylene from reformate:** The stabilized reformate is then taken for fractionation (ortho-meta splitter). The bottom product is a mixture of *o*-Xylene and C_9 aromatics. The top product is a mixture of Ethyl benzene, *o*-, *m*-, *p*-Xylene is sent to PAREX for separation of *p*-Xylene isomer.

(*d*) **Separation and purification of desired xylene isomer:** Typical mix of xylene isomers from a catalytic reformate stream consists of the following: *m*-xylene (40%), *o*-xylene (24%), *p*-xylene (19%) and ethyl benzene (17%). The demand for mixed xylene is low in comparison to the demand for pure isomers, especially *p*-xylene. Hence, separation of isomers of xylene has to be resorted to. Isolation of individual isomers through conventional distillation is difficult for xylene isomers because of the closeness of their boiling points.

Hence, separation of xylene isomers is generally based on differences in adsorptive properties. In PAREX unit, *p*-Xylene is selectively adsorbed on the adsorbent. After the bed is saturated, *p*-Xylene is desorbed by using another organic such as toluene or *p*-diethyl benzene. The PAREX process adopted at IPCL, Vadodara and RIL Patalganga is developed by UOP and separates the *p*-xylene isomer using molecular sieves.

(*e*) **Isomerisation:** The paraxylene lean stream from adsorber (PAREX unit) is heated to reaction temperature and treated with hydrogen to convert ethyl benzene and *m*-xylene into *p*-xylene. Top product, which is a mixture of benzene, toluene and other light hydrocarbons is separated in the stabilizer sent to Gujarat refinery as gasoline blending stock and bottom product is sent back to ortho-meta-splitter.

XYLENE PRODUCTION FROM PYROLYSIS GASOLINE (PYGAS)

Generally, xylene content in pyrolysis gasoline is very low and hence only Petrochemical complexes having very large capacity and/or wanting xylene for downstream uses make xylene from pyrolysis Gasoline. In India, only RIL-Hazira which has a 7,50,000 TPA naphtha cracker makes xylene from pyrolysis gasoline.

A flowsheet of xylene production at Indian petrochemical Co. Ltd. (IPCL), Vadodara is given below:

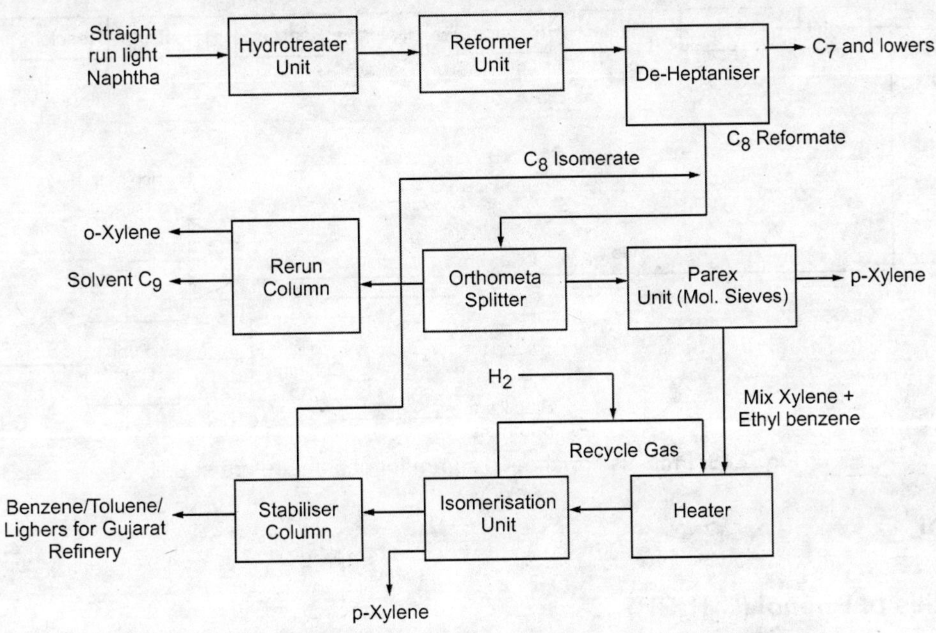

Fig. 2.41. Flowsheet of Xylene Production Process (IPCL, Vadodara)

The pyrolysis gasoline (pygas) is a stream produced by the high temperature, short residence time cracking of naphtha in petrochemical plants. According to cracking severity, the pygas contains typically 20-40% benzene, 16-20% toluene, 10-13% higher aromatics (e.g., ethyl benzene, xylenes, styrenes etc.) with a balance of non-aromatic hydrocarbons (mainly oleflns and diolefins)

The xylene extraction from pyrolysis gasoline (pygas) is carried out in the following reaction steps:

- Hydro-treating
- Solvent extraction
- Secondary hydrogenation
- Fractionation

The details are already explained earlier.

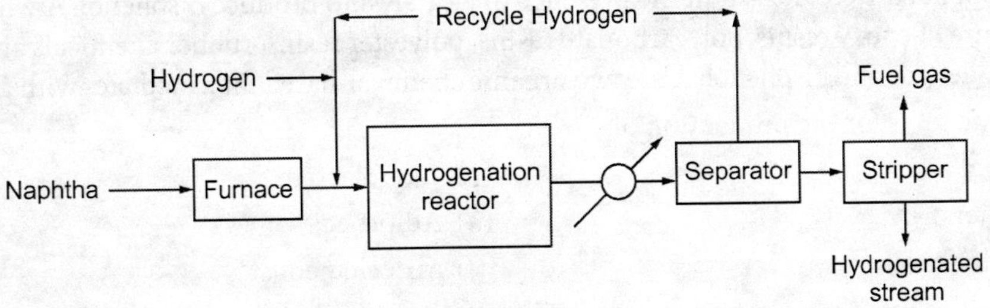

Fig. 2.42. Process flowsheet diagram for hydro-treating of naphtha to improve its quality

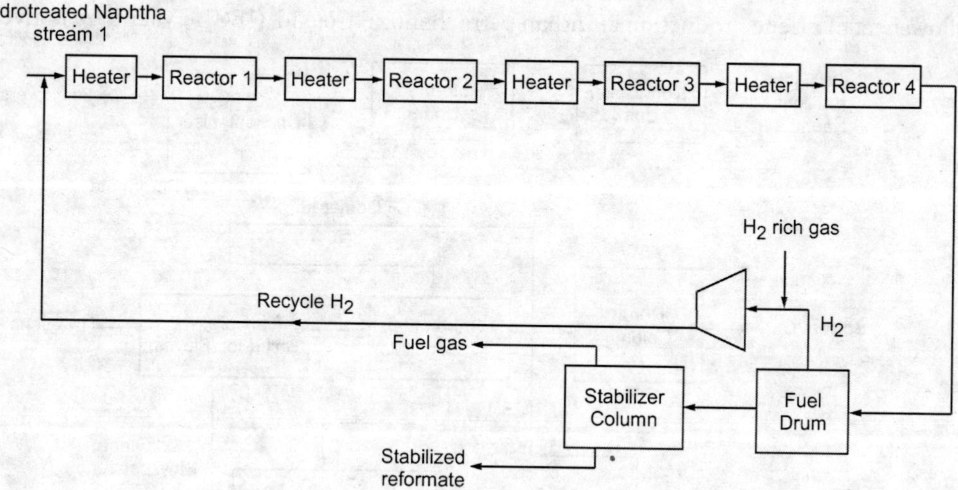

Fig. 2.43. Process flowsheet diagram for naphtha reformer unit

PHENOL

Properties of Phenol (C_6H_5OH)

B.P.: 181.4°C

Sp. gr. @ 25°C: 1.07

Molecular weight: 94

The largest user of phenol is the plastics industry with the chief end-product being cheap thermosetting resins made with formaldehyde. Chemical intermediates for herbicides, insecticides, pharmaceuticals and dyestuffs form the next largest usage with petroleum refining as a solvent being next.

Phenol finds extensive use in the production of phenol formaldehyde resin (bakelite) which is a thermoset resin. It is used in industrial and decorative laminates, molding powders for electrical appliances, plywood and particle board, abrasives, foundry molds, textile auxiliaries and varnishes. Both acetone and phenol are condensed in the presence of a catalyst to produce bisphenol-A which is used in the manufacture of epoxy resins, polycarbonate resins, polyester resins, rubber chemicals and fungicides. Phenol formaldehyde and bisphenol-A are two organic chemical-based intermediates with good demand.

Uses of phenol is for the production of:

(*a*) Phenol resins

(*b*) Caprolactam

(*c*) Bisphenol-A

(*d*) Adipic acid

(*e*) Alkylphenols

(*f*) Miscellaneous*

**e.g.*, aniline, chlorophenols, plasticizers, antioxidants.

Methods of Phenol Production

There are few chemicals which can be made by so many alternate and commercially feasible routes as is the case of synthetic phenol. When one considers that less than 15% of all phenol production comes as natural phenol from coal carbonization plants, the importance of synthetic phenol is apparent.

Seven processes for synthetic phenol are available of which six are still competitive.

(*a*) **Competitive processes:**

- Cumene peroxidation-hydrolysis
- Toluene two-stage oxidation
- **Raschig:** Vapor phase hydrochlorination and hydrolysis
- Chlorobenzene-caustic hydrolysis
- Benzene sulfonate-caustic fusion
- Benzene-direct oxidation

(*b*) **Process not commercially feasible:**

- Chlorobenzene-soda ash hydrolysis

Cumene (iso propyl benzene) Process for Phenol Production

Chemical reactions involved:

(*a*) **Peroxidation:** (emulsification agent is NaOH)

$$C_6H_{11}-CH(CH_3)_2 + \text{Air} \xrightarrow[\text{Emulsion Normal T \& P}]{\text{Aqueous}} C_6H_{11}-C(CH_3)_2-O-OH$$

Cumene orisopropyl benzene → Cumene hydroperoxide

(*b*) **Hydrolysis:**

$$C_6H_{11}-C(CH_3)_2-O-OH + H_2SO_4\ (aq.)\ (10\text{-}25\%) \xrightarrow[\text{1 atm.}]{60°C} C_6H_{11}-OH + (CH_3)_2CO$$

Phenol

Raw Materials Requirements

Basis: 1 ton of phenol (92% yield) + 0.6 ton acetone

Cumene: 1.40 tons

Air: 1.53 tons

Small quantities of sulfuric acid and emulsifying agents.

Co-product: Acetone

Process Description

Cumene (isopropyl benzene) made by alkylation of benzene with propylene is mixed with recycle cumene (1 : 4 ratio) and sent to the hydrogenator. Unsaturates and methyl styrene are converted to saturated materials to avoid undesirable decomposition of the peroxide during the oxidation step. Hydrogen over nickel catalyst at 100°C in a batch reactor is the usual process for purification.

Oxidation is carried out in the presence of air in an aqueous emulsion stabilized by an alkali such as sodium hydroxide in the 8.5-10.5 pH range. Vent gases are passed through a condenser to recover hydrocarbons.

A flowsheet of phenol production from cumene (isopropyl benzene) is given below:

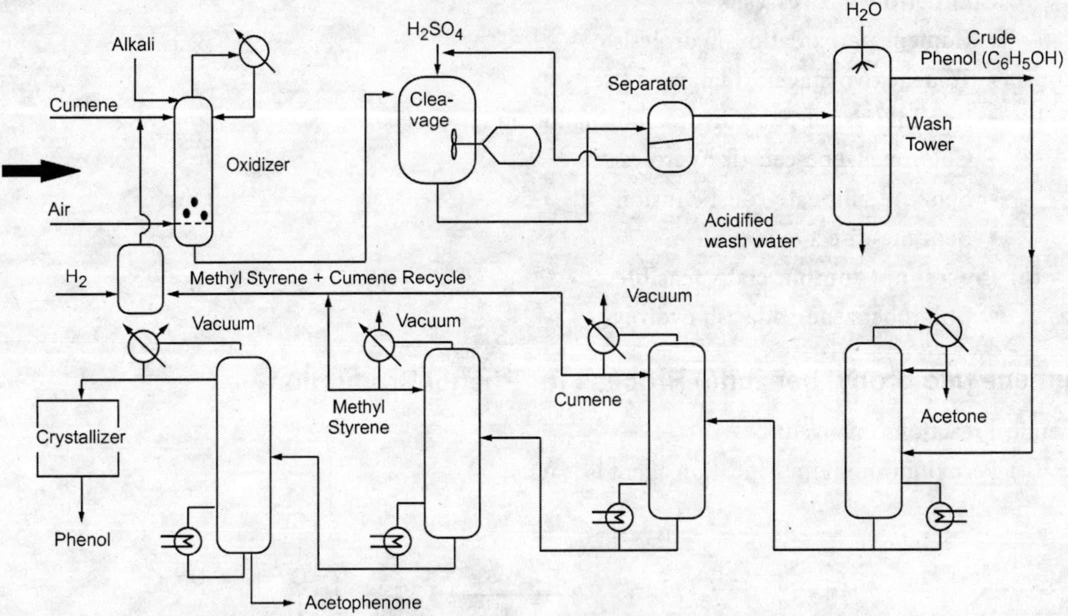

Fig. 2.44. Flowsheet of Phenol production from cumene

MANUFACTURE OF PHENOL FROM TOLUENE OXIDATION

Toluene Oxidation Process for Phenol Production uses air for oxidation of toluene.

Chemical Reactions Involved

(*a*) Oxidation of toluene to benzoic acid

$$\underset{\text{Toluene}}{C_6H_5CH_3} + 1 - 1/2O_2 \xrightarrow[\text{Cobalt Naphthenate}]{150°C} \underset{\text{Benzoic acid}}{C_6H_5COOH} + H_2O$$

(*b*) Oxidation of benzoic acid to phenol

$$\underset{\text{Benzoic acid}}{C_6H_5COOH} + 1/2O_2 \xrightarrow[\text{Cupric benzoate}]{220°C} \underset{\text{Phenol}}{C_6H_5OH} + H_2O$$

Raw Materials Requirements

Basis: 1 ton of phenol (80% yield)

Toluene: 1.34 tons

Air: 8.8 tons

Small quantities of cobalt naphthenate and cupric benzoate catalyst

Process Description

A two-stage air oxidation process is used. In the first stage, fresh plus recycle toluene are mixed with a small quantity of cobalt naphthenate catalyst and charged to the reactor which is a liquid-filled tower through which air is sparged. Cooling tubes are provided to remove the exothermic heat of reaction.

The reactor is run at 150°C and 3 atms. Excess air is used, but toluene conversion is limited to 40% to avoid excessive side reactions. These give by-products such as benzaldehyde, benzyl alcohol, benzyl benzoate, CO and CO_2. With conversion of toluene at 40%, the ultimate yield of benzoic acid is about 90%.

Off-gases from the reactor are vented through a water-cooled condenser to remove water and to allow return of toluene. Liquid from the reactor continuously passes to a distillation column which strips the toluene and other volatile by-products from the acid fraction in the bottoms. Purified benzoic acid is separated by extracting the bottoms with hot water, then crystallizing and filtering the crude benzoic acid. The latter can be recrystallized to meet specifications as a market outlet for benzoic acid.

To make phenol, the crude acid is melted, mixed with cupric benzoate catalyst, then charged to an air-sparged tower containing cooling tubes and mechanical agitation. Reactor conditions are 220°C and 1.3-1.7 atms. Excess air is again necessary to get a 70-80% conversion of benzoic acid with a yield of 90% phenol. The overall process yield for the two steps is about 80%.

Phenol product is obtained by continuously distilling the reactor liquor into a fractionating column where unreacted benzoic acid is returned to the reactor. Non-condensibles such as N_2, O_2 and CO_2 are vented through a condenser along with the condensable fraction of phenol-water. Phenol is withdrawn as the bottom layer in a separator. This crude phenol is again fractionated with purified phenol coming off as bottoms and the overhead phenol-water azeotrope sent to another column for splitting.

The heavy ends in the benzoic acid oxidation tower are water-extracted to recover phenol and benzoic acid which are then recycled, after concentration, to the second stage oxidation tower.

A flowsheet of phenol manufacture from toluene oxidation is shown below:

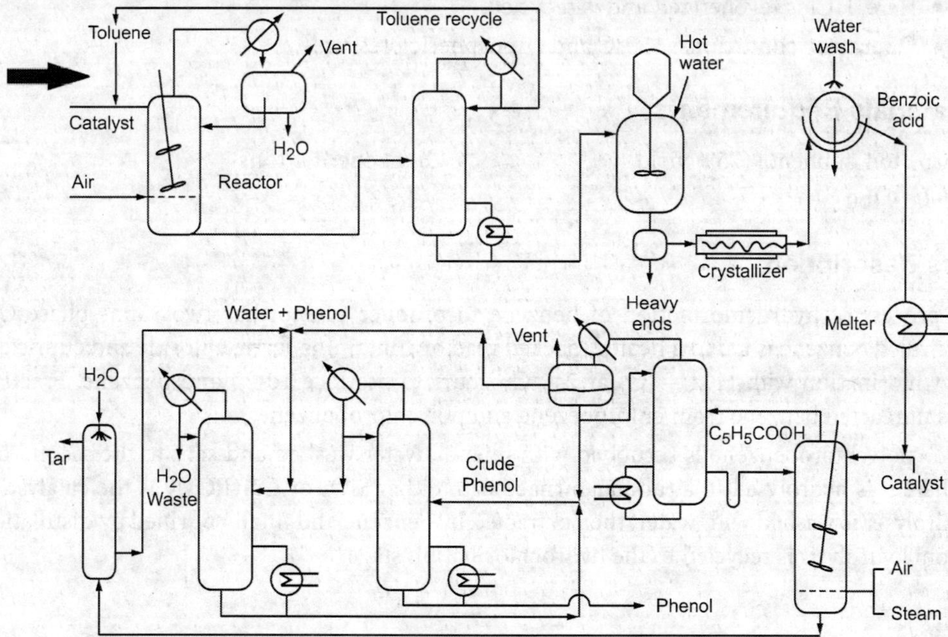

Fig. 2.45. Flowsheet of Phenol manufacture from Toluene Oxidation

MANUFACTURE OF PHENOL FROM HYDROCHLORINATION OF BENZENE

Phenol can be manufactured from Benzene using several ways as given below:

- Benzene hydrochlorination to form Benzyl chloride followed by hydrolysis of benzyl chloride to form phenol.
- Benzene chlorination to form benzyl chloride which is transformed to sodium benzoate and eventually to phenol using NaOH and HCl.
- Benzene sulfonate process: In this process, benzene is converted to benzene sulfonate using sulphuric acid and eventually through neutralization, fusion and acidification, the benzene sulfonate is gradually transformed to phenol.
- However manufacture of phenol from following two processes are important:
  - — Benzene hydrochlorination route.
  - — Benzene from chlorobenzene route.

Manufacture of Phenol using Hydro Chlorination Route of Benzene

Reactions involved:

(*a*) **Hydrochlorination:**

- Benzene + HCl + Oxygen ⟶ Benzyl chloride + Water
- **Catalyst:** $FeCl_3$ + $CuCl_2$
- **Operating conditions:** 240°C and atmospheric pressure

(*b*) **Hydrolysis of benzyl chloride:**

- Benzyl chloride + Water ⟶ Phenol + HCl
- **Catalyst:** SiO_2
- Here, HCl is regenerated and is recycled.
- **Operating conditions:** 350°C and atmospheric pressure

Raw Materials Requirements

Basis: 1 ton of phenol (75% yield)

Benzene: 1.1 tons

HCl: 0.19 ton

Air: 2.4 tons

Process Description

Raschig process of hydrochlorination of benzene to produce phenol has two vapor phase Catalytic stages. Purified benzene is fed to a heated, packed reactor containing ferric chloride and cupric chloride catalyst. Chlorination with HCl — O_2 at 240°C occurs to produce 10-20 and benzene. Fractionation separates unreacted benzene from chlorobezene and polychlorobenzene.

Crude polychlorobenzene is scrubbed with phenol, water washed and sent to the second catalytic stage, where it is hydrolysed in a tubular furnace with either SiO_2 or $Ca_3(PO_4)_2$ as the catalyst. phenol from hydrolyser is washed with water, then extracted by benzene and finally purified by distillation. HCl vapor from hydrolysis is recycled to the hydrochlorination stage.

A Flowsheet of manufacture of phenol using hydrochlorination of benzene is given below:

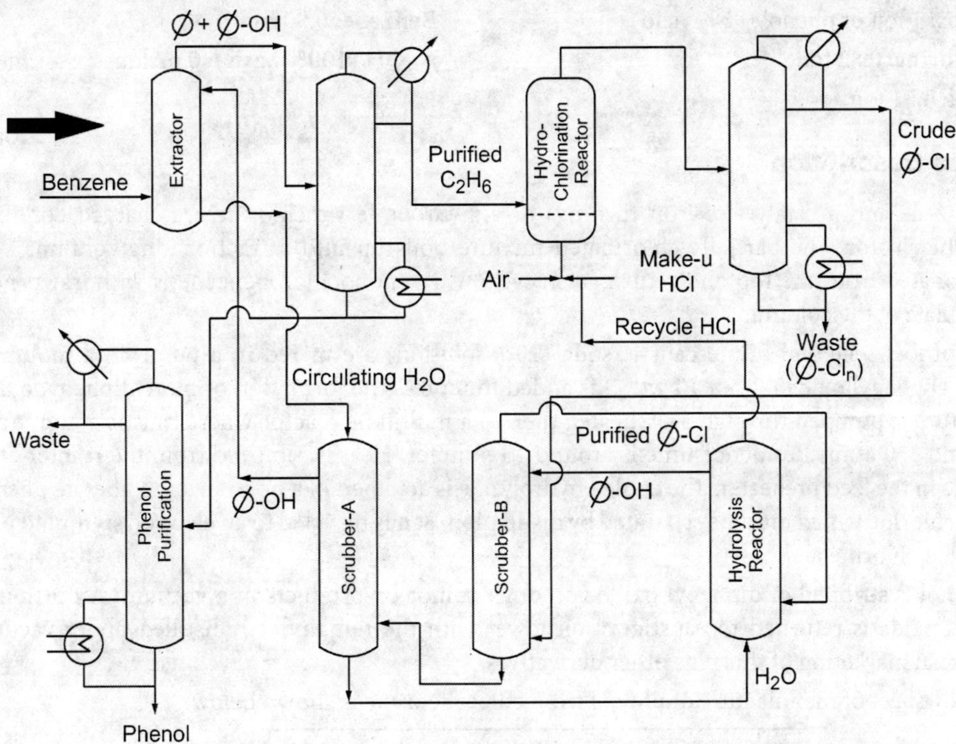

Fig. 2.46. Flowsheet of manufacture of phenol using hydrochlorination of benzene

MANUFACTURE OF PHENOL FROM CHLOROBENZENE

Reactions Involved

There are three reactions to convert benzene to phenol using chlorination route:

(*a*) **Chlorination:**

- Benzene + Cl_2 $\longrightarrow$ monochloro benzene
- Operating temperature: 85°C
- **Catalyst:** Fe or $FeCl_3$ catalyst

(*b*) **Causticization:**

- Benzyl chloride + NaOH $\longrightarrow$ sodium benzoate
- NaOH is in aqueous media
- **Operating conditions:** 425°C and 350 atms
- Exothermic reaction

(*c*) **Hydrolysis:**

- Sodium benzoate + HCl (aq) $\longrightarrow$ Phenol + NaCl (aq)
- **Operating conditions:** Nothing specific

Raw Materials Requirements

Basis: 1 ton of phenol (95% yield)

Benzene: 0.90 ton

Chlorine: 0.82 ton

NaOH (100% basis): 0.67 ton

HCl: 0.51 ton

Process Description

Dry benzene and a catalyst of iron turnings or anhydrous ferric chloride are charged continuously into a chlorinator. The partially chlorinated mixture boils up into a fractionating column. Benzene is fractionated from the top and returns as recycle while monochlorobenzene is withdrawn near the bottom plate of the column.

Chlorobenzene and dilute caustic soda (10% solution) are mixed in a pump in a mole ratio of 1 : 1.25. Diphenyl oxide (about 10 wt.%) is added to repress the formation of more diphenyl oxide and the mixture is pumped through a preheater, then to a multitube reactor where causticization occurs at 425°C and 350 atms. Residence time is around 15 minutes. Heat is removed from the reactor efflux by exchange in the feed preheater. The cooled hydrolyzate is acidified in a neutralizer to liberate phenol and sodium chloride which must be separated by distillation. Salt is recycled to an electrolysis unit to produce the needed chlorine and caustic.

Phenol is stripped of diphenyl oxide and other minor by-products in a vacuum. A portion of the diphenyl oxide is returned to causticization tower with the remainder redistilled under vacuum for commercial marketing of this plus other derivatives.

A flowsheet of manufacture of phenol from chlorobenzene is shown below:

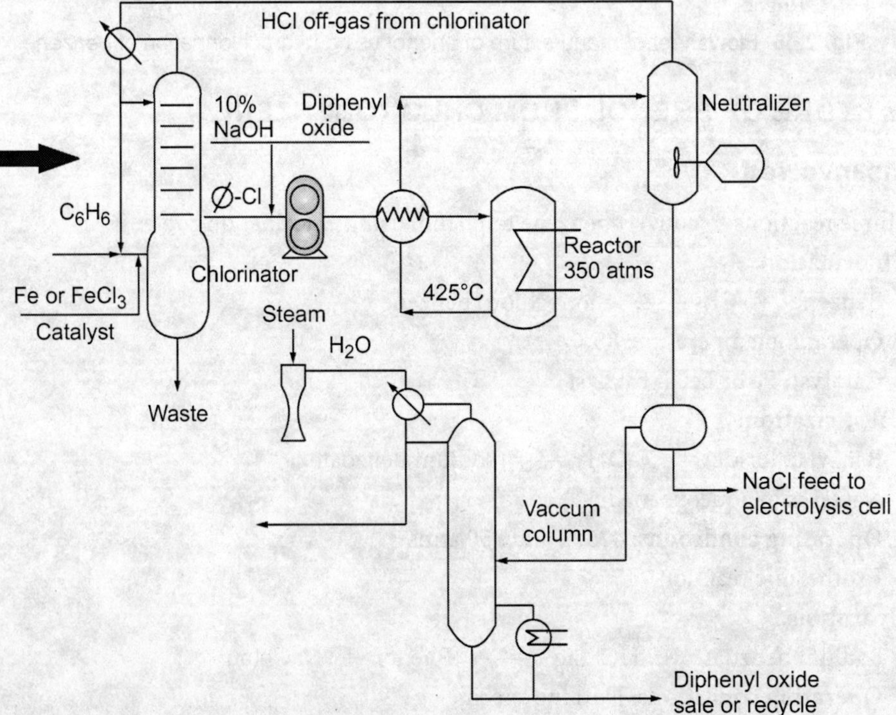

Fig. 2.47. Flowsheet of manufacture of phenol from chlorobenzene

STYRENE

Styrene is produced from benzene via the ethylbenzene route followed by its dehydrogenation.

Reactions Involved

(*a*) Alkylation of Benzene:

- Benzene + ethylene ⟶ Ethyl benzene
- **Catalyst:** $AlCl_3$ granules
- C_2H_5Cl provides hydrogen and chlorine free radicals
- **Operating conditions:** 95°C and 1 atm pressure
- Reaction is exothermic

(*b*) Dehydrogenation of ethylbenzene:

- Ethylbenzene ⟶ Styrene + Hydrogen
- Reaction is endothermic
- **Catalyst:** SnO or FeO
- **Operating conditions:** 800°C

Properties of Styrene ($C_6H_5{\bullet}CH{\bullet}CH_2$)

Molecular weight: 104

M.P.: –30.6°C

B.P.: 145.2°C

Sp. gr. (at 20°C): 0.903

Explosive limit: Lower = 1.1 volume% in air
Upper = 6.1 volume% in air

Toxicity limit: 400 ppm

The impetus to growth of styrene production was the synthetic rubber requirements of World War-II. The copolymerized styrene-butadiene rubber (SBR) is developed for such process. Demands for styrene in the elastomer field continue to rise. Polystyrene molding resin in solidified form and foamed for insulation is of greater importance.

Uses of Styrene

Styrene is used for the production of:

(*a*) Polystyrene
(*b*) Styrene-butadiene rubber (SBR)
(*c*) Acrylonitrile-butadiene-styrene (ABS)
(*d*) Styrene-acrylonitrile (SAN)
(*e*) Polyester resins (unsaturated)

Methods of Production of Styrene

- Dehydrogenation of ethylbenzene.
- Hydrogenation-dehydration of acetophenone
- Dehydrogenation of ethylbenzene is the preferred method of styrene production which is described below.

Raw Materials Requirements

Basis: 1 ton of styrene (86% yield)

Benzene: 0.87 ton

Ethylene: 0.32 ton

$AlCl_3$: 10-11 kg

Process Description of Styrene Production from Benzene

Benzene is alkylated with ethylene by Friedel-Craft method using aluminum chloride or other acid-type catalyst. The resulting ethylbenzene is catalytically dehydrogenated in steam or excess benzene atmosphere to yield styrene.

Recent refinery operation has accomplished the separation of ethylbenzene (with a 2.1°C lower boiling point than *p*-xylene) from the xylene fractions in petroleum reformates. The recovery of this high purity ethylbenzene overhead fraction avoids alkylation procedures. The bulk of the plants, however, use the latter since ethylene and benzene are more plentiful.

The alkylation operation must be done under very dry conditions with high purity feedstock to avoid activity loss of the catalyst. All benzene feed must be dried by azeotropic distillation. Ethyl chloride, which serves ultimately as a source of hydrogen and chlorine free radicals for catalyst promotion, is added to ethylene which is fed continuously along with benzene to the alkylation tower operating at 95°C and 1 atm. pressure. Granular $AlCl_3$ catalyst is fed continuously at the top of the reactor. The reactant mole ratio is 0.6 C_2H_2/1.0 C_6H_6 and no ethylene is recycled.

The alkylation tower is water-cooled to control the exothermic reaction. The alkylated products are pumped to a cooler at 40°C where the aluminum chloride complex is separated and a split stream is fed to the alkylator. The bleed-off fraction is pumped to a high temperature (200°C) dealkylator to break down the polyethyl benzenes to benzene and ethylbenzene which are returned to the system. The tar residue plus aluminum chloride is water-extracted to recover 80-85% of $AlCl_3$.

Crude acidic ethylbenzene from the coolers is neutralized with 50% NaOH solution, stripped to remove polyethylbenzenes and the overhead sent to the benzene column which separates wet benzene from ethylbenzene. A final distillation followed by 20% caustic wash, then drying by percolation in a flake caustic bed, produces 99% ethylbenzene for dehydrogenation in yields averaging 95%.

The polyalkylbenzenes must be removed as bottoms in the atmospheric stripper to avoid formation of polyfunctional, highly reactive products during the dehydrogenation step. The stripper bottoms fraction is distilled at 50 mm Hg pressure with the overhead combined with bottoms from the ethylbenzene column and returned to the alkylation unit for dealkylation of these higher molecular weight compounds. The highest molecular weight polybenzenes as bottoms from the vacuum still can only be split in the high temperature dealkylator.

Dehydrogenation of ethylbenzene produces styrene. Steam is injected with the hydrocarbon in a mole ratio of 15 H_2O/1 $C_6K_5C_2H_5$. The mixed feed passes through the preheater to achieve an input temperature of 500°C to the reactor which operates at 800°C. The dehydrogenation catalyst is promoted zinc, chromium, iron, or MgO on activated carbon, alumina, or bauxite. Conversion is 35-40% with an average yield of 91%.

Reaction product is cooled in the feed preheater, then by steam quenching. A water- cooled condenser and separator operation yields crude styrene-ethylbenzene with impurities of 1% toluene, 0.7% benzene, and 0.3% tar. Sulfur stabilizer is added at this point and the hydrocarbon mixture is passed through a series of vacuum distillation columns to allow the separation of impurities at low enough temperature to avoid styrene polymerization. Benzene and toluene are removed at 160 mm Hg and a reboiler temperature of 90°C. Temperatures higher than this cause autopolymerization of benzene. The second column at 35 mm Hg and 90°C reboiler temperature separates styrene from ethylbenzene. The styrene, stabilized at this point with *p*-tertiary catechol, is vacuum distilled again to remove traces of tar and sulfur. The final overhead (99.8% styrene) is refrigerated at 10°C or less and either stored in insulated tanks or used immediately for polymerization.

Polystyrene is mainly used for making plastic toys and model kits. Moreover, housing for machines as well as refrigerator doors and air conditioner cases are made of polystyrene (a polymer of styrene).

A flowsheet of styrene production from benzene is given below in figure.

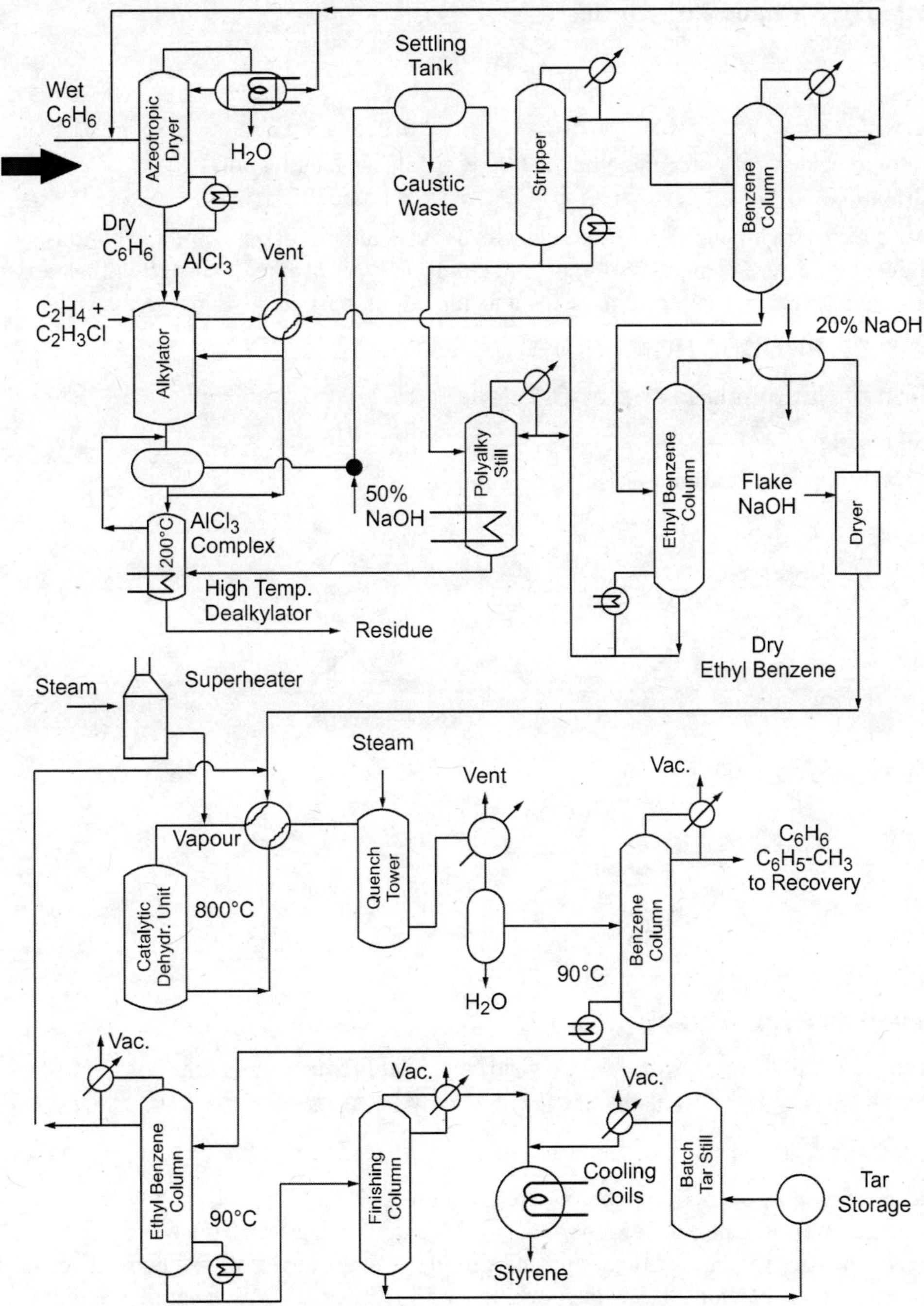

Fig. 2.48. Flowsheet of manufacture of styrene from benzene

PHTHALIC ANHYDRIDE

Properties of Phthalic Anhydride

Mol. wt.: 148.1 **M.P.:** 130.6°C **B.P.:** 284.5 °C **Sp. gr. @ 4°C:** 1.53

Slightly soluble in hot water and ether; sublimes below the melting point.

Ignition temperature: 588°C

Major outlets for phthalic anhydride are in alkyd resins and plasticizers. The alkyd industry has been given a boost by the development of solution and emulsion types of alkyd coatings. Phthalate derivatives continue as a standard plasticizer in many resin formulations, particularly vinyl resins.

METHOD OF PRODUCTION

Oxidation of Naphthalene and/or O-xylene

Chemical reactions:

(*a*) **Naphthalene oxidation:**

$$\text{Naphthalene } (C_{10}H_8) + 4\tfrac{1}{2}\,O_2 \text{ (Air)} \xrightarrow[400°C]{V_2O_5} \text{Phthalic anhydride (PA)} + 2\,CO_2 + 2\,H_2O\,(g)$$

(*b*) ***o*-xyleneoxidation:**

$$\text{o-xylene} + 3\,O_2 \text{ (Air)} \xrightarrow[480\text{-}600°C]{V_2O_5} \text{Phthalic anhydride} + 3\,H_2O$$

Raw Materials Requirements

Basis: 1 ton of phthalic anhydride (75% yield) **Naphthalene:** 1.15 tons

Air: 22.6 tons **Xylene:** 0.96 ton **Air:** 20.1 tons

Process Description

Naphthalene, *o*-xylene, or mixtures thereof are melted and vaporized by bubbling primary preheated air through the molten material. Secondary air to make up the 22 kg air/kg feed requirement is added downstream and the reacting mixture is passed through a preheater to a fixed bed converter. This is a multitubular catalytic reactor filled with supported V_2O_5. The exothermic heat of reaction is removed from the outside of the tubes by a molten salt circulating system. Conditions for oxidation within the converter are 380-450°C.

Reacted effluent gases are passed through a waste-heat steam boiler and thence to a preheater where the final temperature is dropped to a few degrees above the dew point of phthalic anhydride. Condensation of the solid product can be affected by pairs of switching fin- tube condensers, a cyclone, a fluidized bed condenser, or a water scrubber. The crude solid product is melted, distilled under vacuum, and condensed with a flaking or pelletizing operation completing the production scheme. The major impurity is maleic anhydride in 0.2-0.4% quantities.

Maleic anhydride and an easily converted derivative, fumaric acid, can be recovered in 3-5% yield from the off-gas scrubbing liquor of the crude phthalic anhydride condensers.

A process modification which is competitive is the substitution of a fluid bed catalytic reactor for the fixed bed multitubular reactor. Conditions at the head end are somewhat different. The molten naphthalene and/or *o*-xylene is sprayed directly into the air-fluidized catalytic reactor which operates at 600°C. The reacted gases rise upwards through the dense phase into a wider calming-settling zone and thence through the internal cyclone to dust filters. Condensation is the same as above, but no vacuum distillation is required.

A comparison of reactor conditions for the two cases shows that the fluid bed process operates at 140-220°C higher, a longer contact time of 10 sec versus 0.1 to 0.4 sec, a lower air-feed ratio of 15 kg air/ kg feed. The yields are equivalent (75-80%) and almost no maleic acid is produced, thus giving a very high purity phthalic anhydride product.

Phthalic anhydride is used as a versatile intermediate in organic chemical reactions, mainly because it is bifunctional and is cheaply available. It may also be used in the manufacture of phthalate plasticizers like DOP, DEP etc.

Manufacturing process of phthalic anhydride (PA) by oxidation of naphthalene or o-xylene is given below in figure:

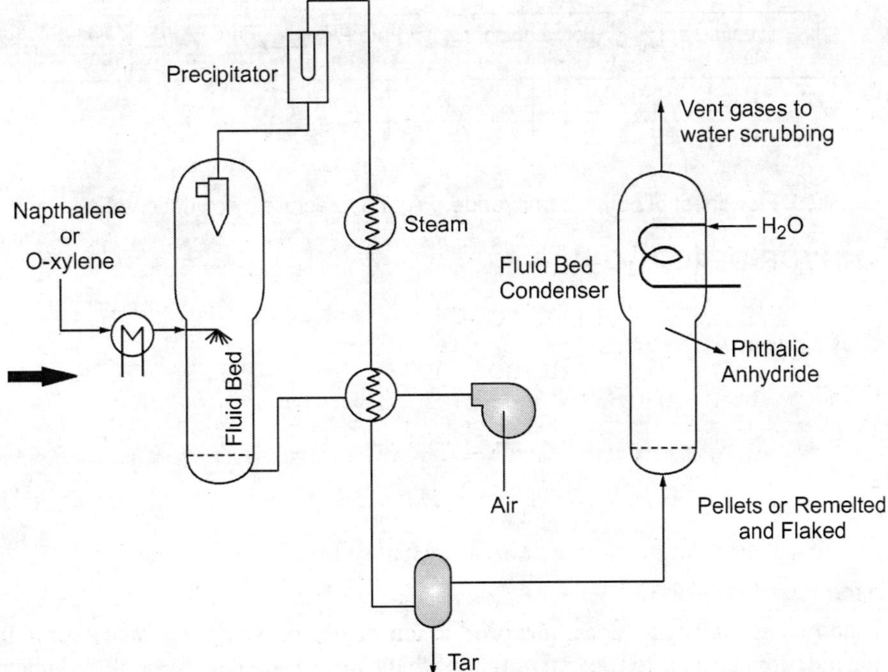

Fig. 2.49. Manufacturing process of phthalic anhydride by oxidation of naphthalene or *o*-xylene

A flowsheet of production of phthalic anhydride by oxidation of *o*-xylene is shown below:

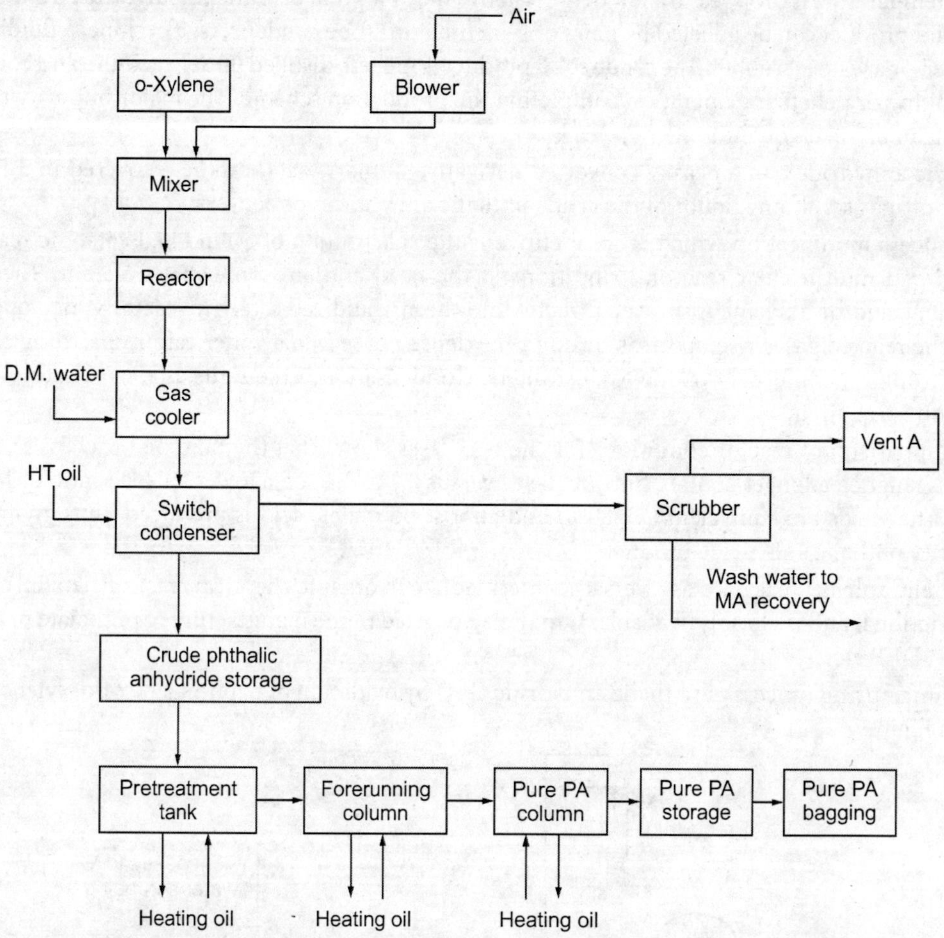

Fig. 2.50. Flowsheet of phthalic anhydride (PA) manufacture by oxidation of *o*-xylene

MALEIC ANHYDRIDE ($C_4H_2O_3$)

Properties of Maleic Anhydride

$$\left[\begin{array}{l} HC\cdot CO \\ \| \\ HC\cdot CO \end{array}\right\rangle O$$

Mol. wt.: 98.1

M.P.: 53.3°C

B.P.: 199.7°C

Sp. gr. @ 20°C: 0.934

Soluble in water (16 wt. % at 30°C), acetone, ether, chloroform, and petroleum.

Ignition temperature: 475 °C

Most of the maleic anhydride goes into production of polyesters. It has made some inroad into alkyd resin formulation, but not to the extent that phthalic anhydride has. Agricultural chemicals such as malathion and soil conditioners consume some maleic anhydride.

Products Made from Maleic Anhydride Include

(*a*) Unsaturated polyesters
(*b*) Fumaric/maleic acid
(*c*) Pesticides
(*d*) Lubricating oil additives

METHOD OF PRODUCTION

Oxidation of Benzene or *n*-Butene

Chemical reactions:

(*a*) **Benzene oxidation:**

$$C_6H_6 + 4.5\,O_2 — V_2O_5 \xrightarrow{400\text{-}500°C} C_4H_2O_3 + 2\,H_2O + 2\,CO_2$$

(*b*) ***n*-Butene oxidation:**

$$C_4H_8 + 3\,O_2 \longrightarrow C_4H_2O_3 + 2\,H_2O$$

Quantitative Requirements

Basis: 1 ton of maleic anhydride (60% yield)
Air: 20-22 tons
Butylene: 1.07 tons

Benzene: 1.33 tons
Basis: 1 ton of maleic anhydride (50% yield)
Air: 38-40 tons

Process Description

This oxidation process is very similar to the fixed bed catalytic operation in making phthalic anhydride. Benzene is vaporized with a large excess of air at 1-1.5 atms, gage in a steam coil to avoid explosive compositions. After preheating, it is fed to a fixed bed converter where the contact time is about 0.1 sec, the temperature is 400-500°C and the pressure 0.8-1 atm. gage.

The exit gases from the reactor are passed successively through a waste heat boiler, feed preheater, tar knock-out drum, and absorbed in water to yield a 40 percent maleic acid solution. This is reconverted to anhydride by a hydrocarbon azeotropic distillation system. The crude product is purified by vacuum distillation.

A flowsheet for the production of maleic anhydride from benzene or *n*-butene is given below:

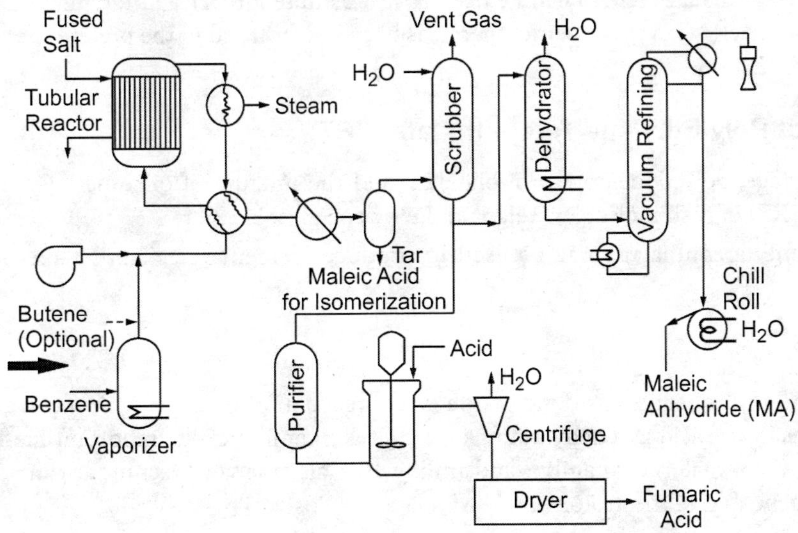

Fig. 2.51. Flowsheet for manufacture of maleic anhydride from benzene or n-butene

Maleic anhydride can be used as a highly reactive and versatile raw material. It can be used in the manufacture of alkyd resins, which in turn are used for making paints and coatings. It can also be used in making agricultural chemicals like herbicides, pesticides and plant growth regulators.

TEREPHTHALIC ACID (TPA) AND DIMETHYL TEREPHTHALATE (DMT)

Terephthalic acid and the ester dimethyl terephthalate are the basic materials for polyester films manufacture

Production of Terephthalic Acid (TPA)

TPA is produced by the air oxidation of paraxylene in the liquid phase. Making polyester polymer can be thought of as occurring in two stages. First is the production of bis-(β-hydroxyethyl) terephthalate (bis-HET) by causing ethylene glycol to esterify with TPA or DMT:

$$ROOC{-}C_6H_4{-}COOR + 2\,HOCH_2{-}CH_2OH \rightleftharpoons HOCH_2CH_2OOC{-}C_6H_4{-}COOCH_2CH_2OH + 2\,ROH$$

$R = H \cdot CH_3$

bis-HET

A number of oligomers with varying values of n are also formed:

$$n\;HOCH_2CH_2OOC{-}C_6H_4{-}COOCH_2CH_2OH \rightleftharpoons \left[{-}OOC{-}C_6H_4{-}COOCH_2CH_2{-}\right]_n + n\,HOCH_2{-}CH_2OH$$

bis-HET PET EG

Dimethyl terephthalate (DMT) can be used as a substitute for purified terephthalic acid (TPA) as polyester raw material. All over the world, there has been a clear trend in the preference of TPA to DMT for polyester production.

Production of Poly Ethylene Terephthalate (PET)

In the second stage, polycondensation of bis-HET and the mixture of oligomers yield poly ethylene terephthalate (PET) of high molecular weight and ethylene glycol.

Poly ethylene terephthalate may be used to produce filament yarns and staple fiber for textile applications.

Nitrobenzene

The production of nitrobenzene is done by the direct reaction of benzene with nitric acid (usually in the presence of sulfuric acid as a dehydrating agent). It is a commercially important method of making nitrobenzene as a raw material for aniline and aniline dyes; also the same technique can be applied to the nitration of toluene to give trinitrotoluene (TNT) as an explosive.

Nitrobenzene is used as a chemical intermediate in the manufacture of aniline, benzidine and many dyestuffs; it is also used widely as an industrial solvent in polishes, etc.

The reaction involved in nitrobenzene manufacture is:

$$C_6H_6 + HNO_3(H_2SO_4) \longrightarrow C_6H_5NO_2 + H_2O(H_2SO_4)$$

Production of Di-methyl Terephthalate (DMT) in an Indian Petrochemical Plant

Di-methyl terephthalate is produced by the reaction of paraxylene with oxygen and then with methanol. The oxidation reaction takes place in the presence of catalyst. The un-reacted paraxylene from oxidation products are stripped completely with steam. The oxidate mixture is then reacted with methanol to produce esters (DMT and Methyl Para toluate). The product esters are then distilled to separate low boiling residue and dibasic acids from DMT. The DMT is further purified by crystallisation to remove impurities like isomers and aldehydes. The purified DMT in the form of pellets is sent to storage.

Chemical Reaction

$$\underset{\text{P-xylene}}{CH_3-C_6H_4-CH_3} + O_2 \longrightarrow \underset{\text{PTA}}{COOH-C_6H_4-COOH} + 2\,CH_2OH \longrightarrow \underset{\text{DMT}}{COOCH_3-C_6H_4-COOCH_3}$$

A flowsheet for manufacture of dimethyl terephthalate (DMT) from oxidation of *p*-xylene is given below:

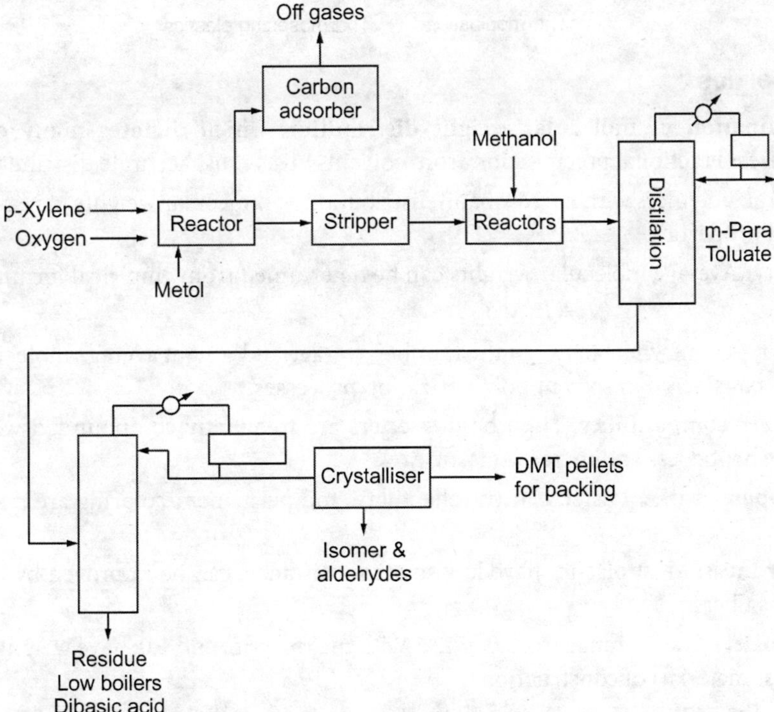

Fig. 2.52. Flowsheet of Manufacturing Process of DMT (Dimethyl Terephthalate)

Properties of Polymers

PROPERTIES OF POLYMERS

In comparison to other manufacturing materials (metals, glass and ceramics), polymeric materials have low density, **low thermal conductivity**, **low electrical conductivity** and are **non-corrosive**. However, they are **flammable** materials and generally have **lower stiffness and strength**. Polymers as per their observed behavior are categorised as follows:

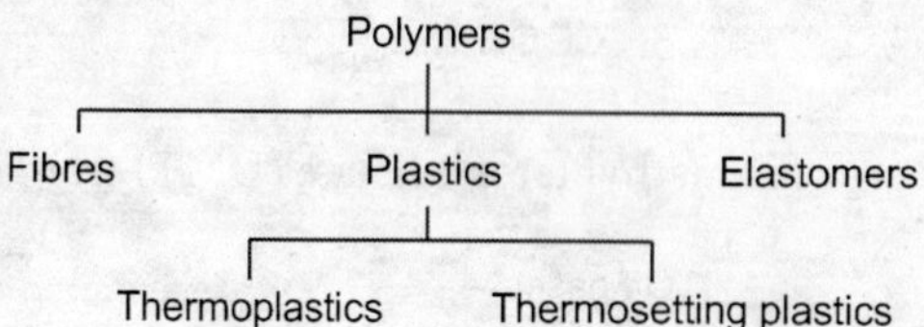

Solvent Properties

(*a*) **Determination of molecular weight distribution which dictates many of the physical properties:** Fractional precipitation from solvents gives most accurate distribution curves.

In general, polymers with narrow band distribution of molecular weights have more predictable physical properties.

(*b*) **Viscosity:** Average molecular weights can be determined from empirical formula:

$$N = K(M_v)_u$$

where M_v is somewhere between the number average and weight average molecular weight; this must be considered in solvent polymerization processes.

(*c*) **Plasticizer compatibility:** High boiling esters are frequently compounded with polymers to decrease brittleness and increase plastic flow.

(*i*) **Properties like:** Luster, clarity, reflectivity, and permanent coloring are real advantages of polymers.

(*ii*) **Hardness:** Most plastics have low scratch resistance, can be improved by lamination with glass fibers.

(*iii*) **Density:** Gravity range from 0.85-2.5 with majority around 1.0; have weight advantage over most materials of construction.

(*iv*) **Weatherability:** Varies, but in general most plastics will darken when exposed to sunlight; moisture resistance excellent.

Mechanical Properties

Plastics exhibit a wide variety of mechanical properties and generalities cannot be used safely to predict performance. Polymers should be tested under expected conditions to guarantee performance. It is not correct to say that Plastics are not as strong as metals and much more sensitive to temperature changes. Improper mechanical design using plastics is largely responsible for the poor service record of polymers.

Polymers, like all materials, respond to an applied stress (force acting on unit area) by changing shape (deformation, expressed as strain = relative change in dimensions).

Linear chain polymers (amorphous thermoplastics) can accommodate applied stresses by three different mechanisms of deformation:

1. Bending/twisting/stretching of bonds in the backbone of the chains
   Classical Elasticity (Immediate strain proportional to the applied stress; immediate recovery when stress is removed)
2. Uncoiling of chains (Segmental movement)
   Rubber like Elasticity (Time-dependent deformation; time dependent recovery)
3. Chains slipping past each other
   Viscous Flow (Time-dependent deformation; no recovery)

Polymers are described as viscoelastic materials, the relative contributions of the three mechanisms depending on:

- temperature
- polymer state (glass, rubbery state, melt)
- repeat unit
- time
- polymer structure
- intermolecular attractions

Amorphous Thermoplastics

Linear Chains

At very low temperatures (glassy state, below T_g) the only mechanism open to chains is the bond bending/twisting/stretching which requires a high stress to achieve deformation.

A high **modulus** (ratio of stress: strain) implies rigidity and hardness. Eventually at high stresses the polymer will be unable to respond further without falling apart and will break in a catastrophic manner.

At temperatures around T_g the polymer gains the additional freedom of rotation about the backbone and the rubberlike elastic mechanism will begin to predominate.

High strains will be achieved without high applied stresses (low modulus = low rigidity or hardness). At a particular strain level segmental movement and viscous flow combine to produce additional deformation without additional stress (**Yield**).

Above T_g the polymer becomes softer, more elastic and, beyond the yield point, high strains can be achieved (**Necking**) as the chains extend and slip over each other. Before final rupture the fully extended chains begin to show more resistance and the stress will increase (**Strain Hardening**).

At much higher temperatures the viscous flow mechanism will dominate, giving very low modulus (very soft) and very high extensibility in the melt state.

Semi-crystalline Thermoplastics

Linear Chains with Secondary Crosslinks

The inclusion of secondary crosslinks (forces of crystallisation) results in restricted deformation at all temperatures up to T_m, giving increased modulus (stiffness). This is particularly evident in the rubbery state. Increasing levels of crystallinity will make the polymer stiffer, stronger but more brittle (lower strain at failure).

Crosslinked Elastomers

Linear Chains with a Few Primary Crosslinks

The inclusion of a few primary crosslinks will extend the rubbery state from below room temperature to high temperatures. The crosslinks limit the extent of the viscous flow mechanism and there is no melt region. The crosslinks also ensure rapid recovery of the rubberlike deformation. Increasing the density of primary crosslinks will increase the stiffness and limit the extensibility before failure.

Thermoset Plastics

3-D Network

With little opportunity of capitalising on segmental movement and chain slippage (due to the high concentration of primary crosslinks) thermoset plastics are very hard, have high stiffness but show brittle failure behaviour over a wide temperature range up to decomposition temperatures.

Stress vs. Strain (in Tension)

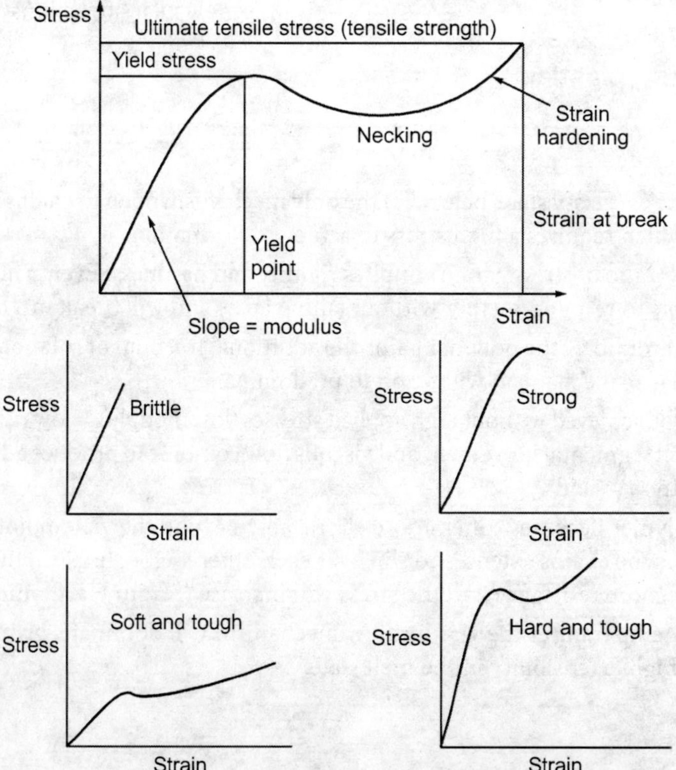

Fig. 3.1. Stress vs. strain curve of polymer plastics

Molecular Weight (MW)

Because of varying chain length, M.W. of most polymers is not a fixed value, but varies over a certain range (10^3-10^7).

Vinyl polymers: 10^5 to 10^6

Very polar functional groups: 15,000-20,000

Crystallinity

Because of varying chain length and high M.W., most polymers are amorphous and only semi-crystalline polymers with higher crystallinity are generally tougher, stiffer, more opaque, more resistant to solvents, and of higher density than polymers of low crystallinity. Also, have more sharply defined melting points. Higher the degree of crystallinity, more pronounced the properties.

The superior mechanical properties of highly crystalline polymers are a reflection of greater cohesive strength arising from more effective intermolecular forces among the closely packed molecules. Opaqueness arises from scattering of light by the crystallites.

Polymer crystallinity is increased by:

- Highly stereoregular structure
- Little or no chain branching
- Highly polar groups

Impact Resistance

A polymer with a high energy to failure (high area under the stress-strain curve) is designated tough. This is important in impact stresses when chains have little time to accommodate the stresses.

Above the T_g, polymer systems show good toughness and impact resistance. Increasing concentration of crosslinks (primary and secondary) will reduce the impact resistance.

Brittle behaviour (failure below the yield point, with a low energy to failure) is encouraged by:

- decreasing temperature
- increasing speed of the applied stress
- inclusion of crystalline regions
- inclusion of primary crosslinks
- introduction of flaws or cracks

Creep Resistance

At extended time scales under stress the rubberlike elastic and viscous flow mechanisms will combine to produce steadily increasing strain with time.

The extent of creep is reduced by:

- decreasing temperature
- increasing crosslink density

Thermal Properties

Thermal properties relevant to the processing and end-use of plastics are:

- Heat Capacity (Specific Heat)
- Thermal Conductivity
- Thermal Expansion
- Glass Transition Temperature (T_g)
- Melt Transition Temperature (T_m)
- Heat Distortion Temperature
- Continuous Use Temperature

The Heat Capacity (Specific Heat) is the heat flux required to raise the temperature of unit mass (1 kg) by one degree Celsius. Plastics have higher heat capacities than metals. Specific Enthalpy represents the heat required to raise 1 kg from room temperature to the melt processing temperature. Semi-crystalline thermoplastics have much higher specific enthalpies due to the latent heat associated with the melting stage

With a poor mechanism of transmitting energy by molecular vibrations plastics have very low Thermal Conductivities compared to metals. This is a useful property in end applications but can cause problems in processing, during heating and cooling stages.

Coefficients of Thermal Expansion are greater for plastics, which accounts for some of the dimensional stability associated with plastics. The greater freedom above the glass transition temperature (T_g) results in higher values of thermal expansion (elastomers). Because of the orderly close packing in semi-crystalline thermoplastics there is a greater expansion at the melt transition (T_m) compared to amorphous thermoplastics (and a greater shrinkage cooling down from the melt state). The addition of fillers with low thermal expansion reduces the overall expansion values.

Softening of amorphous thermoplastics beyond acceptable functional levels is generally associated with the glass transition temperature (T_g).

Semi-crystalline thermoplastics can retain functional rigidity up to the melt transition temperature (T_m) but may be deformed under high loads at lower temperatures. To convey design information on heat softening, the Heat Distortion Temperature (Heat Deflection Under Load) is generally quoted for two different stress (load) levels to recognize the range of possible service conditions.

As temperatures increase the rate of thermal decomposition of polymers also increases, with subsequent deterioration in mechanical properties. A short time at a higher temperature has the same effect as a long time at lower temperature. The continuous use temperature combines heat softening and decomposition.

Unless otherwise specified the Continuous Use Temperature (CUT) refers to temperatures at which a 50% reduction in tensile strength occurs in 20,000 hours

Glass Transition Temperature (T_g)

At sufficiently low temperatures, even amorphous polymers are hard and brittle (glass-like) As temperature is increased kinetic energy of molecules increases, but motion is still restricted to short-range vibrations and rotations, so long as the polymer retains its glass-like structure. At a certain temperature, called the Glass Transition Temperature T_g, the polymer loses its glass-like properties and becomes softer and more elastomeric, but does not melt. Above T_g, molecular motions are more long-range, involving segmental motion of 20–50 atom chains. If heating continues further, polymer will eventually lose its elastomeric properties and melt to a flowable liquid. This melting point (M.P) is usually not sharply defined.

Thermal diffusivity: Low, giving good thermal insulation characteristics, particularly when gas-foamed.

Heat resistance: Poor, ranging from 35-260°C in continuous service.

Flammability: Resistance to ignition and burning poorer than for metals and ceramics.

Optical Properties

Organic polymers usually do not contain chemical groups (chromophores) that absorb in the visible region of the spectrum. Hence, most polymers are colourless in their natural state. Colour in polymeric products is supplied by additives (dyes or pigments).

Semi-crystalline thermoplastics, consisting of crystalline regions of higher density (and hence higher refractive index) and amorphous regions of lower density, will result in scattering of incident light and loss of transmission (haze).

If the crystallites are:

- oriented in one direction;
- or if they are of a similar density to the amorphous regions
- or if the crystallites have dimensions smaller than the wavelength of the light

the amount of scattering will be greatly reduced. Amorphous thermoplastics will have high transparency.

Oriented plastics, eg in biaxially oriented polypropylene film or stretch-blow-moulded polyethylene terephthalate (PET) bottles, will show high transparency even though highly crystalline. Additives of different refractive index from the polymer can also lead to light scattering giving translucent or opaque products.

Electrical Properties

Organic polymers normally do not contain any charge carrying species, such as free electrons as found in metals or ions in electrolytes, and are useful as electrical insulators—insulators with high dielectric strength, low dielectric constant and low power factor. The last two minimize power losses; polyethylene, Teflon and polystyrene are superior to glass and ceramics in UHF insulation applications.

At very high voltages (high electrical stress) polymers will break down catastrophically. In very high frequency electrical fields (kHz to MHz) the dipoles in some polymers are unable to keep up with the alternating field and electrical energy is transformed into heat. Although this can be a problem in the telecommunications field the phenomenon can be used to heat weld certain plastics (e.g., plasticised PVC). Rubbing two polymers together can create static electrical charges which are slow to leak away to earth and cause dust collection and other problems.

Environmental Properties of Polymers

Environmental properties relate to the changes taking place in structure and properties of polymers when exposed to a range of external agencies:

Effect of Heat

All organic materials break down (decompose) when subjected to heat energy through thermal dissociation of chemical bonds. Usually the rate of decomposition increases with increasing temperature. A long time at a low temperature gives much the same result as a short time at a high temperature.

Thermal decomposition is more noticeable in polymers than in other organic chemicals because even less than 0.1% decomposition can result in a 50% reduction in molecular weight, with disastrous consequences for mechanical properties that depend on molecular size. The harmful effects of thermal decomposition can be minimised using heat stabilizer additives.

Effect of Radiation

Most forms of radiative energy can lead to decomposition of polymers but the most serious forms are ultra-violet (UV) radiation (in sunlight) and high energy radiation (X-rays, beta-rays and gamma rays). Visible light has insufficient energy to dissociate the typical bonds found in polymers but additives can act as sensitizers.

Decomposition of polymers by radiation is essentially a surface effect and has little effect on bulk properties such as stiffness but can dramatically alter properties such as impact strength. One of the main outcomes of UV decomposition is discolouration (usually yellowing). Light stabilisers operate as screens for UV light or interfere with the associated chemical reactions.

Effect of Chemicals

About chemical resistance of polymers, generalization cannot be made for polymers. They can be found for each application, but make due allowance for other properties. The most resistant plastic is Teflon (Polytetrafluorethylene) which is not attacked by highly corrosive chemicals or strong solvents in addition to having the highest heat resistance, but the poorest forming properties.

The environmental effects of chemicals can either be a reversible physical process (solvent attack, moisture absorption, permeability) or an irreversible chemical reaction. As with most chemical and physical processes the effect is usually accelerated as the temperature and concentration of the chemical is increased.

Solvent Resistance

Polymers are most strongly attacked by organic solvents that are chemically similar (similar polarity, similar solubility parameter).

Because of the size of polymer molecules true solution is a slow process, accelerated by heating. Attacks take the form of swelling and mechanical weakening. In some cases, additives can be selectively extracted. Crosslinked polymers and semi-crystalline thermoplastics show the best resistance to organic solvents.

Water Resistance

Some polymers, eg polyamides (nylons) can absorb up to 10% water, depending on humidity and temperature, with corresponding increase in dimensions, reduction in modulus but improvement in impact resistance.

The presence of moisture at melt processing temperatures can lead to breakdown of the polymer (hydrolysis) with deterioration in mechanical properties. Hence certain polymers, e.g., polyamides and polycarbonates, must be thoroughly dried before processing.

Oxidation

Atmospheric oxygen can cause serious chemical damage, particularly in hydrocarbon polymers. The chain reaction can lead to discolouration and reduction in molecular weight (degradation). Oxidation can be accelerated in the presence of UV light. Incorporation of antioxidants can minimise the effects of oxidation.

Effect of Biological Organisms

Most synthetic polymers have a high resistance to biological organisms (bacteria, enzymes, fungus, mould) although some additives can be attacked. Naturally occurring polymers (cellulose, natural rubber) are more prone to biodegradation. The concept of modifying synthetic polymers to make them more biodegradable and hence be disposed of in a manner similar to vegetation and waste foodstuffs has not found much support because of the uncertainty of speed of breakdown and the release of unknown chemicals into the food chain.

Weathering of Polymers

Weathering of polymers involves a combination of heat, uv radiation, oxygen and moisture, the net effect on the polymer properties depending on the relative contribution of each.

Combustion: The high temperature oxidation of polymers is a complex process and considered one of the major weaknesses of polymers as manufacturing materials. Fire performance is assessed by a number of factors like:

- Ignitability
- Heat production
- Spread of flame
- Smoke production
- Toxic gas production

All of the above factors can be influenced by the structure of the polymer and the presence of fire performance additives.

Properties of common thermoplastics and thermosets are tabulated below.

S. No.	Polymer	Structural state	Tensile strength (MPa)	Impact strength $(J\ m^{-1})^a$	Density $(g\ cm^{-3})$	T_s $(°C)^b$	T_{max} $(°C)^c$	Applications
1.	Thermoplastics: ABS	Glassy copolymer with rubbery domains	40	200-600	1.03	—	71-93	Refrigerator linings, lawn and garden equipment
2.	Polyamides: (e.g. nylon 6,6)	Highly crystalline, spherulitic structure, H-bonding	80	100	1.15	57	82-150	Bearings, gears, cams, bushings, handles, wire coatings
3.	Polycarbonate	Stiff molecular structure	62	650-850	1.20	150	120	Machine parts, car parts propellers, instrument panels, CDs
4.	LD polyethylene[d]	Branched (60% C)[r]	14	50-700	0.92	–110	82-100	Flexible sheet, bottle, toys
5.	HD polyethylene[c]	Linear (95% C)[r]	28	25-100	0.95	–90	80-120	Batteries, tanks, ice trays
6.	Polypropylene	Isotactic (amorphous)	34	25-100	0.91	–10	107-150	Bottles, packing, TV cabinets
7.	Polystyrene	Atactic (amorphous)	50	12-20	1.09	100	104	Styrofoam, wall tile battery cases, toys, lighting
8.	Polyvinyl chloride (PVC)	Largely amorphous	40	50-125	1.4–1.5	105	110	Pipe, garden hose electrical insulation
9.	Polyethylene terephthalate (PET)	Crystalline	54	12-35	1.15	70-80	80-120	Fiber, film, magnetic carpet, clothing, tires
10.	Thermosets: Phenolic	Amorphous, wood and glass reinforced	52	12-900	1.3-1.9	150-288		Electrical equipment, motor housings, insulators, brake and transmission parts
11.	Polyester	Amorphous, glass reinforced (fiberglass)	28-138	400-800	1.7-2.3	150-177		Car panels and body parts, boat hulls, indoor construction panels, pipes, tanks, ducts

[a] Based on notched Izod test.
[b] Glass to liquid transition temperature.
[c] Maximum use temperature (no load).
[d] LD = low density.
[e] HD = high density.
[f] C = crystallinity.

Table. 3.1. Properties of Commodity Polymer

	Polyethylene			Polystyrene			PVC	
Property	**LDPE**	**HDPE**	**Polypropylene**	**GP-PS**	**HIPS**	**SAN**	**Rigid**	**Flexible**
Specific gravity	0.912-0.94	0.941-0.965	0.902-905	1.04-1.065	1.03-1.06	1.07-1.08	1.30-158	1.16-135
Crystallinity (%)	50-70	80-95	40-68	—	—	—	—	—
Melting temperature (°C)	98-120	127-135	165-174	—	—	—	—	—
Tensile strength (MPa)	15.2-78.6	17.9-33.1	29.3-38.6	36.5-54.5	22.1-33.8	62.0-82.7	41.4-51.7	10.3-24.1
Tensile modulus (MPa)	55.1-172	413-1034	1032-1720	24.3-3378	1792-3240	2758-3861	2413-4136	—
Elongation at break (%)	150-600	20-130	500-900	1-2	13-50	1-4	40-80	200-450
Impact strength (Izod) (ft-lb/in. notch)	716	0.8-14	0.4-6.0	0.25-0.40	0.5-11	0.35-0.50	0.4-20	—
Heat deflection	38-49	60-88	107-121	75-100	77-93	—	57-82	—
Temperature (°C at 66 psi)								

Table. 3.2. Typical Properties of Some Engineering Polymers

Property	ABS[a]	Acetal	Nylon 6	Nylon 6,6	PC	PPO	PSF	PPS	Polyamide
Specific gravity	1.01-1.04	1.42	1.12-1.14	1.13-1.15	1.2	1.06	1.24	1.3	1.36-1.43
Tensile strength (MPa)	33.1-43.4	65.5-82.7	68.9	75.8	65.5	66.2	70.3	65.5	72.4-117.9
Tensile modulus (MPa)	1586-2275	3585	689	—	2378	2447	2482	3309	2068
Flexural strength (MPa)	55.2-75.8	96.5	34.5	42.1	93.1	93.1	106.2	96.5	131-199
Flexural modulus (MPa)	1724-2413	2620-2964	965	1275	2344	2482-2758	2689	3792	3102-3447
Impact strength, Izod (ft-lb/in notch)	3.0-12	1.3-2.3	3.9	2.1	16	1.8-5.0	1.2	1.8-5.0	1.5
Elongation at break	5-70	25-75	300	300	110	20-60	50-100	20-60	8-10
Heat deflection temperature (°C at 455 kPa)									

[a] High impact grade

Table 3.3. Polymer properties used to characterize plastics materials

Property	Definition	Units	Typical Values
1. Physical property			
1.1 Density	Mass per unit volume at 23°C (Water = 1000 kg m^{-3})	kg m^{-3}, g.cm^{-3}	Solid polymers 800-1500 kg m^{-3} 0.8-1.5 g.cm^{-3} Cellular polymers 20-750 kg m^{-3}
2. Mechanical properties			
2.1 Stiffness (Rigidity)	Modulus of Elasticity (stress/strain) [Stress = force acting on unit cross sectional area; Strain = relative change in dimension e.g., length] Modulus of Elasticity in tension Modulus of Elasticity in compression Flexural Modulus (in bend)	N mm^{-2}, GN m^{-2}, GPa Stress: N mm^{-2}, MN m^{-2}, MPa Strain: No units	0.4-8.0 GNm^{-2} 0.1-10 GNm^{-2}
2.2 Yield Stress	Stress at Yield Point	Nmm^{-2}, MNm^{-2}, MPa	10-100 MN m^{-2}
2.3 Tensile Strength	Stress at failure under tensile load	Nmm^{-2}, MNm^{-2}, MPa	10-100 MN m^{-2}
2.4 Compressive Strength	Stress at failure under compressive load (also stress at 0.1 % compression)	Nmm^{-2}, MNm^{-2}, MPa	4-300 MNm^{-2}
2.5 Flexural Strength	Stress at failure under bending load	MNm^{-2}, MPa	10-130 MNm^{-2}
2.6 Elongation at Break	Percentage increase in length at break under tensile load	%	0.4-600%
2.7 Creep	Slow change in dimensions over a long period of time under load Creep strain	No units	—
2.8 Impact strength	Energy to cause failure under an impact force (a) Charpy (notched and unnotched) (b) Izod (notched and unnotched) (c) Falling weight	mJ mm^{-2} (kJ m^{-2}) J/m notch	1-16 kJ m^{-2} 10-800 J/m notch
2.9 Hardness	Indentation of a ball indentor under a static load Shore D – hard plastics Shore A – soft plastics and rubber Rockwell – various scales	Arbitrary scale 100 = infinitely hard 0 = infinitely soft	R 20 – M130
2.10 Abrasion resistance	Loss of material under set abrasive conditions	mg/cycle	—
3. Thermal properties			
3.1 Coefficient of linear expansion	Proportional increase in length for 1°C rise in temperature	K^{-1} (°C^{-1})	2-20 × 10^{-5} K^{-1}
3.2 Thermal conductivity	Rate of heat transfer through unit thickness to maintain 1°C temperature difference	Wm^{-1} K	0.6-5.2 × 10^{-3} Wm^{-1} K
3.3 Melt temperature	Temperature at which the polymer melts (Semi-crystalline thermoplastics only)	°C	30-380°C

3.4 Softening point	Temperature at which the modulus (Stiffness, rigidity, hardness) Drops below an arbitrary acceptable level (a) Vicat Softening temperature (b) Heat Deflection under load (0.45 MPa 03 1.8 MPa)	°C °C	50-300°C 30-250°C
3.5 Continuous Use Temperature	Maximum temperature to give life of x hours under service conditions	°C	50-250°C
3.6 Oxygen Index	Minimum fraction of oxygen in a nitrogen/ oxygen mixture required to sustain burning	No units	0.15-0.80
3.7 Flammability	Behaviour towards standard fire performance tests		—
4. Electrical properties			
4.1 Volume Resistivity	Resistance between opposite faces of a unit cube	ohm cm (or log ohm cm)	10^{10}-10^{18} ohm cm
4.2 Surface Resistivity	Resistance between electrodes on surface		(10-18) ohm
4.3 Relative Permittivity	Ability to store electricity in a capacitor	No units (measured at different frequencies)	2.0-7.0 at 108 Hz
4.4 Dissipation Factor	Indication of heat production in a high frequency field	No units (measured at different frequencies)	0.0002-0.35
4.5 Dielectric Strength (Electric Strength)	Indication of ability to stand up to high voltages		10-30 kV mm^{-1}
5. Chemical Resistance			
5.1 Water Absorption	Percentage increase in weight (or volume) after exposure to water e.g., 1 day total immersion at 70°C 7 days at 50 % relative humidity	%	0-3.3 % (24 hours at 23°C)
5.2 Chemical Resistance	Measured as volume (or weight) increase after exposure to chemical Or effect on mechanical properties		—
6. Processability			
6.1 Drying	Time and temperature to remove moisture	hours at x °C	1-16 hours at 60-120°C
6.2 Processing Temperature	Temperature for extrusion (die temp) or injection moulding (nozzle temp)	°C	180-350°C
6.3 Mould Temperature	Recommended mould temperature	°C	10-120°C
6.4 Melt Index (Melt Flow Rate)	Mass of polymer extruded in 10 minutes from a standard piston extruder	g/10 min	0.1-50 g/10 min
6.5 Melt Viscosity	Ratio of shear stress to shear strain (varies with shear strain rate)	Pa s	—
6.6 Mould Shrinkage	Difference between moulding dimension and corresponding mould dimension, relative to mould dimension	No units but expressed as %	0.4-3.0 %

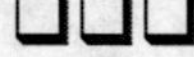

Polymers Processing Technologies

POLYMERS PROCESSING TECHNOLOGIES

All the processing stages between manufacture of the crude polymer and the final molding step are called the compounding processes. These include storage, transportation, metering, mixing, plastication and granulation. Sometimes additives may be added to the polymer mix to allow production on an industrial scale. Addition of plasticizers renders flexibility and ductility to rigid and brittle polymers. In a similar way, fillers can enable to enlarge the volume and also has influence upon the polymer properties.

The various compounding processes are:

(*a*) **Mixing:** Mixing allows obtaining a homogeneous mixture of polymer and additives. The additive is distributed uniformly throughout the crude polymer, without any unnecessary stressing. Mixing can occur under the action of gravitational forces in a tumble mixer ; or in a ribbon mixer, where materials are mixed in a horizontal barrel in a spiral manner. Such mixers are collectively known as cold mixers.

Another class of mixers are the hot mixers, wherein the polymers are heated by contact with heated metal components. Such mixers are also known as high – speed or turbo mixers. This method is widely used in the plastic industry, especially in the compounding of poly vinyl chloride (PVC).

(*b*) **Rolling and Kneading:** Before processing, the plastics are compacted, melted, homogenized and modified with additives. These include rolling, in which the plastic is passed between two counter rotating, heated rolls wherein it is pressed into a thin layer. It is an open process in which additives can be added, and other scrap materials may be reintroduced. PVC is one plastic which extensively employs this process. Moreover, the material temperature can be controlled precisely and cleaning of the machine is easy.

(*c*) **Pelletizing:** It is the method of producing granules of equal size and shape. This enables the optimum feed behaviour on the processing equipment. Pelletizing can be done in hot or cold conditions. Using a pelletizing die, strands, ribbons or sheets are formed ; and after solidification, they are chopped into the desired shapes. This is one procedure strictly followed in cold pelletizing. However, in hot pelletizing, the plasticated material is sent through a die, and a blade that rotates on the surface of the die plate chops the exiting strands. The melt is then cooled; and subsequently the water is removed.

(*d*) **Shredding and Grinding:** In thermoplastics processing, a variety of semi finished and finished products are generated. These products, in order to be made suitable for processing, must be cut to produce a pellet form. A range of particle sizes is obtained in the process.

Regranulating is achieved through cutting mills. The material is added through a feed system wherein a rotor revolves with high momentum. The system is further equipped with multiple cutting heads that operate against stationary cutting edges. The rotating cutting heads are divided and displaced with respect to each other in some other machines. As soon as the desired particle size is obtained, the material is ejected through a sieve, the pores of which determine the size of the particles. The housing and rotor may be equipped with water cooling apparatus to remove the excessive heat generated.

(*e*) **Storage and Transportation:** The storage and transportation of base polymers and intermediate products are achieved through automatic equipments. These equipments have an edge over the manually operated ones as they ensure improved working conditions, lower work place, lower contamination of materials and subsequent reduction losses. Moreover, accident hazards are also reduced. Large external silos are used for the storage of polymers, either in powder form, or in the pellet form. By incorporating vibrating frames, bunker cushions, slotted shelves or similar techniques as ventilation, the formation of granule bridges is avoided. The materials are transported via road or rail. Reserve materials and additives are transported in ships in bulk containers, drums or sacks. Pneumatic equipments are generally preferred for conveying powders and pellets. The conveyors used may be high pressure conveyors, medium pressure conveyors or low pressure conveyors. Fill level sensors help in controlling the transport interval.

In order to meter liquids such as plasicizers or other additives for compounding, oscillating displacement pumps are used. Metering of solids can be done using gravimetric metering equipments or volumetric metering equipments.

TECHNIQUES FOR PRESSURELESS PROCESSING OF POLYMERS

The Pressureless Processing Techniques are used for starting materials such as monomers, plastic solutions and dispersions that can flow and that can be formed without application of external forces. The pressureless processing techniques include:

- Casting
- Coating
- Dipping
- Foaming

(*a*) **Casting (Fig. 4.1):** By polymerizing monomers in molds, thick – walled articles, embeddings, sheets and blocks can be produced. These are preferentially made up of polystyrene, polyamides and polymethylmethacrylate. In order to avoid thermal stresses, adequate temperature regulation is required. The mold material choice is dependent upon the use and surface quality of the moldings.

Film casting is mainly used in the photographic industry for the manufacture of cellulose acetate films. In this method of casting, polymer solution, melts or dispersions are cast directly into a metal drum or a rotating belt; and from a slot die into precipitation bath.

Slush molding enables the production of hollow articles which are made from PVC pastes. These are made using two part metal molds, which are heated and filled with the paste. The paste gels on the outermost edges forming a skin ; and subsequently the excess paste is poured off. The remaining layer is gelled in an oven and after cooling, a hollow mold is obtained.

Slush molding can be further modified into Rotational casting for the production of hollow articles, made from PVC. The only difference is that in rotational casting, the amount of PVC paste is metered to the mold, which makes the removal of excess paste and post gelation techniques redundant. The mold is rotated about two perpendicular axes, while heating, and this enables uniform wetting of the cavity surface. Rotational casting has the advantage of low investment cost over slush molding. In the manufacture of thick-walled, symmetrical structures such as pipes and sockets, centrifugal casting may be used. A partially filled mold is rotated rapidly about an central axis. Centrifugal force press the material outward in the mold, and in the process, the material gets compacted.

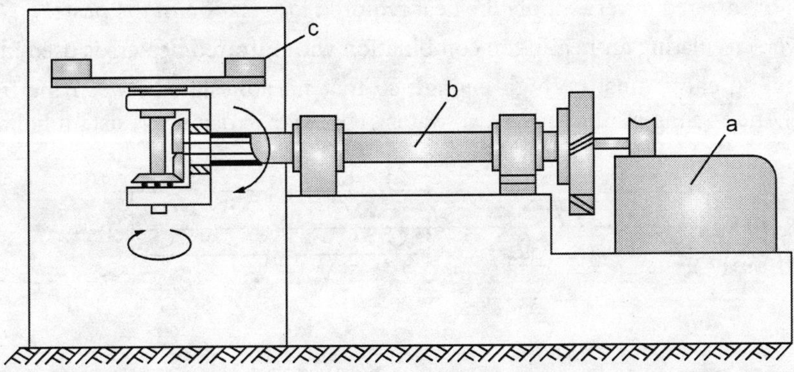

Fig. 4.1. Casting Machine (a) Motor, (b) Drive shaft and (c) Molds

(*b*) **Dipping (Fig. 4.2):** Sometimes, molds or articles are dipped into solutions, dispersions, pastes, melts or powders. Later on, the adhering material on the surface is solidified by heat treatment. For manufacturing articles such as gloves, boots etc. in which one side remains open, appositive mold is immersed and subsequently the casting is gelled. A series of identical molds mounted on dipping frames are used for the mass production of such articles. During the dipping process, degassed PVC paste placed in a vessel is raised and lowered. Sometimes, the dipping frame is moved, while the vessel is stationary. The immersion and removal stages are carried out slowly to allow the material to flow homogeneously from the mold and to prevent the introduction of air into the paste. The gelation is usually carried out in circulating air oven, and during the process the paste temporarily becomes fluid. The molds containing the gelled coating are immersed in water for cooling. It is cooled to around 50°C and at this temperature, the moldings can be easily removed from the molds. Cables and wires are extensively coated using this process.

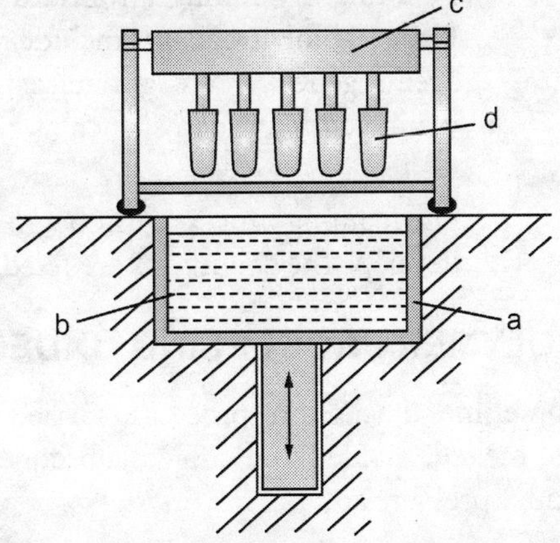

Fig. 4.2. Dipping Unit. (a) Vessel, (b) PVC paste, (c) Dipping frame and (d) Molds

(*c*) **Coating (Fig. 4.3):** This process is used in the production of self-supporting films by coating sheet materials such as paper and textiles. The most important materials that are used forme production of coatings are PVC pastes.

The equipment for coating consists of an unwinding station, an expander, coating head, gelling channel, cooling system, and a rewind station. The disk brakes slows down the unwind station, and the rewind station enables the product to be held under tension. A smooth finish to the sheet is provided by the expanders. With the help of the coating head, the PVC paste is spread upon the moving sheet manually or mechanically. Thereafter, a doctor blade is employed to spread the sheet to required thickness. A revolving roller supports the sheet, due to which an exactly defined contact area is maintained. The rolls distribute the escaping paste and form a film.

The rotation speeds of the rolls can be adjusted.

Hot air or infrared dryers supply the heat required for gelation of the paste.

However, circulating air heating in combination with infrared heaters is used in some cases.

The cooling effect must be high enough so that no adhesion between the individual layers occurs. The cooling is attained by two water - cooled, corrosion – resistant hollow drums.

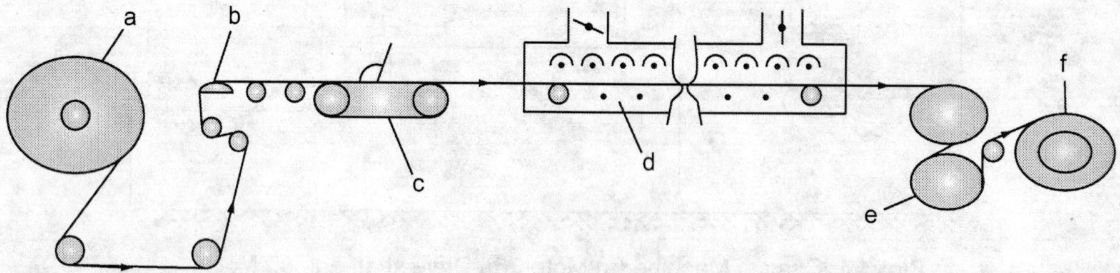

Fig. 4.3. Coating Unit. (a) Unwinding Station, (b) Expander, (c) Coating head, (d) Gelling Tunnel, (e) Cooling Roll, (f) Rewind Station

(*d*) **Foaming:** By adding physical or chemical blowing agents to a polymer, foams are generated. Physical blowing agents include gases and low-boiling compounds, whereas ; chemical blowing agents generally are gas releasing compounds. Only a few materials are employed in the manufacture of foams, such as polystyrene, PVC, LDPE as well as phenolic and polyurethane resins.

The foaming process must begin when the polymer is in a flowable form and allows bubbles to develop. The conditions are fixed, once the bubbles reach the optimum size.

POLYMER PROCESSING UNDER PRESSURE

Sometimes the molding process is carried under high pressure in which the solid thermoplastic materials are melted, molded and cooled subsequently. Some of the commonly used techniques for processing under pressure are:

- Compression molding
- Extrusion
- Injection Molding
- Rolling and Calendering
- Blow Molding

(*a*) **Compression Molding:** For the manufacture of thick-walled components and those that cannot be plasticized in extruders, compression molding is used. It plays an important role in the processing of elastomers and thermosets.

A press consisting of fixed and moving platens is used. The plunger and cavity components are mounted on these plates, and the parting plane is kept horizontal. The cavity is filled with the cold plastic and is melted by heating the mold under low pressure. When the melting is complete, the mold is closed and pressure applied. The molding is removed by releasing the press once the ejection temperature is reached.

Compression molding is uneconomical due to the fact that the process has a long cycle time ; as the material in the mold is heated and cooled by conduction via the mold surface.

(*b*) **Rolling and Calendering (Fig. 4.4):** The rolling mills consist of two rolls that are used exclusively for the compounding of thermoplastics. The product that results from rolling is a sheet that can be granulated after cooling or fed in plasticated form directly into the processing equipment. PVC and rubber are processed using this process.

When the rolls are more than two in number, and the polymer is to be shaped using these rolls; then the machine is called a calendar. The number of rolls in a calendar can be upto seven ; however, most commonly, four rolls are used. The rolls are made of polished cast iron, and are driven individually and heated. The heating medium is usually oil or water. The temperature and speed increases from one roll to the next. The roll mill must be able to withstand the extreme forces to which it is exposed, and hence, be made extremely rigid. The calenders are one of the most expensive types of processing equipments and are employed for the manufacture of a few special products like floor coverings and films made of plasticised PVC.

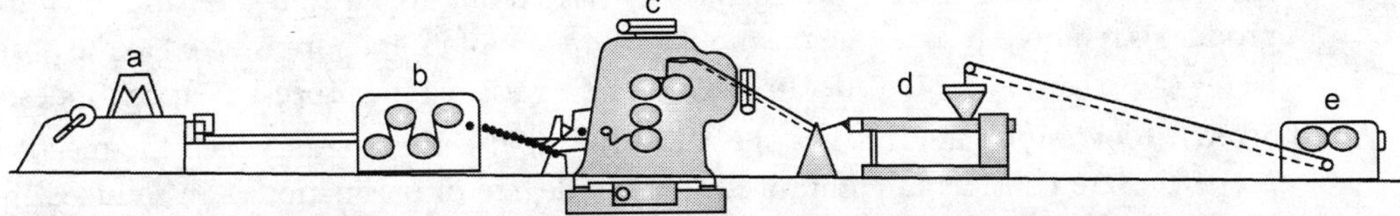

Fig. 4.4. Calendering line (a) Winder and Edge Cutter, (b) Cooling Rolls, (c) Four-roll calendar, (d) Extruder, (e) Mixing Roll Mill.

(*c*) **Extrusion (Fig. 4.5):** This method enables the attainment of a wide range of film thickness, and large scaling processing. Moreover, there is no necessity for expensive solvents in the recovery systems.

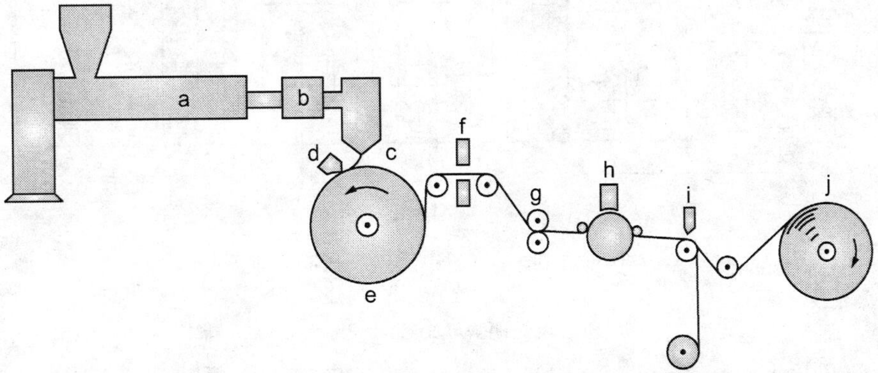

Fig. 4.5. Extrusion Molding. (a) Extruder, (b) Filter, (c) Slot Die, (d) Air Knife, (e) Cooling Roll, (f) Device for measuring film thickness, (g) Tension rolls, (h) Corona Treatment, (i) Edge Treatment, (j) Winding.

A device known as extruder is used to melt the polymer, which is given the specific shape by a die. Later on, a cooling system is used to solidify it.

At the onset, the raw material is dried using heated air or vacuum, before it is melted in the extruder. The extruders may be single – screw extruders, or twin screw extruders, the single screw extruder being the most common. The melt is filtered depending upon the desired purity, and then the filtered raw materials are sent through the slot die. The width and thickness of the film are determined by the width of the die and the die gap respectively. The die gap can be further adjusted during production.

The melt film emerging from the die is cast on to a highly polished roll. A vacuum device prevents the entrapment of air between the melt and the roller surface. The final film that is obtained is wound onto a roll after trimming off the edges. In a separate process, the film is cut into the designate width.

(*d*) **Blow Molding (Fig. 4.6):** This is the most common method for manufacturing hollow articles from thermoplastics. The most simple products and articles that are made by blow molding are bottles, ventilation ducts, suit cases and fuel tanks for automobiles. Two processes are generally followed for blow molding:

(*i*) **Extrusion Blow Molding:** An extruded tube or parison is elongated to reach the required length. The parison is then enclosed by the mold, which is then cut by a blade. It is then conveyed to the blowing station. In the blowing station, a blow mandrel is inserted into the mold and the actual process is subsequently carried out. It is then cooled and the finished product is removed by opening the mold. The open mold is then transferred to the position below the extrusion die in order to receive a new parison. The process then continues.

(*ii*) **Stretch Blow Molding:** This is a special form of blow molding in which the mechanical properties are considerably improved. A high degree of orientation is introduced in the plastics by drawing near the glass transition temperature or crystalline melting points. The molding material is drawn not only in the circumferential direction, but also in the machine direction. A mechanical ram is used for drawing in the longitudinal direction. The process is carried out under low temperature with the application of high deformation forces.

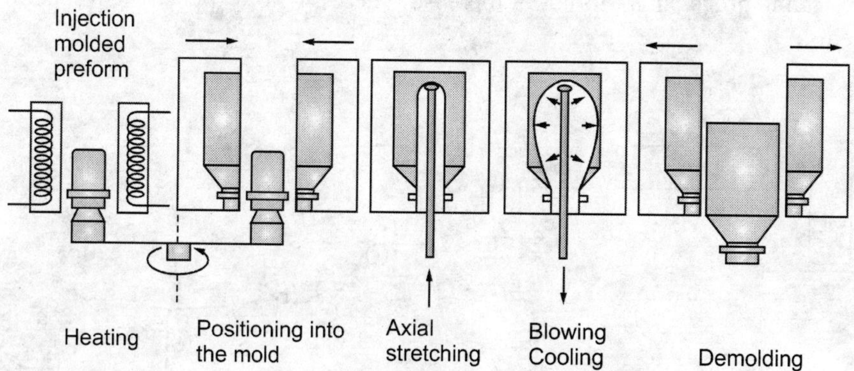

Fig. 4.6. Blow Molding Process

PVC: Polypropylene and poly ethyl terephthalate (PET) are some of the materials for use with this process.

(*e*) **Injection Molding (Fig. 4.7):** The Injection Molding technique is significant because of its ability to manufacture complex molding geometries in a single stage operation. The process is suitable for mass production operations because of the high degree of automation.

The Injection molding technique is a discontinuous process. The plastic pellets or granules are melted and injected under pressure into the mold cavity, wherein it is thermally cross- linked or solidified by cooling.

Through a runner system the hot melt prepared in the injection cylinder is injected into the cavity of the mold. The pressure is increased towards the end of the filling stage because of the increasing length of the flow path. After the injection phase, the holding pressure stage follows. It reduces the volume losses due to cooling of the melt, and thus, avoids marks and voids in the molding. After this, the molding is cooled sufficiently so that it is dimensionally stable for ejection. When the holding pressure phase is finished, the plastic is prepared for the next cycle by the plastication unit by rotating the screw. The melt is transported to the area in the front of the screw by the helical backward movement of the screw in the plastication cylinder against a back pressure. After this the molding is cooled and is mechanically removed from the mold via a handling equipment.

The entire process is fully automated and is coordinated by a control unit.

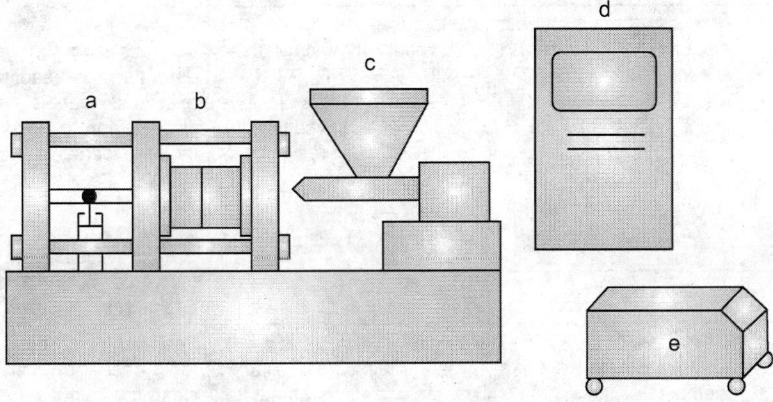

Fig. 4.7. Injection Molding Machine. (a) Clamping unit, (b) Mold, (c) Plastication Unit, (d) Control Unit, (e) Temperature Control Unit.

Table 4.1. Characteristics and typical products produced by various polymer processing processes.

Process	Characteristics	Resin Employed	Typical Products
Extrusion	A process for making indeterminate length of thermoplastics with constant cross-section	Most thermoplastics including PE, PP, PVC, ABS, PS	Pipes and tubing used for soda straws, garden hose, drains and vents, control cable housing, gas and water pipes, profiles, e.g., auto trim, home siding, storm windows, sheets used for window glazing, refrigerator liners, signs, plates, lighting, films for bags, covering laminates, packaging, fibers or filaments for brush bristles, rope, fishing line, insulated wire for homes, automobiles, appliances, telephone and electric power distribution, coated paper for milk cartons, meat packaging.

Injection molding	Versatile process, most suitable for high speed, low cost molding of intricate plastic parts required in high volume	Virtually all thermoplastics and some thermoset, most common are commodity plastics such as PVC, PE, PP and PS, others include ABS, nylon, cellulosics, acrylics	Automobile parts, appliance housing camera cases, knobs, gears, grilles, fan blades, bowls, spoon, lenses, flowers, wastebaskets and garbage cans.
Blow molding	Process used for making bottles and other hollow plastic parts having relatively thin walls	Several thermoplastics with PE (particularly ND, PE) having the largest volume, others include PVC, PP, PS, ABS, acrylics, nylons, acrylonitrile, acetates and PC	Bottles, watering cans, hollow toys, gas tanks for automobiles and trucks.
Rotational molding	Economical process for the production of hollow seamless parts with heavy walls and/or complex shapes	PE, (highest volume) PP, PVC together account for almost all plastics used, others include a number of engineering thermoplastics, including ABS, acetal copolymers, nylon (6 and 11), polycarbonate	Hollow balls, squeeze toys, storage and feed tanks, automobile dashboards, door liners and gearshift covers, industrial storage tanks and shipping containers, whirlpool tubs, recreational boats, canoes and camper tops, hobby horses, heater ducts, auto armrest skins, athletic balls, portable toilets.
Thermo forming	Process of forming moderately complex shapes that are not readily amenable to injection molding	Almost all thermoplastics but most commonly used include ABS, PP, PS, PVC polyesters, others include acrylics, polycarbonate, cellulosic, nitrile resins	Automobile, airline and mass transportation industries for such uses as auto headliners, fender walls, overhead panels, aircraft canoples, construction industry for exterior and interior paneling, bathtubs, shower stalls, outdoor signs, appliances, e.g., refrigerator liners, freezer panels, packaging trays for meat packing, egg cartons, fast-food disposables and carryouts, blister packages, suitcases, tote boxes, cups and containers.
Compression and transfer molding	Most widely used techniques for molding thermosets	Phenolic (largest volume), urea, melamines, epoxy, rubber, diallyl phthalate, alkyds	Pot handles, electrical connectors, radio cases, television cabinets, bottle closures, buttons, dinnerware, knobs, handles, replacement for metal parts in electrical automotive, aircraft industries.

Casting: Process for converting Polyesters, nylons,polyurethanes, silicones, epoxies, liquid resins into rigid objects of desired shape e.g., phenolics acrylics etc.

Tooling and Metal-Forming Industries

Cast epoxy dies used to produce airplane and missile skins, automobile panels and truck parts; epoxies used by artists and architects for outdoor sculpture, churches, homes and commercial buildings, and encapsulation in electronic industry; cast acrylic sheets used in airplanes, helicopters, schools.

MANUFACTURE, PROPERTIES & APPLICATIONS OF IMPORTANT PLASTICS

Polythylene (Polythene)

(POLYTHYLENE(POLYTHENE)

The manufacture of polyethylene follows addition polymerization reaction involving catalysis of purified ethylene. Its molecular formula is $—(CH_2 — CH_2)_n —$ with a molecular weight of 1,500 to 100,000. Its melting point is 85-110°C. Its density is 0.91-0.93, when produced by high pressure process(which produces low density polyethylene) and 0.96, when produced by low pressure process (which produces high density polyethylene).

Manufacture:

Poly(ethene) is made by several methods by addition polymerization.

There are three processes by which polyethylene is manufactured.

(*a*) **High Pressure Process (for producing LDPE):** This process was developed in the UK by ICI. It uses peroxide catalyst at 100-300°C and produces low density randomly oriented polymer which have a low melting point. The process is run at pressure of 1000-2500 atm. This process yields Low Density Polyethylene (LDPE). LDPE is generally amorphous and transparent with about 50% crystallinity. The branches prevent the molecules fitting closely together and so it has low density.

(*b*) **Intermediate Pressure Process:** This process was developed in the USA by Phillips Petroleum Co. for preparing high density polymer with increased rigidity, crystallinity, tensile strength and softening point. The process uses MoO_3 and Cr_2O_3 on alumina as catalyst and is operated at 30-100 atms.

(*c*) **Low Pressure Process (Ziegler process for producing HDPE):** This process was originally developed in Germany for preparing high density polyethylene (HDPE). The catalyst used in this process consists of aluminium triethyl activated with heavy metal derivatives such as $TiCl_4$.

Two types of catalyst are used principally in the manufacture of HDPE:

- a Ziegler-Natta organometallic catalyst (titanium compounds with an aluminium alkyl).
- an inorganic compound, known as a Phillips-type catalyst. A well-known example is chromium(VI) oxide on silica, which is prepared by roasting a chromium(III) compound at 1000 K in oxygen and then storing prior to use, under nitrogen.

Polymerisation reaction is as follows:

$$nC_2H_4 \longrightarrow \left[CH_2 - CH_2 \right]_n \qquad (\Delta H^\phi = -92 \text{ kJ mol}^{-1})$$

HDPE is produced by three types of processes. All operate at relatively low pressures (10-80 atm) in the presence of a Ziegler-Natta or inorganic catalyst. Typical temperatures range between 350-420 K. In all three processes hydrogen is mixed with the ethene to control the chain length of the polymer.

In the **Ziegler Process**, High purity desulphurised ethylene is treated to remove traces of oxygen which deactivates the catalyst. The purified ethylene is pumped to a reactor where it is continuously mixed with the catalyst diluent stream. Solvent polymerisation reaction conditions are maintained at 7 atm. and 70°C. The effluent goes to a series of flash drum to remove the solvent from the catalyst. Water is added to destroy residual catalyst (which is highly flammable in the dry condition).

The flashed solvent is dried, redistilled and recycled to the catalyst make up unit. Polythene slurry is dewatered by centrifuging and water is treated to remove the catalyst before recycling. Polyethylene solids are dried, extruded and finished to required form (pellets, flakes or powder). Yield based on ethylene is 95-98 %. Granules of poly(ethene) are used to make film, extruded into pipes or moulded.

Similar process with modifications in reaction conditions can be used for polypropylene.

In the process flow sheet (Figure below), is the description of Low Pressure Ziegler Process to produce polyethylene (HDPE).

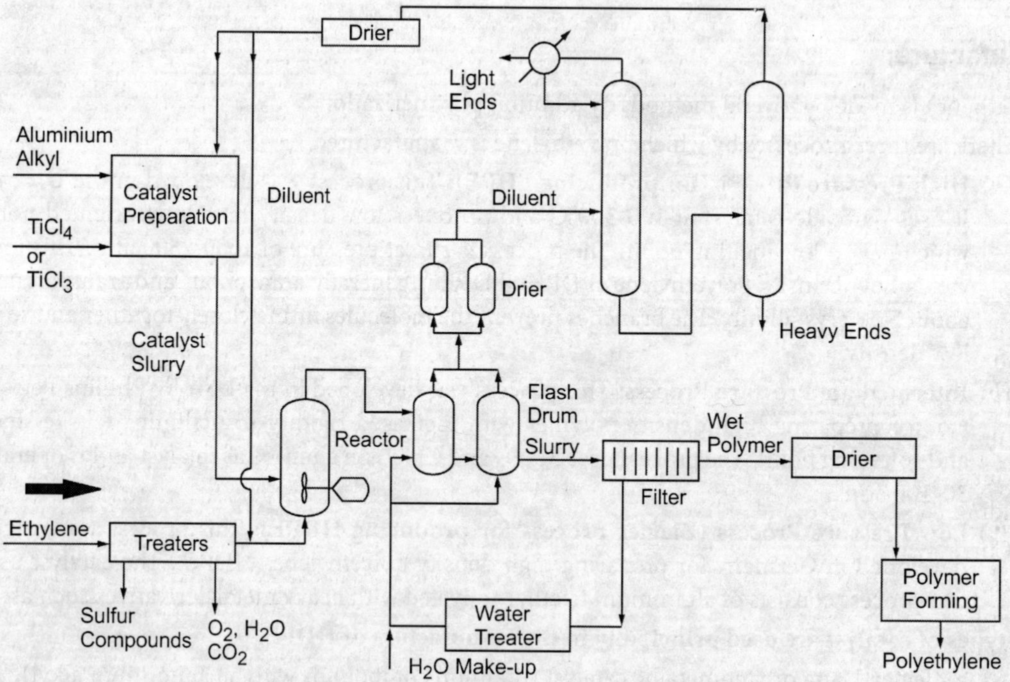

Fig. 5.1. Manufacture of polythene by low pressure Ziegler process to produce high density polyethylene (HDPE)

CLASSIFICATION OF POLYETHYLENE

Polyethylene is broadly classified into High Density Polyethylene (HDPE), Linear Low Density Polyethylene (LLDPE) and Low Density Polyethylene (LDPE).

High Density Polyethylene (HDPE)

High Density Polyethylene is predominantly linear in structure, closely packed with density of 0.960-0.980 gram/cm^3, crystalline as high as 95% and a melting point as high as 138.5°C.

Currently High Density Polyethylene finds its major application in Raffia, Blow Moulding, Film, Pipes and Injection Moulding. Consumption wise it is in order of HD-Film GP*Blow Moulding, Raffia, Pipe, have the major share. Large and Medium Blow Moulding containers, Pipe, General Purpose BM containers are currently the leading growth sectors.

HDPE because of its unique properties and increased availability offers abundant opportunities for growth. There is a huge untapped potential for the sector like for e.g., by the dilution of JPMA, loose to packed packaging in Blow moulding containers, automotive sector, infrastructure, and in installing drinking water and sewerage system etc.

Low Density Polyethylene (LDPE)

LDPE is predominantly a branched polymer. The properties depend upon the frequency and the length of branches. LDPE helps in processability, has excellent transparency and good sealing properties and low glass transition temp. Due to its inherent strength, LD is extensively used as a blend with other polymers.

LDPE finds its application in general purpose film, heavy duty film, liquid packaging, injection moulding, extrusion coating, wire and cables, adhesive lamination.

The major application wise share is in the order of general-purpose film, extrusion coating, milk packaging, heavy duty, injection moulding, wire and cables and others. Flexible Packaging constitutes 84% of the LD consumption. The growing sectors are general purpose film, heavy duty, milk packaging, injection moulding sectors, adhesive laminates.

Linear Low-Density Polyethylene (LLDPE)

LLDPE is linear in structure. The branches are formed due to presence of co monomer. The co-monomer used is Butene (C_4), Hexene (C_6) and Octene (C_8). The low pressure used in manufacture of LLDPE does not allow the formation of large branches.

LLDPE finds its application in mono and multi-layered films, roto-moulding, wire and cables and high flow LLDPE is used in master batch compounds etc.

Currently 59% of LL share comes from Butene film, followed by HAO and Metallocene LL film, roto-moulding, wire and cables, high flow LLDPE grades.

Currently the fast growing sector in LLDPE is Metallocene, followed by wire and cable and high flow grades. Metallocene LLDPE grades are fast moving due to its properties like better strength, high dart, higher line speed, better sealing, downsizing of the film etc. LLDPE is the major constituents of flexible packaging growing at more than 17%. Key end uses of flexible packaging being lamination, edible oil and milk.

Low density poly(ethene) has many uses but the high pressure method of manufacture by which it is produced has high capital costs. However, an elegant technique has been developed by using both Zeigler Natta and inorganic catalysts to produce linear low density poly(ethene) LLDPE, which has even improved properties over LDPE. Any of the three processes, slurry, solution and gas phase, can be used when a Ziegler-Natta catalyst is chosen. The gas phase process is used when the inorganic catalyst is employed.

Small amounts of a co-monomer such as but-1-ene or hex-1-ene are added to the feedstock. The monomers are randomly polymerized and there are small branches made up of a few carbon atoms along the linear chains.

The structure is essentially linear but because of the short chain branching it has a low density. The structure gives the material much better resilience, tear strength and flexibility without the use of plasticizer.. This makes linear low density poly(ethene) an ideal material for the manufacture of film products, such as those used in wrappings.

Metallocene Linear Low Density Polythene (mLLDPE)

This poly(ethene), known as mLLDPE, is produced by a new family of catalysts, the metallocenes. Another name for this family is *single site catalyst*. The benefit is that the mLLDPE is much more homogenous in terms of molecular structure than classical LLDPE produced by Ziegler-Natta catalysts. Each catalyst is a single site catalyst which produces the same PE chain. Chemists have compared the structure of metallocenes to that of a sandwich. There is a transition metal (often zirconium or titanium) 'filling' a hole between layers of organic compounds. The catalysts are even more specific than the original Ziegler-Natta and it is possible to control the polymer's molecular mass as well as its configuration. Either the slurry or solution processes are usually used. Poly(ethene) produced using a metallocene can be used as very thin film which has excellent optical properties and sealing performance, thus making them very effective for wrapping foods. The real plus for the metallocene catalysts are the enhanced mechanical properties of the films made out of mLLDPE.

Co-polymers of Polythene

Ethylene forms co-polymers with propylene which have very useful properties. The properties of the polymer, and hence its uses, can be varied by varying the proportion of ethylene and co-monomer and by using different co-monomers.

Uses of Polyethylene

Polyethylene is produced in three main forms: low density (LDPE) (< 0.930 g cm^{-3}) and linear low density (LLDPE) (*ca* 0.915-0.940 g cm-3) and high density (HDPE) (*ca* 0.940-0.965 g cm^{-3}).

The LDPE or LLDPE form is preferred for film packaging and for electrical insulation. HDPE is blow-moulded to make containers for household chemicals such as washing-up liquids and drums for industrial packaging. It is also extruded as piping.

All forms can be used for products such as buckets, food boxes and washing-up bowls (See Table below).

Process	HDPE	LDPE	LLDPE
Making film	Food packaging Shopping bags	Cling film Milk carton lining	Stretch film
Injection moulding	Dustbins Crates	Buckets Bowls	Food boxes
Blow moulding	Detergent bottles Drums	Squeezable bottles	
Extrusion	Water pipes	Flexible water pipes Cable jacketing	Cable coating

Table 5.1. Important properties of polyethylenes

Polymer	Melting point, °C	Density g/cm^3	Degree of crystallinity %	Stiffness modulus psi × 10^3
Branched, Low density	107–121	0.92	60–65	25–30
Medium density	—	0.935	75	60–65
Linear, High density				
Ziegler type	125–132	0.95	85	90–110
Phillips type	—	0.96	91	130–150

Table 5.2. Typical Applications of Polyethylene

Product Type	Typical Applications
Film extrusion	LDPE films used in packaging (bags and wrappings for frozen and perishable foods, produce and textile products); construction cover (moisture barriers and utility covering material); agriculture (greenhouse; ground cover; tank, pond, and canal liners); garment and stripping bags. HDPE film used for floral wrapping, grocery bags, snack and food packaging; EVA film used as produce bags, heavy-duty shipping bags, disposable protective gloves
Extrusion coating	Laminates of foil, paper and used in milk-type cartons (LDPE) for a variety of foods and drinks. Some HDPE used in extrusion coatings over flexible foil packages either as a glue layer between paper on plastics film and aluminum foil or as a heat-seal layer
Wire and cable insulation	LDPE is used as insulation of high-frequency electrical cables, insulation materials for television, radar, and multicircuit long-distance telephones
Blow molding	Squeeze bottles, toys, housewares, lids, containers made from LDPE. Blow-molded HDPE used as containers for bleaches, liquid detergents, milk and other beverages, automobile and truck gas tanks, crates and pails
Injection molding	Packaging containers, pails and lids, houseware dishpans, and waste baskets, molded furniture seats, medical labware. HDPE is used in molded structural foam pallets, and crates, underground conduits and housing.

Polypropylene

POLYPROPYLENE

Polypropylene (PP) is a major thermoplastic polymer.

Structure of Polypropylene

The propylene molecule is asymmetrical,

$$\mathrm{H_2C{=}CH{-}CH_3}$$

and, when polymerized, can form three basic chain structures dependent on the position of the methyl groups: two are stereoregular (isotactic and syndiotactic) and the third does not have a regular structure and is termed atactic as shown diagrammatically below:

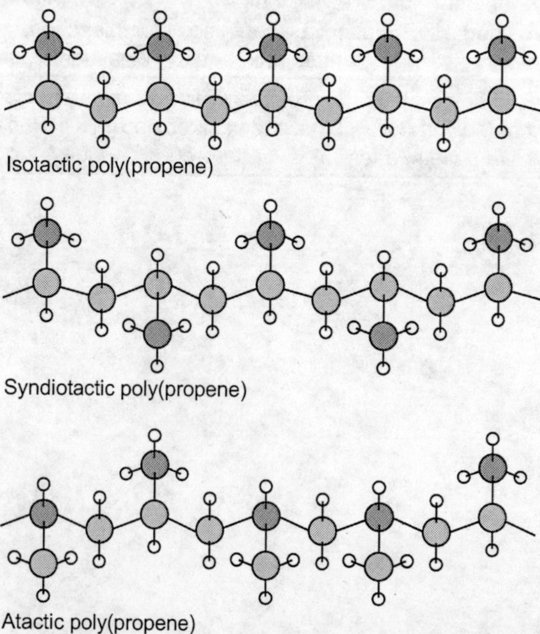

Fig. 6.1. Molecular structures of polypropylene

The 'one handed' structure of isotactic poly(propene) causes the molecules to form helices. This regular form permits the molecules to crystallize to a hard, relatively rigid material, which, in its pure form, melts at 440 K.

The syndiotactic polymer, because of its regular structure, is also crystalline. Atactic chains are completely random in structure and consequently they do not crystallize. High molecular mass atactic poly(propene) is a rubber-like material.

Commercial poly(propene) is a predominantly isotactic polymer containing 1-5% by mass of atactic material.

Manufacture of Polypropylene

Polypropylene is produced from propene. Propylene is produced in large quantities from gas oil, naphtha, ethane and propane. Parallel to this, there are several methods being developed to produce bio-based poly(propene).

(*a*) **Using a Ziegler-Natta catalyst:** Ziegler-Natta catalysts are used in the polymerization process. These are produced by interaction of titanium (IV) chloride and an aluminium alkyl, such as triethyl aluminium. Two main processes are used for making the polymer with these catalysts, although the slurry method is used as well.

(*i*) **The bulk process:** Polymerization takes place in liquid propene, in the absence of a solvent at a temperature of 340- 360 K and pressures of 30-40 atm (to keep the propene as a liquid). After polymerization, solid polymer particles are separated from liquid propene, which is then recycled. The use of liquid propene as a solvent for the polymer as it is formed means that there is no need to use hydrocarbons such as the C_4-C_8 alkanes which are used in the parallel manufacture of poly(ethene).

A mixture of propene and hydrogen is passed over a bed containing the Ziegler-Natta catalyst at temperatures of 320-360 K and a pressure of 8-35 atm.

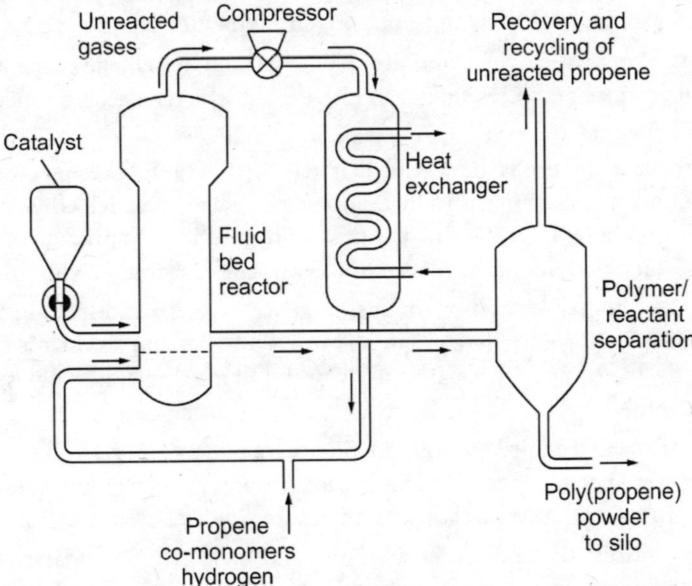

Fig. 6.2. Low pressure gas-phase process of polypropylene manufacture

The polymer is separated from the gaseous propene and hydrogen using cyclones and the unreacted gas is recycled.

Both processes can be operated continuously and use 'stereospecific' Ziegler-Natta catalysts to effect the polymerization. The catalyst remains in the product and needs to be destroyed using water or alcohols, before the polymer is converted into pellets.

Both bulk and gas phase processes have virtually eliminated gaseous and aqueous effluents by the use of high activity catalysts, resulting in low residues in the final polymer.

(*b*) **Using a metallocene as catalyst:** Metallocenes are being increasingly used as catalysts for the production of polythene and polypropylene.

Metallocenes are strictly defined as molecules which have a transition metal atom bonded between two cyclopentadienyl ligands which are in parallel planes. Ferrocene is a particularly well known example.

However, the term is now used more widely to include other ligands related to cyclopentadienyl. One such metallocene is based on zirconium.

As with the Ziegler-Natta catalysts, the bulk or gas phase (described above) can be used. Alternatively, the slurry process is used. Polypropylenes made in this way, mPP, are used in particular to make non-woven fibres and heat- seal films. Metallocenes also catalyse the production of co-polymers of propylene and ethylene.

Co-Polymers of Propylene

There are two main types of co-polymer. The simplest are the random co-polymers produced by polymerizing together ethylene and propylene. Ethene units, usually up to 6% by mass, are incorporated randomly in the polypropylene chains.

The crystallinity and melting point are reduced and the products are more flexible and are optically much clearer. Major uses for these random co-polymers are for medical products (pouches, vials and other containers) and packaging (for example, bottles, CD and video boxes).

Many other co-polymers of ethylene and propylene with higher alkenes such as hexene, are being developed which will produce polymers similar to LLDPE but which have better mechanical and optical properties.

The second type of co-polymers is the so-called 'block' co-polymers. These are made by following the poly(propylene) homopolymerization with a further, separate stage, in which ethylene and propylene are co-polymerized in the gas phase. Thus, these two processes are in series. The products of these two processes form a composite in which nodules of the block co-polymer are distributed with the homopolymer.

The ethylene content of the block co-polymer is larger (between 5 and 15%) than used in randomly alternating co-polymers. It has rubber-like properties and is tougher and less brittle than the random co-polymer. Consequently, the composite is particularly useful in making crates, pipes, furniture and toys, where toughness is required.

When ethene, propene and a third monomer, a diene, are polymerized, a rubber is formed, known as EPDM (**E**thene, **P**ropene, **D**iene, poly**m**ethylene. The ethene and propene molecules polymerize to form very long molecular chains, with several thousand monomer molecules in a chain.

The polymerization is usually effected in solution using a Zeigler-Natta catalyst but more recently, metallocenes have been used very successfully. Usually the ethene content is about 60% and that of the diene varies between 2 and 7%.

As the diene (usually ENB ethylidene norbornene) has two double bonds, one is used in the chain and the other is used to form a three dimensional structure. The reactive sites are pendant (not part of the backbone of the chain) and are joined together in the next part of the process when the polymer is heated with sulfur, the process used to vulcanize rubber. The two dimensional structure shown above becomes three dimensional.

Properties of Polypropylene

The properties of commercial polypropylene vary widely according to the percentage of crystalline isotactic polymer and the degree of polymerization.

Articles made from polypropylene have good electrical and chemical resistance and low water absorption. Its other useful characteristics are its light weight (lowest thermoplastic polymer density), high abrasion resistance, dimensional stability, high impact strength, and no toxicity.

Polypropylene can be extruded into sheets and thermoformed by solid- phase pressure forming into thin-walled containers. Due to its light weight and toughness, polypropylene and its copolymers are extensively used in automobile parts.

Improvements in melt spinning techniques and film filament processes have made polypropylene accessible for fiber production. Low-cost fibers made from polypropylene are replacing those made from sisal and jute.

Materials suitable for a much wider range of applications can be made by compounding poly(propene) with, for example, fillers, pigments and elastomers. Poly(propene) has remarkable properties, making it suitable to replace glass, metals, cartons and other polymers. These properties include:

- low density (weight saving)
- high stiffness
- heat resistance chemical inertness
- steam barrier properties (food protection)
- good transparency
- good impact/rigidity balance
- stretchability (film and fibre applications)
- good hinge property (for example where a lid and box are made together, for DVD boxes)
- high gloss (appearance)
- easy to weld (design) recyclability

The majority (60% of the total produced) of poly(propene) is produced as a homopolymer.

Poly(propene) is one of the lightest thermoplastics (density 0.905 g cm^{-3}). It has a melting point of 440 K and a crystallinity of 50-60%. The polymer, unlike polyethene, is transparent.

Table 6.1. Properties of Polypropylene

Property	Value
Density, g/cm^3	0.90–0.91
Fill temperature, max. °C	130
Tensile strength, psi	3,200–5,000
Water absorption, 24 hr., %	0.01
Elongation, %	3–700
Melting point, T_m °C	176
Thermal expansion, 10^{-5} in./in. °C	5.8–10
Specific volume, cm^3/lb	30.4–30.8

Uses of Polypropylene

Poly(propene) is processed into film, for packaging and into fibres for carpets and clothing. It is also used for injection moulded articles ranging from car bumpers to washing up bowls, and can be extruded into pipe

Use profile of polypropylene is shown below in figure.

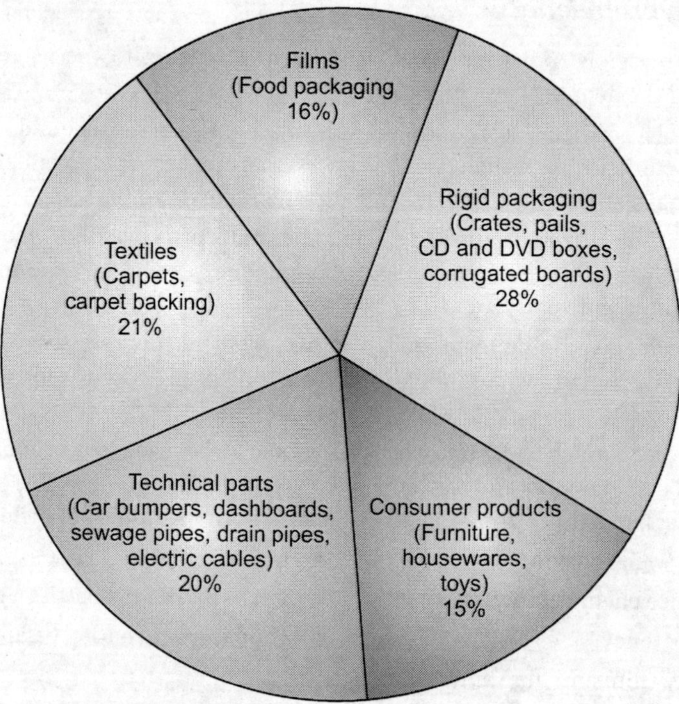

Fig. 6.3. Global Uses of polypropylene in 2012

Table 6.2. Typical Applications of Polypropylene

Product Type	Typical Applications
Injection molded	Automotive and appliance fields: distributor caps, radiator fans, accelerator pedals, battery casings, pillboxes, cabinet doors, bottle crates, jerry cans, cups, plates, file jackets, toys, food and drug containers
Extruded	Yarn for carpets, woven sacks and upholstery fabrics, hoses, drinking straws, hypodermic syringes, reusable towels, overwrap for cigarettes and phonograph records, liners for disposable diapers, furniture dust covers, geotextiles for road stabilization and erosion control

Poly Vinyl Chloride (PVC)

POLY VINLY CHLORIDE (PVC) $-[CH_2 - CHCl]_n-$

Polyvinyl chloride (PVC) is the most versatile plastic and, after poly(ethene), the most widely used.

The basic monomer for manufacture of PVC is vinyl chloride which is made from either acetylene or ethylene. However, ethylene is the chief source of vinyl chloride throughout world. Various sources of ethylene are naphtha gas cracking, acetylene route and alcohol route.

Manufacture of Poly Vinyl Chloride (PVC)

The production of PVC involves several stages:

(*a*) ethylene is converted into 1,2-dichloroethane

(*b*) 1,2-dichloroethane is cracked to chloroethene (vinyl chloride)

(*c*) polymerization of vinyl chloride to produce poly vinyl chloride

(*a*) **Production of 1,2-dichloroethane:** Ethylene is obtained from the cracking of ethane, propane, naphtha and gas oil. In practice, two processes for the overall conversion of ethene to chloroethene are used in parallel.

(*i*) **Direct chlorination of ethylene:** Ethylene is reacted with chlorine in the liquid phase, using excess chloroethene as solvent. The catalyst, iron(III) chloride, is soluble in chloroethene. The reaction is exothermic, and no external heat is needed.

$$\underset{\text{Ethene}}{CH_2 = CH_2\ (g)} + \underset{\text{Chlorine}}{Cl_2\ (g)} \xrightarrow{FeCl_3 \text{ catalyst}} \underset{\text{1,2-dichloroethane}}{ClCH_2CH_2Cl\ (g)}$$

The reaction temperature is maintained at 320-350 K and with a pressure of 4 atm. Higher temperatures give unwanted polychlorinated compounds.

(*ii*) **Oxychlorination route:** The cracking of 1,2-dichloroethane during stage (*b*) of the overall process forms hydrogen chloride as a by-product. It is unwanted as hydrogen chloride is very detrimental to the environment. However, it can be used in this second method of making 1,2-dichloroethane.

Ethene is mixed with hydrogen chloride and air (enriched with oxygen). The gases are passed over a heated solid catalyst in metal tubes, in a fluid bed reactor maintained at 500 K and 5 atm. The catalyst generally used is a mixture of copper(II) and potassium chlorides, deposited on alumina.

As the reaction is highly exothermic, the reactor vessel is cooled to give the optimum reaction temperature and to reduce unwanted by-product formation. Unreacted hydrogen chloride is removed by scrubbing the gases with aqueous sodium hydroxide solution. Impurities, such as unreacted ethylene and unwanted chlorinated hydrocarbons, are removed by distillation. Ethylene is recycled.

(*b*) **Production of vinyl chloride (chloroethene):** 1,2-dichloroethane is then passed rapidly through metal tubes heated to 650 K. It is cracked to eliminate HCl.

$$\underset{\text{1,2-dichloroethane}}{ClCH_2-CH_2Cl\ (g)} \xrightarrow{650\ K} \underset{\text{Chloroethane}}{CH_2=CHCl\ (g)} + HCl\ (g)$$

Thus the above two processes are normally operated in parallel, with the purified 1,2-dichloroethane from both routes cracked to give vinyl chloride. It is separated from hydrogen chloride and unconverted 1,2-dichloroethane in a two-stage distillation process. Hydrogen chloride is removed in the first stage and recycled to the oxychlorination process. The second stage separates **chloroethene (Vinyl chloride)** from unreacted 1,2-dichloroethane, which is recycled to the cracker.

The cracking process has been modified by increasing the volume of the reactor. This increase in volume increases the conversion and reduces the energy required to produce chloroethene.

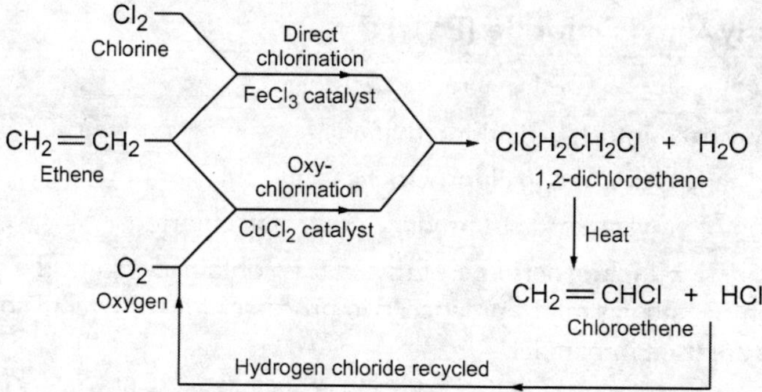

Fig. 7.1. The manufacture of chloroethene(Vinyl Chloride)

The vinyl chloride monomer (VCM) has a boiling point of 13.4°C and is a gas at room temperature and pressure. The vapor pressure of VCM over the typical polymerization temperature range of 50 to 70°C is 800-1250 KPa.

(*c*) **Polymerization of vinyl chloride (chloroethene) to produce polyvinyl chloride(polychloroethene):** This is an example of addition polymerization. PVC is made by free-radical polymerization in suspension. The vinyl chloride monomer (bp 259 K) is polymerized in aqueous dispersion at 325-350 K. Pressure (13 atm) is used to keep the monomer in a liquid phase. For polymerization to be controlled, an initiator is needed. In suspension polymerization an initiator, an organic peroxide is used, which is soluble in chloroethene. After the reaction, excess monomer is removed and the polymer is separated, by centrifuging and drying.

$$\underset{\text{Chloroethene}}{CH_2=CHCl} \xrightarrow[13\ atm]{325\text{-}350\ K} \underset{\text{Poly (chloroethene)}}{-\!\!\left[CH_2-CHCl\right]\!\!-_n}$$

The two most commercially used methods for the manufacture of PVC are Emulsion Polymerization and Suspension Polymerization. The Suspension Polymerization process provides 80 % of the world production. During polymerization, the polymer precipitates out as it is formed, since it is insoluble in the monomer. It is used for the extrusion, injection moulding and film-making processes.

It is estimated that nearly 80% of the PVC polymer is produced by suspension polymerization. Alternatively, vinyl chloride is polymerized as an emulsion in water. Ammonium peroxy disulfate is often used as the catalyst as it is soluble in water. The monomer is in the form of very fine droplets rather similar to paint. On evaporation after the polymerization, a very fine powder is formed. This is used as a coating on, for example, flooring and wallpaper.

A monomer soluble free radical initiator is charged into the reactors. After charging, the reactor temperature is increased to 45 – 75 °C. The heat decomposes some of the initiators to free radicals, and the monomers in these droplets begin to polymerize., the reaction is highly exothermic and the heat is removed via cooling jackets or by boil – off to the condenser. Thereafter, the condensed monomer is returned to the reactor. Although the PVC is insoluble in its monomer, it is swollen by VCM to form a coherent gel. Even in the gel phase, the polymerization continues. The polymerization is rapid at first, but slowly, as the conversion reaches 80 – 85 %, the rate is reduced due to monomer starvation.

At a predetermined pressure, the reaction is ended by adding a chain terminator or by venting the unreacted monomer to a recovery plant. Even after venting, the aqueous slurry contains 2 – 3 % unreacted monomer, which is then removed by stripping in a Stripping Column. The unreacted monomer is recovered and stored for later polymerization stages. The slurry is then passed through a heat exchanger and is passed through a continuous centrifuge to give a wet cake with 20 – 30 % moisture.

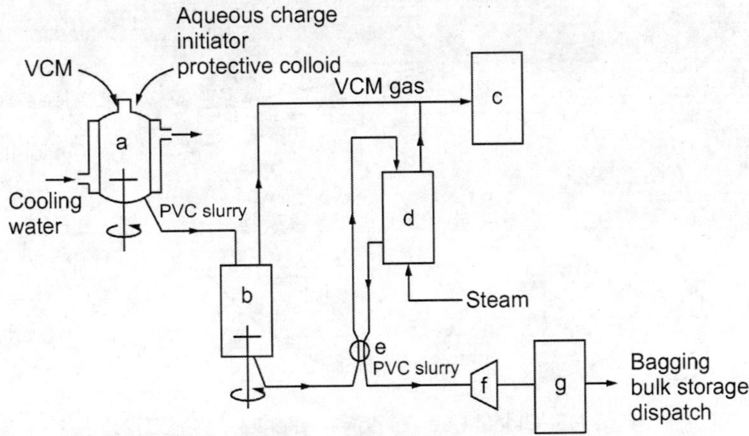

Fig. 7.2. Manufacture of Poly (Vinyl Chloride)
(a) Reactor, (b) Blowdown Vessel, (c) VCM Recovery Plant, (d) Stripping Column, (e) Heat Exchanger, (f) Centrifuge, (g) Driers

Properties and Uses of Polyvinyl Chloride

Polyvinyl chloride (PVC) is one of the most widely used thermoplastics. It can be extruded into sheets and film and blow molded into bottles. It is used in many common items such as garden hoses, shower curtains, irrigation pipes, and paint formulations.

Two types of the homopolymer of vinyl chloride are available, the flexible and the rigid. Both types have excellent chemical and abrasion resistance. The flexible types are produced with high porosity to permit plasticizer absorption. Articles made from the rigid type are hard and cannot be stretched more than 40% of their original length. An important property of PVC is that it is self-extinguishing due to presence of the chlorine atom.

Flexible PVC grades account for approximately 50% of PVC production. They go into such items as tablecloths, shower curtains, furniture, automobile upholstery, and wire and cable insulation.

Many additives are used with PVC polymers such as plasticizers, antioxidants, and impact modifiers. Heat stabilizers, which are particularly important with PVC resins, extend the useful life of the finished product.

Rigid PVC is used in many items such as pipes, fittings roofing, automobile parts, siding, and bottles

PVC is used in *building and construction*, for example in making window and door profiles and in making pipes (from guttering to sewers, and ducting for cables). It is also much used for *packaging*, including film for food and bottles. *Wire* and *cable* insulation and sheathing, for example in telephones, computers as well as for power transmission cables, is another main outlet for the polymer.

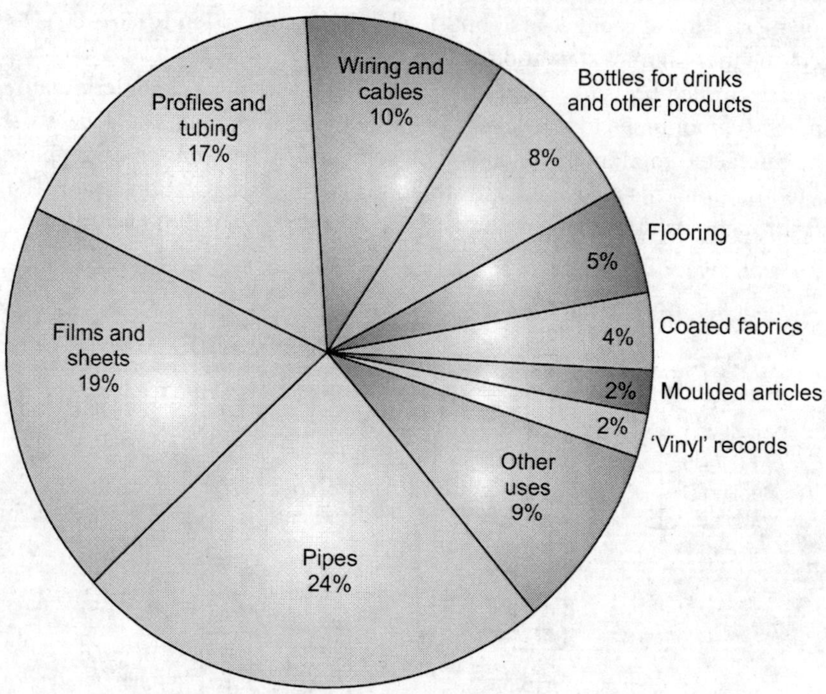

Fig. 7.3. Global Uses of polyvinyl chloride in 2012

Co-polymers of Vinyl Chloride

Vinyl chloride can be copolymerized with many other monomers to improve its properties. Examples of monomers used commercially are vinyl acetate, propylene, ethylene, and vinylidine chloride. The copolymer with ethylene or propylene (T_g = 80°C), which is rigid, is used for blow molding objects. Copolymers with 6-20% vinyl acetate (T_g = 50-80°C) are used for coatings.

Vinyl chloride is also used to produce co-polymers for example with vinyl acetate. Co-polymer (Poly vinyl acetate-PVA) can be processed at much lower temperatures than poly vinyl chloride. It is used for coating metals and wood. The film is flexible, hard wearing, resistant to chemical attack and can be pigmented so that it is particularly suitable, for example, for ships. The co-polymer is also used in adhesives and in inks.

Vinyl chloride is copolymerized with other monomers to improve quality of the resin. When vinyl acetate is used as copolymer, polymer is more stable to light and heat than the homopolymer of vinyl chloride. With vinylidene chloride has a higher melting point than homopolymer and may be spun into a fibre

PVC today is recognized as 'an infrastructure plastic' finding applications in pipes, conduits, ducts, wire and cables, flooring, windows and doors and roofing etc. It is also used in many other sectors including automobiles, medical and healthcare, packaging, sports and leisure. Pipes continues to be the biggest end use sector with a share of ~70%, other sectors being calendered products, wire and cables, films and sheets, profiles, footwear and medical and healthcare.

Green Building concept is increasingly being accepted in the country. Indian Green building Council, TERI Griha, and Bureau of Energy Efficiency are working towards green construction activities with usage of materials that save energy at all stages. PVC is known to save energy during the entire life cycle. Besides; products like windows, doors, partition, wall paneling false ceiling, and furniture when made in PVC, save natural resource like wood which would help in saving fast depleting forests.

Fig. 7.4. Table 7.1. Typical Applications of Poly vinyl chloride

Area	Typical Application
Piping systems	Pressure pipes — water supply and distribution, agricultural irrigation, chemical processing; non pressure pipes — drain, waste and vent pipes, sewer systems, conduits for electrical and telephone cables
Building construction	Siding, window frames, gutters, interior molding and trim, flooring, wire and cable insulation, wall coverings, upholstery, shower curtains, refrigerator gaskets
Transportation	Upholstery, floor mats, auto tops, automotive wire, interior and exterior trim
Consumer products	Footwear, outerwear, phonograph records, sporting goods, toys

Polymethyl Methacrylate (PMMA)

POLY METHYL METHAACRYLATE (PMMA-POLYMETHYL-2-METHYL PROPNOATE)

$$\left[-H_2C-\overset{\overset{\displaystyle CH_3}{|}}{\underset{\underset{\displaystyle C=O}{|}}{C}}- \right]_n \quad \text{with } C=O \text{ bearing } O-CH_3$$

Polymethyl methacrylate(PMMA) is one of the best known polymers, used widely under trade names such as Lucite, Perspex and Altuglas.

It is a polymer of the methyl metha acrylate monomer, which is produced from acrylonitrile (CH_2CHCN) and sulfuric acid at a temperature below 90°C to yield methyl acrylate (CH_2CONH_2) followed by reacting with methanol to convert methyl acrylate to methyl methacrylate ($CH_2CHCOOCH_3$). This methyl metha acrylate monomer is then polymerised to produce PMMA in a stirred tank autoclave under nitrogen pressure. Any of the solution, emulsion, mass or suspension polymerisation methods are used.

Manufacture of Poly Methyl Metha Acrylate (PMMA)

The monomer methyl methacrylate is made by two processes:

(*i*) **From acetone:** Worldwide, over 80% of the monomer is made from acetone by a sequence of steps which begins by reacting acetone with hydrogen cyanide.

(*ii*) **From ethylene, carbon monoxide and methanol:** Another promising route for production of methyl methacrylate, which is now in use, uses a mixture of ethylene, carbon monoxide and methanol in the liquid phase under pressure of about 10 atm at 350 K:

Polymerization of methyl metha acrylate is achieved by a free radical process using an initiator, such as an azo compound or a peroxide. The amount of initiator employed affects both polymerization rate and resulting molecular mass of the polymer. Polymerization is carried out commercially in several ways, i.e. in bulk, solution, suspension and emulsion. The process is an example of addition polymerization.

Acrylic, a polymer called **poly (methyl methacrylate)** (PMMA), is also known by trade names such as Lucite, Perspex and Plexiglas. It serves as a sturdy substitute for glass for such items as aquariums, motorcycle helmet visors, aircraft windows, viewing ports of submersibles, and lenses of exterior lights of automobiles. It is extensively used to make signs, including lettering and logos. In medicine, it is used in bone cement and to replace eye lenses. Acrylic paint consists of PMMA particles suspended in water.

Co-polymers of Methyl Methacrylate

Co-monomers are often used together with the methyl methacrylate. For example, most commercial grades of poly methyl methacrylate used in injection moulding or extrusion applications contain a small amount (4%) of co-monomer, such as methyl acrylate (when casting sheets of the polymer) and ethyl acrylate (when extruding sheets of the polymer).

In these co-polymers, the monomers are randomly arranged. The resulting polymers have increased thermal stability compared to the homopolymer. With butyl acrylate, a co-polymer is produced which is used as a base for emulsion paints.

It is also co-polymerized with ABS to produce a very tough polymer which is both rigid and has excellent clarity. It is used, for example, in medical applications and in cosmetic packaging.

Uses of Polymethyl Methacrylate (PMMA)

PMMA is better known as Lucite, Perspex, plexiglass and Altuglas (when in sheet form) and as Diakon (when in powder form). The cast sheet is used in baths and other sanitary ware, which along with illuminated signs, is the largest use of the polymer. High molecular mass cast sheet (Perspex) is also used as a lightweight replacement for glass. Lower molecular mass products, made by suspension or solution polymerization (Diakon), are used in car lights and domestic lighting. Special grades are used in diverse applications such as false teeth and eyes and as a major component of bone cements. The monomer is used in adhesives, surface coatings and in paints.

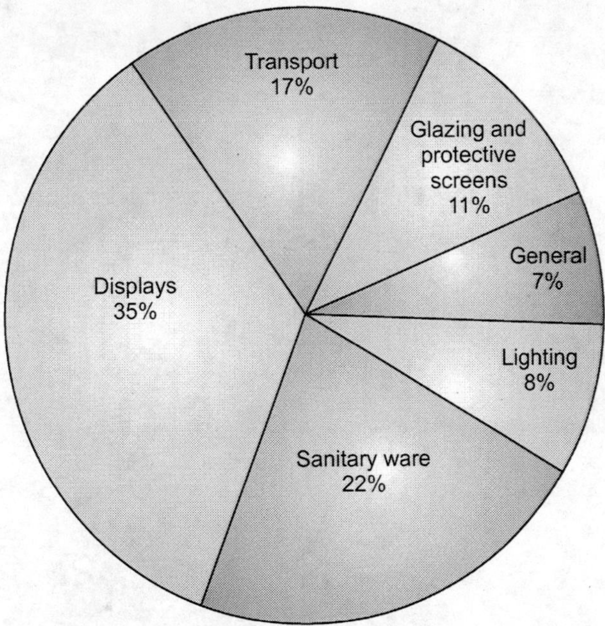

Fig. 8.1. Global Uses of poly methyl methacrylate (PMMA) in 2012

Table 8.1. Typical Applications of Poly methyl methacrylate (PMMA)

Area	Typical Applications
Construction	Enclosures for swimming pools, shopping malls and restaurants, tinted sunscreens to reduce air- conditioning and glare, domed skylights
Lighting	Lighted signs, luminous ceilings, diffusers, lenses and shields
Automotive	Lenses, instrument panels, signals and nameplates
Aviation	Windows, instrument panels, lighting fixture covers, radar plotting boards, canopies
Household	Housings, room dividers, decorating of appliances, furniture, vanities, tubs, counters
Others	Display cabinets and transparent demonstration models in museums, exhibits, and departments stores

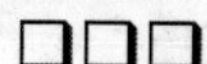

Polystyrene (PS)

POLYSTYRENE (PS)

$$\left[-CH_2 - CH(C_6H_5) - \right]_n$$

Polystyrene, is the third most important polymer, in terms of amount made from ethylene. Its physical properties can be adjusted to suit a range of everyday uses. Techniques have been developed which increase its mechanical strength, its ability to absorb shock and its thermal insulating properties.

Manufacture of Polystyrene

Polystyrene is manufactured from its monomer styrene. Styrene, in turn, is produced from benzene and ethylene *via* ethylbenzene. There are thus following three stages in the manufacture of polystyrene :

(*a*) the manufacture of ethylbenzene from benzene

(*b*) the manufacture styrene

(*c*) the polymerization of styrene

(*a*) **The manufacture of ethylbenzene from benzene:** Benzene vapour and ethene are mixed and passed over an acid catalyst, at 650 K and 20 atm pressure.

$$\underset{\substack{\text{Benzene} \\ C_6H_6}}{C_6H_6\,(g)} + \underset{\text{Ethene}}{H_2C{=}CH_2\,(g)} \xrightarrow[\text{Acid catalyst}]{650\text{ K, 20 atm}} \underset{\text{Ethylbenzene}}{H_3C-CH_2(C_6H_5)\,(g)}$$

(*b*) **The manufacture of styrene (Phenylethene):** Ethyl benzene vapour is mixed with excess steam and passed over heated iron (III) oxide. Other metal oxides used as the catalyst including those of magnesium, chromium (III) and zinc, usually coated on carbon or alumina. It is dehydrogenated:

$$\underset{\text{Ethylbenzene}}{H_3C-CH_2(C_6H_5)\,(g)} \xrightarrow[Fe_2O_3\text{ catalyst}]{850\text{ K}} \underset{\text{Phenylethene}}{H_2C{=}CH(C_6H_5)\,(g)}$$

A small amount of potassium oxide is mixed with the iron (III) oxide (which keeps the catalyst in the iron (III) state).

The steam reduces 'coking' (the formation of soot on the catalyst from the decomposition of ethyl benzene at the high temperatures used).

(*c*) **The polymerization of styrene to produce polystyrene:** The vast majority of all polystyrene produced today is produced via Solution (bulk) polymerization. The common solvents used in this process are the styrene monomer itself and ethyl benzene. The two types of mass polymerization are batch and continuous, of which continuous mass is by far the most popular.

Batch process: Batch mass polymerization consists of a polymerization section containing agitated vessels polymerizing up to 80% conversion in a batch method. The polymerized solution is then pumped to a batch finishing section for either devolatisation for final polymerization and grinding. In a typical batch suspension process, styrene is suspended in water by agitation and use of a stabilizer. The polymer forms beads. The bead/water slurry is separated by centrifugation, dried, and blended with additives.

Continuous process: The most widely used process for polymerization of styrene today is the continuous mass process. This solution is continuously prepared in a holding vessel and will then be injected into the reactor system.

Typical feed to the first reactor would consist of 50 weight percent styrene monomer, 100 ppm water (based on styrene weight), 2000 ppm boron trifluoride (based on styrene weight), with the balance being organic solvent. The polymerization reaction gives off heat that is carried away from the reactors by jacketing them with a heat transfer fluid. The temperature of the reactants should not vary by more than 15°C throughout the reactor series. Temperature control is very important in this reaction because as the reaction temperature increases, the average molecular weight of the polystyrene decreases. The reaction temperature range is 40-70°C. Temperature can also be controlled by intermediate shell and tube heat exchangers. Monomer conversions of up to 85wt% styrene are obtainable in these reactors.

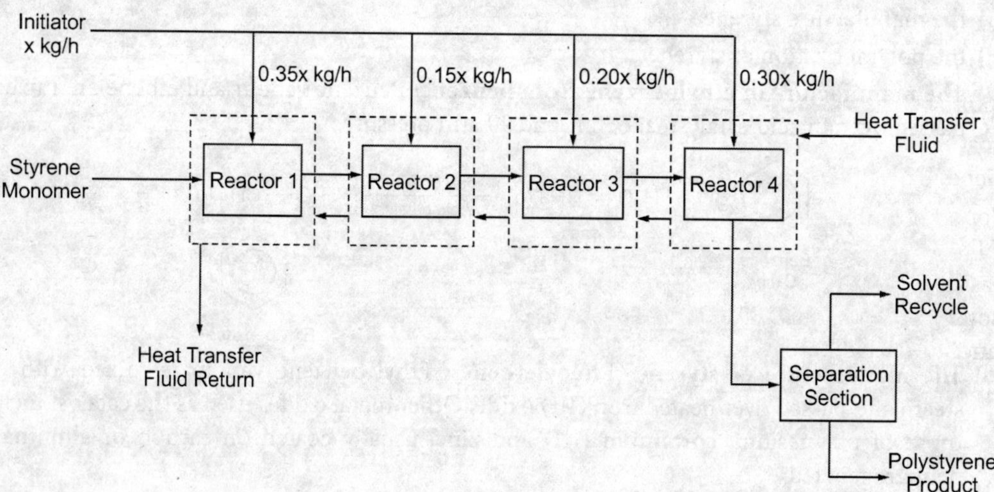

Fig. 9.1. Flow diagram for the manufacture of polystyrene

Classification of Polystyrene

We have three types of polystyrenes with different application - General purpose Polystyrene (GPS), High Impact Polystyrene (HIPS) and Expandable Polystyrene (EPS).

General Purpose polystyrene (GPS): It has excellent transparency, processability and its grades are used to cover a wide variety of applications such as disposable containers, cutlery, CD storage cases, co-extrusion to obtain gloss layer, toys, stationery and refrigeration components.

High Impact polystyrenes (HIPS): Polystyrene is a clear thermoplastic, with good moisture resistance, but is rather brittle. A tougher product is also manufactured by polymerizing styrene containing 5-10% dissolved poly(buta-1,3-diene) rubber. This tougher product - generally known as High Impact Polystyrene (HIPS) - is made exclusively by continuous thermal mass polymerization, in which heat is required to initiate the polymerization reaction. This toughened polymer is translucent. The structure of polystyrene made by these technologies is completely random in structure and is known as an atactic structure. By modification of the polymerization technique - principally by the use of metallocene catalysts - stereoregular (syndiotactic) structures can be obtained. This syndiotactic polymer (sPS) has improved properties - particularly thermal and mechanical. HIPS resins are specifically designed to meet the mechanical properties and productivity requirements of a wide variety of applications both in injection and extrusion/thermoforming processes. Typical applications include electronics, household appliances, disposable containers, toys and refrigeration components.

Expandable Polystyrene (EPS) also known as Polystyrene foam: It is prepared by suspension polymerization of styrene, and then adding foaming agent and white bead-like particles, with relative density of 1.05 are obtained. Thermal conductivity is low. EPS is a rigid foam honeycomb material with good insulation, vibration absorption, high compressive strength, very light weight and resistance to moisture; its uses include building insulation and sound insulation, sidewall, and within the walls of the Cover, packaging materials and disposable containers. Expanded polystyrene is manufactured as beads containing pentane (a liquid at room temperature). When they are heated in steam, the hydrocarbon volatilises and the beads expand. Beads are impregnated with graphite powder to improve further their ability to absorb heat. These are subsequently blown into moulds and fused by further steaming and then cooling. The expanded polystyrene has good thermal insulation and shock absorbing properties.

Copolymers of Styrene

Copolymers with butadiene in a ratio of approximately 1:3 produces **SBR** (Styrene Butadiene Rubber), the most important synthetic rubber. The general purpose rubber (SBR), a block copolymer with 75% butadiene, is produced by anionic polymerization.

Copolymers of styrene-acrylonitrile (**SAN**) have higher tensile strength than styrene homopolymers. SAN ($T_g = 105°C$) is stiffer and has better chemical and heat resistance than the homopolymer. However, it is not as clear as polystyrene, and it is used in articles that do not require optical clarity, such as appliances and houseware materials. SAN (styrene-acrylonitrile) is less flexible, more transparent and has more resistance to heat and chemicals than polystyrene. It is used in car headlamps, cassette covers, syringes and high quality kitchen appliances. Often SAN is blended with ABS to make it an even more rigid polymer.

A copolymer of acrylonitrile, butadiene, and styrene (**ABS**) is an engineering plastic due to its better mechanical properties. ABS has a specific gravity of 1.03 to 1.06 and a tensile strength in the range of 6 to 7.5×10^3 psi. These polymers are tough plastics with outstanding mechanical properties. A wide variety of ABS modifications are available with heat resistance comparable to or better than polysulfones and polycarbonates. Another outstanding property of ABS is its ability to be alloyed with other thermoplastics for improved properties. For example, ABS is alloyed with rigid PVC for a product with better flame resistance.

Among the major applications of ABS are extruded pipes and pipe fittings, appliances such as refrigerator door liners, and in molded automobile bodies.

ABS is tougher, scratch proof and more chemically resistant than rubber-modified polystyrene and is used, for example, in casings for computers, cycle helmets, calculators, telephones, vacuum cleaners and toys.

ABS may be processed into end products by most thermoplastic processing methods such as injection and compression moulding, extrusion, blow moulding and calendering. Other operations such as vacuum forming, vapour metallizing, plating, hot stamping and painting may also be carried out. ABS's light weight and ability to be injection molded and extruded make it useful in manufacturing products such as drain-waste-vent (DWV) pipe systems, musical instruments (recorders, plastic clarinets, and piano movements), golf club heads (because of its good shock absorbance), automotive trim components, automotive bumper bars, medical devices for blood access, enclosures for electrical and electronic assemblies, protective headgear, whitewater canoes, buffer edging for furniture and joinery panels, luggage and protective carrying cases, small kitchen appliances, and toys, Household and consumer goods are the major applications of ABS. Keyboard keycaps are commonly made out of ABS.

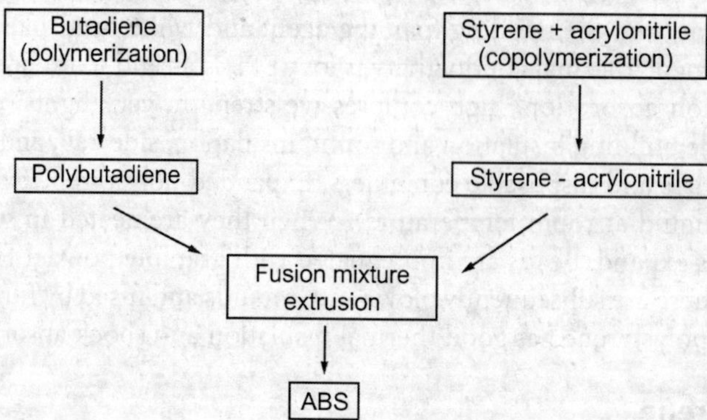

Fig. 9.2. Flow diagram for manufacturing of ABS

Polystyrene accounts for 50% of the polymers produced from styrene and SBR,ABS and SAN, a further 27%, 11% and 1% respectively.

Polystyrene is an important thermoplastic. Polystyrene because of its ease of fabrication, low specific gravity, thermal stability and low cost, find wide applications in consumer durable goods, electronics, packaging toys, structural foams, wall tiles, shoe soles, blister packages, lenses, bottle caps, wire and cable sheathing, small jars, vacuum formed refrigerator liners, containers of all kinds, transparent display boxes and automobile interior parts.

Properties and Uses of Polystyrene

Polystyrene homopolymer produced by free radical initiators is highly amorphous (T_g = 100°C). The most important use of polystyrene is in packaging. Molded polystyrene is used in items such as automobile interior parts, furniture, and home appliances. Packaging uses plus specialized food uses such as containers for carryout food are growth areas. Expanded polystyrene foams, which are produced by polymerizing styrene with a volatile solvent such as pentane, have low densities. They are used extensively in insulation and flotation (life jackets).

Uses of (Polystyrene)

The largest use for polystyrene is for packaging, particularly for foods such as poultry and eggs, for cold drinks and take-away meals. It is also used in making appliances, including refrigerators, microwaves and blenders. It is the leading choice for jewel boxes (cases for CDs and DVDs) and is also widely used for its insulating properties.

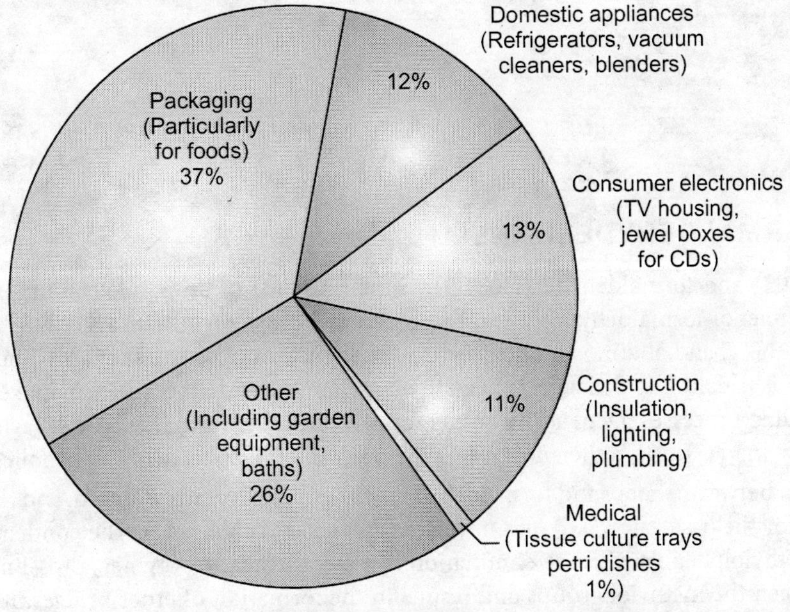

Fig. 9.3. Global Uses of polystyrene in 2012

In India, Polystyrene end use is mainly into the "Electronics/Appliances" sector, which accounts for almost 60 percent of total demand. One reason why PS demand in the Indian Subcontinent has been below expectations is the competition from other polymers, mainly polypropylene in both packaging and durables, primarily due to cost reasons.

Table 9.1. Some Applications of Different Grades of Polystyrene Processing

Methods	Typical products of polystyrene
Extrusion	Containers, tubs and trays for food packaging, mirror and picture frames, refrigerator breaker strips, room dividers, shower doors, glazings, lighting
Thermoforming	Refrigerator door liners, signs, horticultural trays, luggage; furniture panels, glove compartment boxes; meat, poultry, and egg containers; fast-food containers, blister packs, container lids; cookie, candy, pastry and other food packages
Injection molding	Packaging and nonpackaging disposables; air conditioner grilles; refrigerator and freezer components; small appliance housings
Others	Consumer electronics such as cassettes, reels, radio, television; and stereo dust covers; business machine housings, smoke detectors, display racks, toys, disposable tumblers, cutlery, bottles, combs, brush block

Phenol Formaldehyde (Bakelite)

PHENOL FORMALDEHYDE (BAKELITE)

Phenol (C_6H_5OH) and formaldehyde ($HCHO$) in the presence of an aqueous ammoniacal medium condense to a phenol-formaldehyde resin. The reaction is carried out in a kettle-type reactor at a temperature of 160°C and at atmospheric pressure. The reaction time is less than 1 hour. The product from the reactor is usually blended with hexamethylene triamine and other fillers to make finished resins ready for moulding processes. Depending on the ratio of phenol to formaldehyde, a variety of grades of this resin may be made. This is a thermosetting type resin and is also termed as phenolic resin.

The reaction between phenol and formaldehyde is either base or acid catalyzed, and the polymers are termed resols (for the base catalyzed) and novalacs (for the acid catalyzed). The condensation reaction between the methylolphenols with the elimination of water forms the polymer. Crosslinking occurs by a reaction between the methylol groups and results in the formation of ether bridges. It occurs also by the reaction of the methylol groups and the aromatic ring, which forms methylene bridges. The formed polymer is a three-dimensional network thermoset.

Usually, the ratio of formaldehyde to phenol is kept less than unity to produce a linear fusible polymer in the first stage. Crosslinking of the formed polymer can occur by adding more formaldehyde and a small amount of hexamethylene tetramine [hexamine, $(CH_2)_6N_4$]. Hexamine decomposes in the presence of traces of moisture to formaldehyde and ammonia. This results in crosslinking and formation of a thermoset resin.

Manufacture of Phenol Formaldehyde

The formation of phenol formaldehyde follows step-growth polymerization. The process can be either acid-catalysed or base-catalysed.

Phenol and formaldehyde (methanal) are heated in an acidic or alkaline solution. The first step in forming a polymer chain involves substitution of formaldehyde in the phenol ring in the 2- or 4-position:

$$H_2C=O + C_6H_5OH \longrightarrow C_6H_4(OH)CH_2OH$$

Methanal　　Phenol

The product then undergoes a condensation reaction:

When formaldehyde and phenol, in the proportion of 0.5 : 1, are reacted under acidic conditions, a phenolic resin (known as a novalac resin) is produced, where $n = 0$ to 6.

Base-catalysed phenol formaldehyde resins are made with formaldehyde to phenol ratio of around 1.5. Such resins are sometimes called resols, and are found in varied industrial products. Laminates are made by impregnating one or more layers of base materials like paper or cotton with phenolic resin with the application of temperature and pressure. Moreover, the resins act as binding agents in brake pads, brake shoes and various other applications.

The polymerization process for the manufacture of phenol formaldehyde is an exothermic one and is controlled by a batch reactor as the viscosity of the material changes rapidly.

The raw materials for the process – phenol, formaldehyde and the catalyst are mixed in a jacketed autoclave, which is also termed as a resin kettle. Inside the autoclave, the mixture is heated with steam. Water cooling and refluxing remove the excess heat of reaction. During the initial stages of the reaction, the heavy viscous resins settle as the bottom layer, with an aqueous layer on top. A combination of heat and vacuum in the resin kettle (autoclave) enables the dehydration of the reaction mixture.

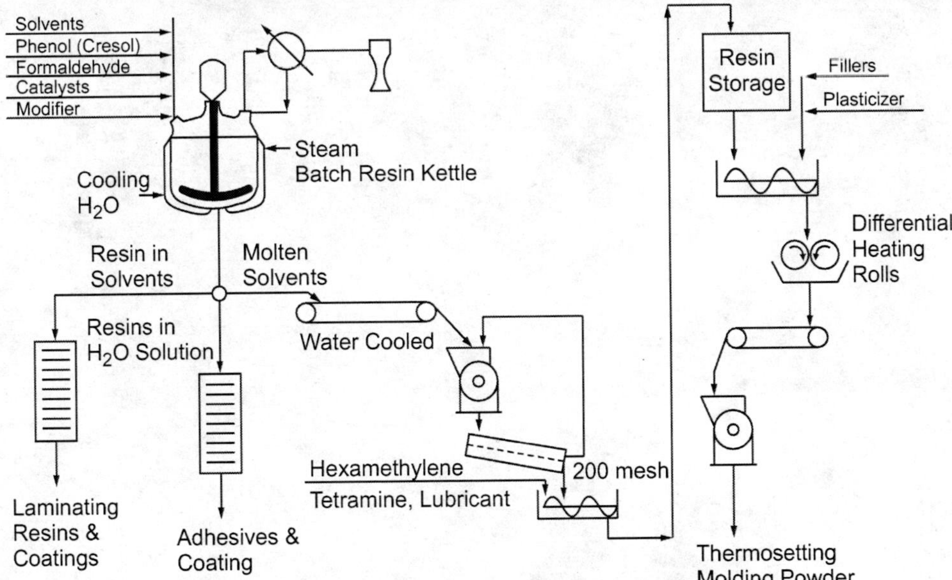

Fig. 10.1. Manufacture of Phenol Formaldehyde

At around 130-150°C, the resin fuses and is removed from the kettle. Then product is then cooled and is ground to form fine powders. This molding powder which forms can be mixed with fillers, coloring agents, lubricants and catalysts in a ball mill. In order to prepare the commercial phenolic molding powder, the final product may be further heated on a pair of differential heating rolls.

Properties and Uses of Phenol-Formaldehyde

The properties of a resin can be altered by using other phenols. For example, if the phenol has an alkyl group with four or more carbon atoms (for example, a butyl group) in the 4-position, the resulting polymer is soluble in oils. Such polymers also have lighter colours. The polymers with giant three-dimensional, cross-linked structures form resins that cannot be melted and reshaped. On heating to high temperatures, they begin to decompose and char. They are also insoluble in a wide range of solvents and are unreactive. Their high electrical and heat resistance means that they are used as an electrical insulator and for the manufacture of brake and clutch linings for vehicles. When they are used with fillers such as wood flour (wood that has been pulverised roughly to the size of a rice grain) or mica, they have major applications in the construction of worktops and printed circuit board insulation. They are also widely used as adhesives, for example in producing ply wood and hard board.

Urea Formaldehyde

UREA FORMALDEHYDE

The addition polymerization reaction of urea(carbamide) with formaldehyde produces mainly monomethylol urea and some dimethylol urea. When the mixture is heated in presence of an acid, condensation occurs, and water is released. This is accompanied by the formation of a crosslinked polymer. Urea $(CONH_2)_2$ and formaldehyde (HCHO) readily react to polymerise into urea formaldehyde resin. Polymerisation is carried out in a glass-lined (or stainless steel) stirred tank reactor in an ammoniacal or weak alkaline medium. The polymerisation reaction takes place at room temperature and atmospheric pressure. This resin is a thermosetting resin.

Urea (Carbamide) is manufactured by heating ammonia and carbon dioxide under pressure. Reaction between formaldehyde(methanol) and carbamide (urea) in alkaline solution gives mono- and di-methylol carbamides:

$O{=}C \langle {N{-}CH_2OH \atop NH_2}$

Mono-methylol carbamide

$O{=}C \langle {N{-}CH_2OH \atop NH{-}CH_2OH}$

di-methylol carbamide

These products condense together in acidic conditions to give a linear polymer. Water is eliminated:

$H, CH_2OH{-}N{-}C{=}O, HN{-}CH_2OH$ (di-)

$H, CH_2OH{-}N{-}C{=}O, H_2N$ (mono-)

$H, CH_2OH{-}N{-}C{=}O, HN{-}CH_2OH$ (di-)

$$\xrightarrow[-H_2O]{H^+,\ heat}$$

$-N(CH_2)-N(CH_2)-N(CH_2)-$ chain, with each N bearing $C{=}O$; the C=O groups bearing $HN{-}CH_2OH$, H_2N, $HN{-}CH_2OH$ respectively.

Further heating (curing) causes further condensation, with the elimination of water between the OH and NH groups, resulting in a cross-linked 3-dimensional structure.

Properties and Uses of Urea Formaldehyde Resin/Plastics

These plastics are often used after incorporating fillers (/polymers such as paper pulp and wood flour. They have the advantage over phenolic resins (phenol formaldehyde) of being colourless, although they are less water resistant and less heat resistant. Urea formaldehyde resin is a non transparent resin and find application in adhesives, molded objects, decorative laminates, textiles, paper, foundry sand molds. Because of its high reactivity and good performance it is widely used as an adhesive for particle board and hard wood plywood.

Their greatest use is as wood adhesives, usually mixed at point of use with filler, extenders and acidic hardeners. They are used in particular as a binder in the production of chipboard. The resin powder or granules are moulded into electrical plugs and sockets, toilet seats and some table wear. Used in solution, the resins wet-strengthen paper and improve the shrink resistance of cotton, wool and rayon where a stiff finish is required.

The key restraint is the emission of toxic emissions of formaldehyde vapour which has been of concern in many countries.

Melamine Formaldehyde

MELAMINE FORMALDEHYDE

Melamine $[C_3N_3(NH_2)_3]$ and formaldehyde (HCHO) combine readily to form melamine-formaldehyde resin, which is a thermosetting resin. The temperature of the reaction is 100°C in a neutral medium, i.e., at a pH of 7 in the aqueous medium at atmospheric pressure.

Manufacture of Melamine Formaldehyde Plastics

Melamine is produced by heating urea(carbamide):

$$6O{=}C(NH_2)_2 \xrightarrow{\text{Heat}} C_3N_3(NH_2)_3 + 6NH_3 + 3CO_2$$

Carbamide (urea) — Melamine

The production of melamine formaldehyde plastics is similar to that for urea formaldehyde polymers, and their structures are analogous. The resin is usually produced as a powder.

Properties and Uses of Melamine Formaldehyde Plastics

Melamine resins combine the best of the properties of phenolic and carbamide (urea) resins. They are stable to heat, light and moisture and are colourless, and can absorb dyes easily. Melamine formaldehyde resin is commonly used as kitchen ware, a melamine formaldehyde resin has low water absorption and better chemical and heat resistance than urea resins. However major application is as bonding and adhesive components.

The powder can be moulded under pressure and most of the resin is used to make laminates for surfaces. Kitchen work tops can be made using a cheaper phenolic resin to form the base, which is then coated with a more expensive melamine surface. Light-weight picnic ware and tableware (e.g., Melaware) can also be made from melamine. To absorb sound, and make commuting more comfortable, the roofs of 25 new subway stations built recently in Seoul, Republic of Korea are constructed of perforated steel behind which is a layer of melamine resin.

Like the urea resins, melamine resins are also used, in solution, to wet-strengthen paper and to improve the shrink resistance of cotton, wool and rayon. Resins are also used as adhesives and as ion-exchange resins to demineralise water. If melamine, formaldehyde and an alcohol (for example, butanol) are heated together, a resin is formed with both $—CH_2OH$ and $—CH_2OR$ groups. The resulting resins are then cross linked with other polymers to form paints for cars.

Polycarbonates

POLYCARBONATES $\left[O - \underset{\underset{O}{\|}}{C} - O - R \right]_n$

The polycarbonates are polymers which have organic functional groups linked together by carbonate groups. The most used is a thermoplastic which has long molecular chains.

Polycarbonates (PC) are condensation thermoplastics used mainly for special engineering purposes. These polymers are considered polyesters of carbonic acid.

Manufacture of Polycarbonates

$HO-C_6H_4-C(CH_3)_2-C_6H_4-OH$ (Bisphenol A)　　$O=CCl_2$ (Carbonyl chloride)　　$(C_6H_5O)_2C=O$ (Diphenyl carbonate)

The polycarbonate most used is manufactured by condensation polymerization between bisphenol A and either carbonyl chloride or diphenyl carbonate.

Bisphenol A is produced by the condensation of phenol with acetone. Carbonyl chloride is produced from carbon monoxide and chlorine:

$$CO\,(g) + Cl_2\,(g) \longrightarrow \underset{\text{Carbonyl chloride}}{COCl_2\,(g)}$$

Diphenyl carbonate is produced from dimethyl carbonate, which is often made from methanol, oxygen and carbon monoxide in the liquid phase in the presence of a copper salt such as copper (II) chloride:

The polymer is usually formed by the reaction of bisphenol A and carbonyl chloride in a basic solution. A solution of bisphenol A in sodium hydroxide (i.e., a solution of the sodium salt of the phenol) is prepared. It is mixed with a solution of carbonyl chloride in an organic solvent (dichloromethane). The polymerization takes place at the interface between the aqueous and organic layers with the help of a catalyst (an amine):

$$n\ {}^{-}O-C_6H_4-C(CH_3)_2-C_6H_4-O^{-} + n\ COCl_2 \longrightarrow \left[-O-C_6H_4-C(CH_3)_2-C_6H_4-O-\overset{O}{\overset{\|}{C}}-\right]_n + 2n\ Cl^{-}$$

Polycarbonate

The polycarbonate is held in solution in the organic layer. This solution is then run off from the aqueous layer and is either evaporated to form granules of the polymer or ethanol is added to precipitate the solid polymer.

However, an increasing proportion of polycarbonates is made *via* diphenyl carbonate, in order to eliminate the use of carbonyl chloride, an extremely poisonous gas.

Bisphenol A and the ester are heated together to form a molten mass of polymer:

$$n\ HO-C_6H_4-C(CH_3)_2-C_6H_4-OH + n\ (C_6H_5O)_2C{=}O \longrightarrow \left[-O-C_6H_4-C(CH_3)_2-C_6H_4-O-\overset{O}{\overset{\|}{C}}-\right]_n + 2n\ C_6H_5OH$$

Diphenyl carbonate

Polycarbonate

Phenol

The phenol and excess reactants are removed by distillation under reduced pressure. The polycarbonate melt is then pressed through fine nozzles to form long ‘spaghetti-like’ threads, which are cooled down and granulated.

Properties and Uses of Polycarbonates

Polycarbonate are transparent and are break resistance. Polycarbonates are thermoplastics used in several applications and have exceptional optical clarity, impact strength, flame resistance, low wrap page, excellent electrical properties and dimensional stability, high heat distortion temperature, lightness, durability.

There are many polycarbonates which vary in properties depending on their molecular mass and structure. As the molecular mass increases, the polymer becomes more rigid. Further, the properties are changed by blending it with other polymers, for example, with ABS and polyesters such as PET. Because of their remarkable properties (they are flame and heat resistant, tough and transparent), the polymers are very widely used. The properties of polycarbonates, their strength and remarkable transparency and their biocompatibility make them ideal materials for medical appliances. They are used in dialysis machines for kidney problems and for oxygenators which take over the work of the heart and lungs during serious operations on them, for example in by-pass surgery. The polycarbonate is sterilized with epoxyethane, radiation or heating in steam.

Polycarbonates, known for their toughness in molded parts, typify the class of polymers known as engineering thermoplastics. These materials, designed to replace metals and glass in applications demanding strength and temperature resistance, offer advantages of light weight, low cost, and ease of fabrication.

Materials made from polycarbonates are transparent, strong, and heat and break resistant. However, these materials are subject to stress cracking and can be attacked by weak alkalies and acids.

Polycarbonates are used in a variety of articles such as laboratory safety shields, street light globes, and safety helmets. The maximum usage temperature for polycarbonate objects is 125°C.

Polycarbonate is a high performance engineering resin. Polycarbonate finds its application in Sheets/film, Automotive (window and non window), optical media, Appliances, Housewares, Electronics/Electrical, medical, ophthalmic, construction, sports, recreational, packaging etc. As blends, it finds usage as auto body.

Among the uses of the polycarbonates and their blends are:

- medical (for example, for dialysis housing and spectacle lenses).
- electro-electronic (for example, sockets, lamp covers.
- fuse-boxes, computer and television housing.
- construction (for example, stadium roofs, signs, skylights).
- optical storage (CDs, DVD, HD-DVDs).
- cars (interior lighting and headlamps, sunroofs, side windows, radiators, grilles, bumpers).
- packaging (for example, large water bottles)

Epoxy Resins

EPOXY RESINS

Epoxy resins are produced by reacting epichlorohydrin and a diphenol. Bisphenol A is the diphenol generally used. The reaction, a ring opening polymerization of the epoxide ring, is catalyzed with strong bases such as sodium hydroxide. The linear polymer formed is cured by cross-linking either with an acid anhydride, which reacts with the —OH groups, or by an amine, which opens the terminal epoxide rings. Cresols and other bisphenols are also used for producing epoxy resins.

Most common epoxy resins are produced from a reaction between epichlorohydrin (ECH) and bisphenol-A (BPA), though the latter may be replaced by other raw materials (such as aliphatic glycols, phenol and o-cresol novolacs) to produce specialty resins. The epoxy resins can be obtained in either liquid or solid states. The two processes are similar. Firstly ECH and BPA are charged into a reactor. A solution of 20-40% caustic soda is added to the reaction vessel as the solution is brought to the boiling point. After the evaporation of unreacted ECH, the two phases are separated by adding an inert solvent such as methyl isobutyl ketone (MIBK). The resin is then washed with water and the solvent is removed by vacuum distillation. The producers will add the specific additives to create a formula that lend special properties such as flexibility, viscosity, color, adhesiveness, and faster curing, depending on a particular application.

In order to convert epoxy resins into a hard, infusible, and rigid material, it is necessary to cure the resin with hardener. Epoxy resins can cure at practically any temperature from 5-150°C depending on the choice of curing agent. Primary and secondary amines are widely used to cure epoxy resins.

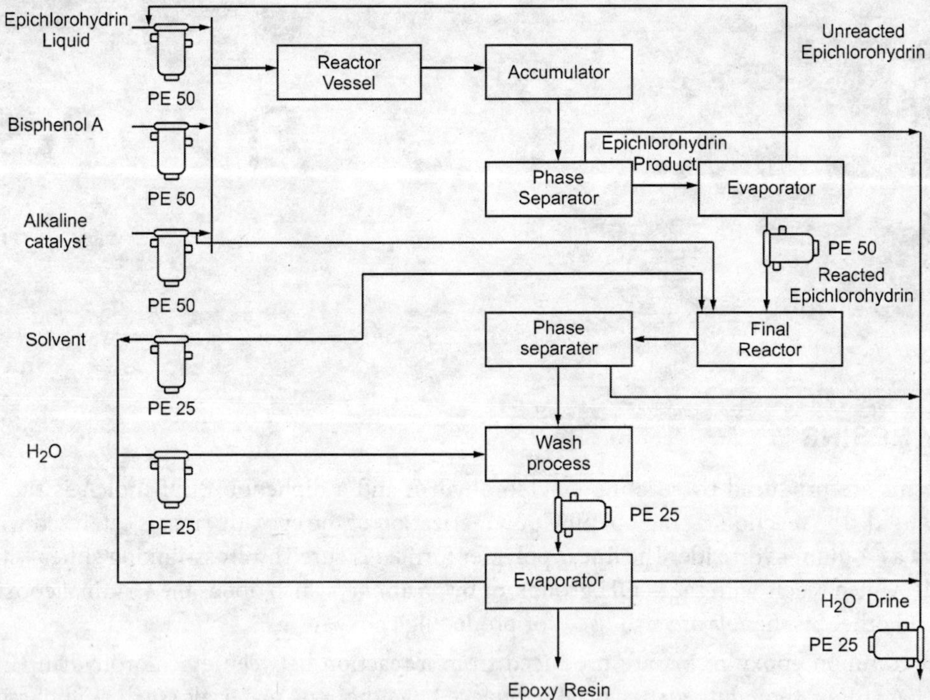

Fig. 14.1. Epoxy resin manufacturing process with flow chart.

Properties and Uses of Epoxy Resins

Epoxy resins have a wide range of molecular weights (1,000-10,000). Those with higher molecular weights, termed phenoxy, are hydrolyzed to transparent resins that do not have the epoxide groups. These could be used in molding purposes, or crosslinked by diisocyanates or by cyclic anhydrides.

Important properties of epoxy resins include their ability to adhere strongly to metal surfaces, their resistance to chemicals, and their high dimensional stability. They can withstand temperatures up to 500°C. Epoxy resins with improved stress cracking properties can be made by using toughening agents, such as carboxyl-terminated butadiene-acrylonitrile liquid polymers. The carboxyl group reacts with the terminal epoxy ring to form an ester. The ester, with its pendant hydroxyl groups, reacts with the remaining epoxide rings, then more crosslinking occurs by forming ether linkages. This material is tougher than epoxy resins and suitable for encapsulating electrical units. Crude epoxies have good chemical and corrosion resistance, high adhesion to a variety of substrates and low shrinkage during occurring. Coatings and laminating binders are the largest users of epoxies. Major uses of epoxy resins are coatings for appliance finishes, auto primers, adhesive, and in coatings for cans and drums. Interior coatings of drums used for chemicals and solvents manifests its chemical resistance.

Polyurethanes

POLYURETHANES

A carbamate functional group is synthesized by the addition reaction of an alcohol with an isocyanate called as urethane. If a polymer chain is linked together by urethane bonds, they are called as polyurethane. Polyurethanes are thermosetting resins obtained by condensation polymerisation of polyol (diols or triols) and a diisocyanate. Examples of monomers are toluene diisocyanate $[(CH_3)(C_6H_4)(NCO)_2]$ and dihydric alcohol or glycol $[(CH_2OH)_2]$. The polymerisation is carried out in a stirred tank autoclave reactor at a temperature of around 200°C under nitrogen pressure. Polyurethane in the form of foams has been used for furniture stuffing, carpet backings and insulations.

Manufacture of Polyurethanes

If the polyol has two hydroxyl groups and is mixed with either TDI (toluene diisocyanate) or MDI (methylene diisocyanate), a linear polymer is produced. For example, a linear polyurethane is produced by reaction with a diisocyanate and the simplest diol, ethane-1,2-diol, a condensation polymerization occurs:

$$\underset{\text{4,4'-MDI}}{nO{=}C{=}N{-}C_6H_4{-}CH_2{-}C_6H_4{-}N{=}C{=}O} + \underset{\text{Ethane-1,2-diol}}{nHO{-}CH_2{-}CH_2{-}OH}$$

$$\rightarrow \left[\overset{O}{\overset{\|}{C}}{-}\underset{H}{\underset{|}{N}}{-}C_6H_4{-}CH_2{-}C_6H_4{-}\underset{H}{\underset{|}{N}}{-}\overset{O}{\overset{\|}{C}}{-}O{-}CH_2{-}CH_2{-}O \right]$$

A polyurethane — The urethane linkage

A different diisocyanate used in polyurethane synthesis is methylene diisocyanate (MDI), which is prepared from aniline and formaldehyde. The diamine product is reacted with phosgene to get MDI.

A much used polyurethane is made from TDI and a polyol derived from epoxypropane:

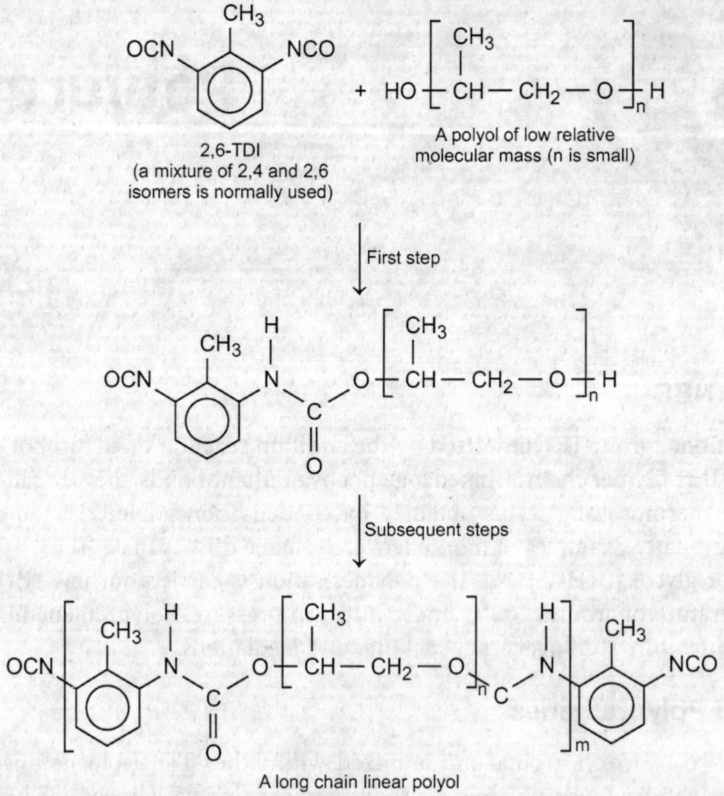

2,6-TDI
(a mixture of 2,4 and 2,6 isomers is normally used)

$$+ \; HO{-}[CH(CH_3){-}CH_2{-}O]_n{-}H$$

A polyol of low relative molecular mass (n is small)

First step

Subsequent steps

A long chain linear polyol

If the polyol has more than two reactive hydroxyl groups, adjacent long-chain molecules become linked at intermediate points. These crosslinks create a stiffer polymer structure with improved mechanical characteristics which is exploited in the development of 'rigid' polyurethanes. Thus a diisocyanate, such as MDI or TDI which reacts with a polyol with three hydroxyl groups, such as one derived from propane-1,2,3-triol and epoxyethane, undergoes crosslinking and forms a rigid thermosetting polymer.

Besides, polyisocyanates and polyols, the manufacture of polyurethanes needs a variety of other chemicals to control the polyurethane-forming reactions and to create exactly the right properties in the end-product. All practical polyurethane manufacturing systems include some, but not necessarily all additives, as described in table below.

Table 15.1. Additives used in the manufacture of polyurethanes

Additives	Reasons for use
Catalysts	To speed up the reaction between polyol and polyisocyanate
Cross-linking and chain-extending agents	To modify the structure of the polyurethane molecules and to provide mechanical reinforcement to improve physical properties (for example, adding a polyisocyanate or polyol with more functional groups)
Blowing agents	To create polyurethane as a foam
surfactants	To control the bubble formation during the reaction and, hence, the cell structure of the foam

Pigments	To create coloured polyurethanes For identification and aesthetic reasons
Fillers	To improve properties such as stiffness and to reduce overall costs
Flame retardants	To reduce flammability of the end product
Smoke suppressants	To reduce the rate at which smoke is generated if the polyurethane is burnt
Plasticisers	To reduce the hardness of the product

Manufacturing Processes for Polyurethane Products

There is a fundamental difference between the manufacture of most polyurethanes and the manufacture of many other plastics. Polymers such as poly(ethene) and poly(propene) are produced in chemical plants and sold as granules, powders or films. Products are subsequently made from them by heating the polymer, shaping it under pressure and cooling it. The properties of such end-products are almost completely dependent on those of the original polymer.

Polyurethanes, on the other hand are usually made directly into the final product. Much of the polyurethanes produced are in the form of large blocks of foam, which are cut up for use in cushions, or for thermal insulation. The chemical reaction can also take place in moulds, leading to, for example, a car bumper, a computer casing or a building panel. It may occur as the liquid reactants are sprayed onto a building surface or coated on a fabric.

No other plastic allows itself to be made in the same way as a polyurethane. Foams can be flexible or rigid, resistant to cold or particularly kind to skin. It all comes down to the way in which the polyurethane 'building blocks' are blended.

As an example, consider the manufacture of a moulded item that might otherwise be made from a thermoplastic polymer by injection moulding. To make it of polyurethane, it is necessary to mix exactly the right masses of the two major components (polyisocyanate and polyol), which must be liquids. The reaction starts immediately and gives the solid polymer. Depending on the formulation, the catalysts used and the application, the reaction is typically completed in between a few seconds and some minutes. In this time, therefore, it is essential to dispense the reacting liquid mixture into the mould and also to clean the combined 'mixing and dispensing' equipment ready for the next operation. The exothermic chemical reaction is completed within the mould and the manufactured article can be taken from the mould immediately.

Foamed Polyurethanes

When the two liquids react, a solid polymer is formed. The polymer may be elastic or rigid. However, it may also contain bubbles of gas so that it is cellular - a foam.

$$OCN—R—NCO(l) + 2H_2O(l) \longrightarrow H_2N—R—NH_2(l) + 2CO_2(g)$$

When producing a foamed polyurethane, there are two possible ways to generate a gas inside the reacting liquid mixture. The so called chemical blowing uses water that may have been added to the polyol which reacts with some of the polyisocyanate to create carbon dioxide:

Alternatively (physical blowing), a liquid with a low boiling point, for example pentane, is mixed into the polyol. The reaction is exothermic and so, as it proceeds, the mixture warms up and the pentane vaporizes.

A tiny amount of air is dispersed through the mixture of polyisocyanate and polyol. This provides nucleation seeds for the multitude of gas bubbles that are produced throughout the polymer. Heat makes the bubbles expand until the chemical reaction changes the liquid to solid polymer, and the available gas pressure cannot create any further expansion.

A shoe sole, for example, may be 'blown' to double the volume of solid polymer. This process is so versatile that the expansion can take much further. In low-density foams for upholstery or thermal insulation less than 3% of the total volume is polyurethane. The gas has expanded the original volume occupied by the liquid by 30 to 40 times. In the case of cushions, only just enough solid polymer is needed to ensure that we can sit comfortably.

In thermal insulation, it is the gas trapped in the cells which insulates. The polymer that encloses the cells reduces the insulation efficiency, so it makes sense to have as little of it as possible.

Adhesive

In the final stages of the polyurethane-forming reaction, the mixture becomes a gel with very effective surface adhesion. Hence polyurethanes can be used as adhesives. Equally important is the fact that polyurethanes, which are being created as, for example, cushioning or insulation materials, can be bonded to surface materials without the introduction of separate adhesives.

Flexible foam and fabric can create a composite cushion or rigid foam and sheet building materials (e.g., plasterboard, steel sheet, plywood) can provide composite building insulation panels.

Polyurethane polymers are either rigid or flexible, depending on the type of the polyol used. For example, triols derived from glycerol and propylene oxide are used for producing block slab foams.

Improved polyurethane can be produced by copolymerization. Block copolymers of polyurethanes connected with segments of isobutylenes exhibit high temperature properties, hydrolytic stability, and barrier characteristics. The hard segments of polyurethane block polymers consist of $—(RNHCOO)_n—$, where R usually contains an aromatic moiety.

Properties and Uses of Polyurethane

The physical properties of polyurethanes vary with the ratio of the polyol to the diisocyanate. For example, tensile strength can be adjusted within a range of 1,200-600 psi; elongation, between 150-800%.Polyurethane find application in manufacture of flexible, high resilience foam seating; rigid foam insulation panels, microcellular foam seals and gaskets; durable elastomeric wheels tires; automotive suspension bushings, electrical potting compounds; high performance adhesives; surface coatings and surface sealants; synthetic fibre.

The major use of polyurethanes is to produce foam. The density as well as the mechanical strength of the rigid and the flexible types varies widely with polyol type and reaction conditions. For example, polyurethanes could have densities ranging between 1-6 lb/ft^3 for the flexible types and 1-50 lb/ft^3 for the rigid types. Polyurethane foams have good abrasion resistance, low thermal conductance and good load bearing characteristics. However, they have moderate resistance to organic solvents and are attacked by strong acids. Flame retardency of polyurethanes could be improved by using special additives, spraying a coating material such as magnesium oxychloride, or by grafting a halogen phosphorous moiety to the polyol. Trichloro butylene oxide is sometimes copolymerized with ethylene and propylene oxides to produce the polyol.

The combined effects of controlling the polymer properties and the density lead to the existence of a very wide range of different materials so that polyurethanes are used in many applications (see table below). Some examples of the main reasons for choosing polyurethanes for various uses is tabulated below.

Table 15.2. Main reasons for choosing polyurethanes for various uses

Uses	Reasons
Cushioning	Low density, flexibility, resistance to fatigue
Shoe soles	Flexibility, resistance to abrasion, strength, durability
Building panels	Thermal insulation, strength, long life
Artificial heart valves	Flexibility and biostability
Electrical equipment	Electrical insulation, toughness, resistance to oils

Polyurethanes can be rigid or rubbery at any density between, say 10 kg m^{-3} and 100 kg m^{-3}. The overall range of properties available to the designer and the manufacturer is clearly very wide and this is reflected in the many, very different, uses to which polyurethanes are put.

Uses of Polyurethanes

Major markets for polyurethanes are furniture, transportation, and building and construction. Other uses include carpet underlay, textural laminates and coatings, footwear, packaging, toys, and fibers.

The largest use for rigid polyurethane is in construction and industrial insulation due to its high insulating property.

Molded urethanes are used in items such as bumpers, steering wheels, instrument panels, and body panels. Elastomers from polyurethanes are characterized by toughness and resistance to oils, oxidation, and abrasion. They are produced using short chain polyols such as polytetramethylene glycol from 1,4-butanediol. Polyurethanes are also used to produce fibers. Spandex (trade name) is a copolymer of polyurethane (85%) and polyesters.

Polyurethane networks based on triisocyante and diisocyanate connected by segments consisting of polyisobutylene are rubbery and exhibit high temperature properties, hydrolytic stability and barrier characteristics.

Polyurethane Foam: Polyurethane rigid foam are characterized by good structural strength, excellent adhesion to most substance, processing flexibility and long life. Rigid polyurethane foams are most widely preferred insulation and find application in refrigerator, manufacture of thermo-ware, cold store panel, refrigerated trucks and wagons. A rigid polyurethane foam is a cellular polymer in which the individual small cells are filled with a gaseous blowing agent which imparts the remarkably low thermal conductivity to these foams.

Polyurethane Rubber: Polyurethane rubber has good abrasion resistance besides oil, solvent, temperature and ozone resistance. It finds wide application in solid tyres for industrial trucks, seals and boots, calendar sheet, potting and sealing of electronic components, general engineering mechanical goods, shoe heels and soles, elastic threads, insulation, mattresses, vibration damping.

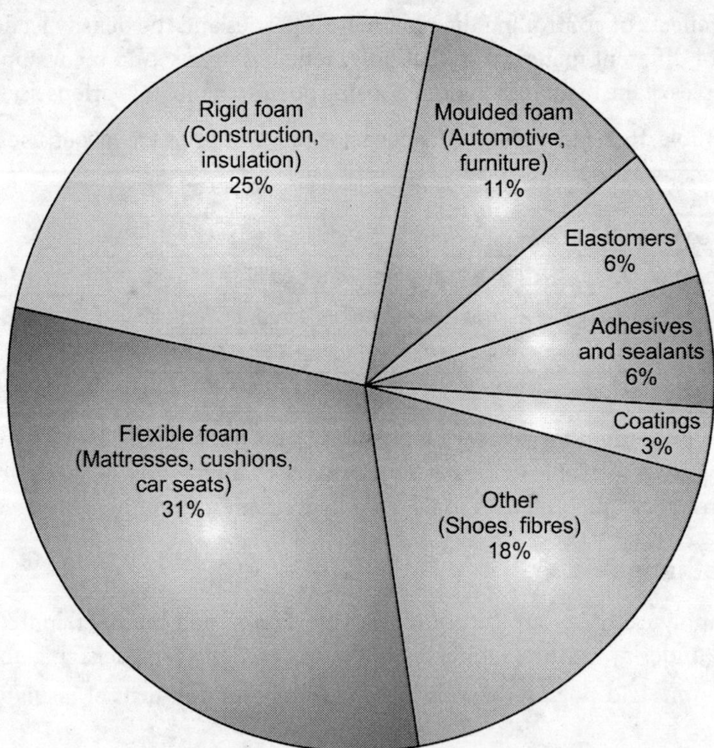

Fig. 15.1. Global uses of polyurethanes in 2012

❑❑❑

Polyethylene Terephthalate (PET)

POLYETHYLENE TEREPHTHALATE (PET)

Polyethylene terephthalate (PET) is produced by esterifying terephthalic acid (TPA) and ethylene glycol or, more commonly, by the trans-esterification of dimethyl terephthalate (DMT) and ethylene glycol (EG). This route is favored because the free acid is not soluble in many organic solvents. The reaction occurs in two stages. Methanol is released in the first stage at approximately 200°C with the formation of bis(2-hydroxyethyl) terephthalate. In the second stage, polycondensation occurs, and excess ethylene glycol is driven away at approximately 280°C and at lower pressures ($\approx$ 0.01 atm):

Using excess ethylene glycol is the usual practice because it drives the equilibrium to near completion and terminates the acid end groups. This results in a polymer with terminal —OH. When the free acid is used (esterification), the reaction is self catalyzed. However, an acid catalyst is used to compensate for the decrease in terephthalic acid as the esterification nears completion. In addition to the catalyst and terminator, other additives are used such as color improvers and dulling agents. For example, PET is delustred by the addition of titanium dioxide.

The molecular weight of the polymer is a function of the extent of polymerization and could be monitored through the melt viscosity. The final polymer may be directly extruded or transformed to chips, which are stored. Batch polymerization is still used. However, most new processes use continuous polymerization and direct spinning.

An alternative route to PET is by the direct reaction of terephthalic acid (TPA) and ethylene oxide. The product bis(2-hydroxyethyl) terephthalate reacts in a second step with TPA to form a dimer and ethylene glycol, which is released under reduced pressure at approximately 300°C.

This process differs from the direct esterification and the transesterification routes in that only ethylene glycol is released. In the former two routes, water or methanol are coproduced and the excess glycol released.

PET is a condensation polymer made by continuous melt-phase polymerization process followed by Solid State Polymerization (SSP). In the past, it was manufactured either by DMT or PTA route:

DMT (Di Methyl Terepthalate) Route: DMT + MEG = PET Polymer

PTA (Purified Terephthalic acid) Route: PTA + MEG = PET Polymer

DMT is seldom used for PET now a days. It is manufactured using PTA to help achieve better quality and consistency.

Properties and Uses of PET

Polyethylene Terephthalate (PET) is a thermo plastic polyester. PET, the material used for the manufacturing of the plastic beverage bottles is of high strength and high barrier material. Pepsi-cola was the first company to commercially introduce its beverages products in PET. PET containers when properly manufactured, offer excellent clarity and gloss, toughness and impact resistance, chemical resistance, low gas permeation, good process ability and good dimensional stability.

PET is used for the production of bottles, sheet, strapping and injection moulded products. PET bottles are used for packaging wide range of consumer products such as all types of Beverages including alcoholic, Edible Oil, Pharmaceutical formulations, Agro chemicals, Personal and Home Care etc. PET is also used for non-bottle applications such as thin/thick thermoformable sheet, dual-oven able crystallised PET (CPET) containers, injection moulded components, fastening straps etc.

Injection moulding of PET is becoming more prevalent in today's market place. PET can be best injection moulded into parts with wall thickness of up to 4.0 mm.

The stretch blow moulding process is used to convert the injection moulded preform which is of amorphous structure to a specific bottle design of bi-axially oriented structure. To achieve the best bottle performance from a preform, it must be stretched by its natural amount both in longitudinal and transverse direction. This means for any desired bottle shape, preform must be specifically designed. The natural amount of stretching which need to be applied to amorphous PET is of the order of 10-12 times, resulting in a thickness reduction of 10 times between preform and bottle blown at 35-40 bar under controlled temperature conditions..

The concern for Recycled PET (RPET) is continuously growing in recent times. PET bottles, which form the major market of PET packaging resin (94%), are the major concern from the point of recycling. More than 90% of PET is consumed in food packaging with drinks/ beverages forming almost 80%. Since drinks and beverages are consumed mostly in residential houses, restaurants, stadia, railway stations, airports, entertainment venues and other public places, the importance of organized collection and recycling of post-consumer PET bottles need not be over-emphasized.

The PET bottle is 100% recyclable and most eco-friendly in nature. The need of the hour is proper handling system at all stages. The end-consumers should take care to remove the cap and label from the used PET bottle before disposing it off and also should not throw the empty bottles in a public place. In order to ease the handling at different stages, the PET bottle should be 'flattened' and baled before sending it to the intended area of usage.

Recycled PET is becoming alternative for PET packaging options today and is getting serious consideration by the packaging world. There is a growing awareness among consumers towards their environment obligation in individual capacity and a marked preference thus for recycled content packaging. Predominantly PET consumption in India is driven by beverages, health care, personal care, agro chemicals, cooking oil, confectioneries and non-bottle segments.

In domestic market, beverages mostly covered mineral water, fortified water, flavored water, sparkling water, juices, Coke and Pepsi Cola and other alcoholic drinks.

Amino Resins (Aminoplasts)

AMINO RESINS (AMINOPLASTS)

Amino resins are condensation thermosetting polymers of formaldehyde with either urea or melamine. Melamine is a condensation product of three urea molecules. It is also prepared from cyanamide at high pressures and temperatures.

$$3H_2N-C\equiv N \longrightarrow H_2N-C\begin{array}{l} \diagup N=C-NH_2 \\ \qquad\qquad \diagdown \\ \qquad\qquad\quad N \\ \qquad\qquad \diagup\!\!\diagup \\ \diagdown\!\!\diagdown N-C-NH_2 \end{array}$$

Melamine

Urea-melamine resins: A similar reaction occurs between melamine and formaldehyde and produces methylolmelamines. A variety of methylols are possible due to the availability of six hydrogens in melamine. As with urea formaldehyde resins, polymerization occurs by a condensation reaction and the release of water.

Properties and Uses of Amino Resins (Aminoplasts)

Amino resins are characterized by being more clear and harder (tensile strength) than phenolic resins. However, their impact strength (breakability) and heat resistance are lower. Melamine resins have better heat and moisture resistance and better hardness than their urea analogs.

The most important use of amino resins is the production of adhesives for particle board and hardwood plywood. Compression and injection molding are used with amino resins to produce articles such as radio cabinets, buttons, and cover plates. Because melamine resins have lower water absorption and better chemical and heat resistance than urea resins, they are used to produce dinnerware and laminates used to cover furniture. Almost all molded objects use fillers such as cellulose, asbestos, glass, wood flour, glass fiber and paper.

Polytetrafluoroethylene (PTFE-Teflon)

POLYTETRAFLUOROETHYLENE (PTFE-TEFLON)

Teflon is the brand name given by DuPont Corp. for a polymer called polytetrafluoroethylene (PTFE), which belongs to a class of thermosetting plastics known as fluoropolymers. It is famous as a coating for non-stick cookware. Being chemically inert, it is used in making containers and pipes that come in contact with reactive chemicals. It is also used as a lubricant to reduce wear from friction between sliding parts, such as gears, bearings and bushings.

Polytetrafluoroethylene (PTFE) or *Teflon* is a polymer of the tetrafluoroethylene (C_2F_4) monomer. This monomer is prepared from methanol chlorination followed by fluorination. The reactions are carried out in three reactors. Trichloromethane produced in the chlorinator is reacted with hydrofluoric acid in the first reactor at about 65°C in the presence of $SbCl_5$ as the catalyst and is converted to produce mono-chloro-difluoro- methane (CHF_2Cl) in the presence of $AlCl_3$ as the catalyst in the second reactor. In the third reactor, CHF_2Cl is then converted to C_2F_4 by catalytic pyrolysis at a temperature of 650-800°C in the presence of platinum (Pt) as the catalyst. Polymerisation of C_2F_4 is then carried out in a batch reactor preferably by the suspension polymerisation method at a temperature of 200°C and at a pressure of 70 kg/cm^2. The reaction time is about 1 hour.

Manufacture of Polytetrafluoroethylene (PTFE)

PTFE is made from methane in a series of reactions:

(*a*) Production of trichloromethane (chloroform)
(*b*) Production of chlorodifluoromethane
(*c*) Production of tetrafluoroethylene (TFE)
(*d*) Polymerization of tetrafluoroethylene to produce PTFE

(*a*) **Production of trichloromethane (chloroform):**

$$CH_4\,(g) \;+\; 3\,Cl_2\,(g) \longrightarrow CHCl_3\,(g) \;+\; 3\,HCl\,(g)$$

Trichloromethane is one of the products formed by the reaction of methane and a mixture of chlorine and hydrogen chloride. This can be performed in the liquid phase at 370-420 K using a zinc chloride catalyst. Alternatively, the reaction is carried out in the vapour phase, using alumina gel or zinc oxide on silica as a catalyst at 620-720 K.

(*b*) **Production of chlorodifluoromethane:**

$$CHCl_5\,(g) + 2HF(g) \longrightarrow \underset{\text{Chlorodifluoromethane}}{CHClF_2\,(g)} + 2\,HCl\,(g)$$

Trichloromethane is reacted with anhydrous hydrogen fluoride in the presence of antimony (III) and antimony (V) chlorofluoride to give chlorodifluoromethane:

(*c*) **Production of tetrafluoroethene (tetrafluoroethylene, TFE):** Since TFE is an explosive gas (bp 197 K), it is usually made when and where required for polymerization so that there is minimum storage time of the monomer between its production and its polymerization. Chlorodifluoromethane is heated in the absence of air, a process known as *pyrolysis*:

$$2\,CHClF_2\,(g) \rightleftharpoons F_2C = CF_2\,(g) + 2\,HCl\,(g) \qquad \Delta H^{\phi} = +127.6\ kJ\ mol^{-1}$$

Low pressures (atmospheric) and high temperatures (940-1070 K) favour the reaction. Steam, preheated to 1220 K, and chlorodifluoromethane, at 670 K, are fed into a reactor. Steam is used to dilute the reaction mixture and hence reduce the reactant partial pressure, and thus the formation of carbon and toxic by-products. The steam also supplies all the heat required by this endothermic reaction. Very little hydrolysis of reactant and product occurs.

$$F_2C = CF_2\,(g) \rightleftharpoons CF_4\,(g) + C\,(s) \qquad \Delta H^{\phi} = -282\ kJ\ mol^{-1}$$

Once formed, the product must be rapidly cooled to 770 K to prevent the reverse reaction occurring and the explosive decomposition of TFE:

The cooling is done by passing the vapour through a water-cooled heat exchanger, made of graphite to resist chemical attack and thermal shock. Reactor residence time is 1 second.

(*d*) **Polymerization of tetrafluoroethylene (TFE) to produce PTFE:** The monomer TFE is transformed into the polymer, PTFE, by radical polymerization. The reaction is carried out by passing TFE into water containing a radical initiator, e.g. ammonium persulfate, $(NH_4)_2S_2O_8$, at 310-350 K and a pressure of 10-20 atm.

Two different procedures are used:

- Granular polymerization gives a suspension of string-like PTFE particles up to 1 cm long in water. These are milled to produce fine powders (30 pm) used for moulding. The fine powders are also agglomerated to larger particles (500 pm) to give better flow. Unlike other thermoplastics, such as PVC, PTFE cannot be processed by melt extrusion. The powder is therefore, moulded into rods for extrusion and heating at temperatures above 530 K to force the particles to stick together.
- Dispersion polymerization can be used to obtain a colloidal dispersion of PTFE particles (0.25 pm) in water. The dispersion can be concentrated and used for dip coating or spraying articles. The dispersion can also be coagulated and dried to give a powder, which, in turn, is made into a paste and extruded on to wire.

Co-polymers of Tetrafluoroethylene (TFE)

There is a group of polymers which are formed by the co-polymerization of tetrafluoroethylene and other unsaturated organic compounds such as ethylene, hexafluoropropene and perfluoropropylvinyl ether. As described above, these co- polymers are used in many of the examples given for PTFE.

$$\underset{\text{Hexafluoropropene}}{F_3C - CF = CF_2} \qquad \underset{\text{Perfluoropropylvinyl ether}}{F_3C - CF_2 - CF_2 - O - CF = CF_2}$$

The co-polymer produced from ethylene and tetrafluroethylene is an alternating co-polymer usually known by its trivial name, ethylene tetrafluoroethylene (ETFE): It is used, in particular as a lining for containers as it is stable to attack by concentrated solutions of acids and alkalis, and because of its good electrical properties of insulation and its strength, it is used as a coating for wires and cables. Its most spectacular use is as a roofing material in buildings and Stadium. The roofs are made up of 2-5 layers of large cushions of ETFE. It is also used as an outer skin of large buildings.

$$\left[CH_2 - CH_2 - CF_2 - CF_2 \right]_n$$

ETFE

Properties and Uses of Polytetrafluoroethylene (PTFE)

The uses of poly(tetrafluoroethene) (PTFE) are a function of its resistance to chemical attack, unreactivity (even above 500 K), low friction, non-stick properties and high electrical resistance.

In many applications, tetrafluoroethene (TFE) is co-polymerized with other fluorinated monomers, such as hexafluoropropene and perfluoropropyl vinyl ether, and also with ethylene.

Polytetrafluoroethylene (PTFE) and its co-polymers are used in:

- Cable insulation
- Reactor and plant equipment linings, when reactants or products are highly corrosive to ordinary materials such as steel.
- Semi-permeable membranes in chlor-alkali cells and fuel cells.
- Bearings and components in mechanical devices such as small electrical motors and pumps permeable membrane (for clothing and shoes, which allows water vapour to diffuse away from the skin but prevents liquid water (rain) from soaking in.
- Non-stick domestic utensils, e.g., frying pans
  - Medical - catheter tubing
  - Hose and tubing
  - Solid lubricants
  - Combinations with magnesium and aluminium as an igniter for explosives

The table below summarises list of consumer end products produced from five commodity plastics in India.

Table 18.1. Polymers and corresponding plastic products

LDPE/LLDPE	HDPE	PP	PVC	PS
Cost and stretch cling films	Raffia Mono filaments	Raffia/FIBC	Irrigation pipes and fittings	Thermocol
Mono/Multi layer films	Crates	Fish nets	Wire and cables	Foam
Wires and Cables	Multi layer films	Buckets, Tiffin boxes	Floorings/Tiles	Refrigerator trays
Water storage tanks	Water Conveyance pipes	Automobile bumpers	Shower curtains	CD/DVD eases

LDPE/LLDPE	HDPE	PP	PVC	PS
House wares	Pesticide bottles	BOPP films	Blister films	Structural insulation
Carrier bags	Woven sacks	Washing machine tubs	Profiles and Builders' ware	Bottles
Furniture	Caps *and* closures	Leno bags	Footwear	Disposable cups/ cutleries
Milk Packaging	Pallets	Fibres and Filaments	IV bottles and blood bags	Smoke detectors
Detergent Packaging	Thin wall moldings	Syringes	Bag liners	Films

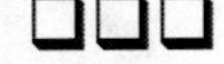

MANUFACTURE, PROPERTIES & APPLICATIONS OF IMPORTANT ELASTOMERS (RUBBERS)

Overview of Rubbers (Elastomers)

OVERVIEW OF RUBBERS (ELASTOMERS)

Rubbers(natural or synthetic) also known as elastomers is an elastic high polymer, which returns to its original dimension when stretched considerably from its original shape. It has elastic properties in excess of 200% ; that means it must elongate at least 200 % and return to its original dimensions rapidly and forcibly.

Generally, the elastomer (rubber) material is an amorphous polymer, that is stable in three dimensional structures (network) in which large molecules chains motion depends on the material's glass transition temperature (T_g). All polymers above glass transition temperature (T_g), act as rubbers due to entanglements. However, when the entanglements break, the polymer will begin to flow. The glass transition temperatures of all rubber are quite low, usually below room temperature, therefore many polymers at room temperatures act as rubbers (elastomers) with many useful properties. The elastomers are used in many applications because of their unusual properties, which are unmatched by other types of materials.

The elastomer materials can be modified after curing by heat, irradiation, and vulcanization with chemical materials by chemical reaction. Elastomer have the property of viscoelasticity (having both viscosity and elasticity), also dissipates energy because of its viscoelastic nature. However, the vulcanization reduces the viscosity; increases the elasticity, also elastomer would be insoluble in boiling solvent and becomes more resilience.

The elastomer is often used interchangeably with the term rubber. The difference between an elastomer and a rubber is the term originally used for natural rubber (NR), which is naturally derived from organic materials that occurred naturally in nature. While, the term elastomer is used for materials which is derived from by- products of petroleum or natural gas, and produced synthetically by human, to give the synthetic rubber.

Characteristics of Elastomers:

- Elastomers are amorphous solids.
- Elastomers have linear but highly coiled structure which imparts it the property of elasticity.
- Inter chain cohesive forces are negligible because these are hydrocarbons.
- They have high degree of elasticity.
- They can retract rapidly.

Properties of rubber: Important properties of rubber products include tensile strength, hardness, tear resistance, resilience, flex failure, abrasion, repeated stress failure, permanent set, aging, ozone and oxygen reactivity, oil and solvent resistance and thermal and electrical properties. These properties are dependent not only on the original molecular structure of the polymers but also on additional processing methods which include:

Compounding: An essential aspect of the production of a finished part from any elastomer (whether it be natural rubber, SBR or silicone rubber) is the addition of compounding ingredients like fillers, coloring agents, plasticisers, anti oxidants,vulcanizing agents etc. Since pure rubber is as useless as pure gold, the desirable properties of plasticity, elasticity, hardness, softness, abrasion resistance permeability etc. and their various combinations are achieved by the rubber compounders. The most important compounding ingredient is carbon black which is used to improve wearing qualities of both natural and synthetic rubber by imparting toughness.

Mixing: Homogenous distribution of compounding ingredients into solid elastomeric polymer for which rubber mill is used which has one of the two rolls running at slower speed, giving kneading action.

Vulcanisation: The properties of natural raw rubber can be modified by introducing —S—S— polysulphide crosslinks in its structure. This process of introducing —S—S— crosslinks in the structure of natural rubber by heating with sulphur at 110-140°C is called vulcanisation of rubber. Vulcanisation is carried out by adding sulphur (3-5%) and zinc oxide to the rubber, and then heating the object at about 110°C for about 20-30 minutes. Zinc oxide accelerates the rate of vulcanisation. Vulcanisation introduces polysulphide (—S—S—) bonds between the adjacent chains. These crosslinks tend to limit the motion of chains relative to each other.

The degree of vulcanization depends on the amount of sulfur used. Tyre industry uses 3-5% sulfur. If the amount of sulfur is increased to 30%, a hard and rigid rubber called "ebonite" is produced.

Table 19.1. Comparison of some properties of natural rubber and vulcanized rubber.

S. No.	Natural Rubber	Vulcanized Rubber
1.	Natural rubber is soft and sticky.	Vulcanized rubber is hard and non-sticky.
2.	It has low tensile strength.	It has high tensile strength
3.	It has low elasticity.	It has high elasticity.
4.	It can be used over a narrow range of temperature (from 10° to 60°C).	It can be used over a wide range of temperature (–40° to 100°C).
5.	It has low wear and tear resistance.	It has high wear and tear resistance.
6.	It is soluble in solvents like ether, carbon tetrachloride, petrol, etc.	It is insoluble in most of the common solvents.

Advantages of Vulcanization

- The tensile strength of vulcanized rubber is very good. It is 10 times the tensile strength of raw rubber.
- It has excellent resilience i.e., articles made from it returns to the original shape when the deforming load is removed.
- It has better resistance to moisture, oxidation, abrasion.
- It has much higher resistance to wear and tear compared to raw rubber.
- It has broader useful temperature range (–40 to 100°C) compared to raw rubber's useful temperature range (10-60°C).

- It is a better electrical insulator. **Ex:** Ebonite
- It is resistant to (*i*) Organic solvents like petrol, benzene, CCl_4, (*ii*) Fats and oils, but it swells in them.
- It has only slight tackiness.
- It has low elasticity. The property depends on the extent of vulcanization. Ebonite has practically no elasticity.
- It is very easy to manipulate the vulcanized rubber to produce the desired shapes.

Forming: Production of final product shapes by moulding,calendaring, extrusion etc.

General Properties of Elastomers

- Relatively soft.
- **High elasticity (flexibility):** Elasticity is the property of elastomers that allows them to being compressed, stretched or bended, after force is applied and return to their original shape when the force is removed.
- **High resiliency:** Resiliency is the property of elastomers that refers to the speed of return (recovery) to the original shape after a deforming force (external force) is removed. The deforming force is applied as input energy to the elastomer, when the elastomer returns to its original shape, part of these input energy is not returned but is dissipated as heat energy within the elastomer. The ratio of the returned energy to the input energy is defined as the material's resilience.
- High durability.
- Oil and fuel resistant.
- Swell in the certain solvents.
- Low permeability to gases, water and steam.
- Good electrical and thermal insulation.
- Weathering resistance.
- Deteriorate by oxidation.
- High abrasion resistance.
- Low creep resistance under load.
- Low modulus of elasticity.
- High strength especially under conditions of shear and compression.

Classification of Elastomers

Depending on the polymer structure and degree of the chemical bonds in the polymers, elastomeric materials can have properties similar to thermosetting or thermoplastics, so elastomeric materials can be divided broadly into two types:

- **Thermoplastics Elastomers:** are those elastomer materials having a linear structure, and can be dissolved in suitable solvents but do not swell, which melt when heated, usually that do not require vulcanization, such as polystyrene rubber, polyethylene rubber. Some elastomers are thermoplastic, so these materials do not require curing and the molecules chains are joined by the physical bonds, this is weaker bonds such as (hydrogen bonds, Vander walls bonds or dipole-

dipole interactions) to aggregation of the molecules into hard domains. These materials melt to a liquid state when heated, and return to the solid state when cooled sufficiently, therefore, can be repeatedly remelted, remolded and recycled.

- **Thermosetting Elastomers:** are those elastomer materials having a three dimensional (cross link) structure, and can be swelling but do not dissolve in solvents, which do not melt when heated, usually that require vulcanization, such as styrene butadiene rubber and butadiene rubber. Thermosetting elastomers are classified as general purpose elastomers and specialty elastomers. General purpose elastomers are used in some applications, such as styrene butadiene rubber (SBR), butadiene rubber (BR), and natural rubber (NR). Whereas general elastomer are unsuitable in many applications and when required higher performance, the specialty elastomers are used.These materials are more costly and hence are produced in smaller volumes, such as silicon rubber, polyacrylic rubber.

Most elastomers are thermosets, these materials require curing. In the curing process, the long molecules chains are held together by chemical bonds (covalent bonds), then material becomes stronger, and it cannot be remelted, remolded, recycled, and reprocessed.

The thermosetting elastomers classification is shown below in Figure.

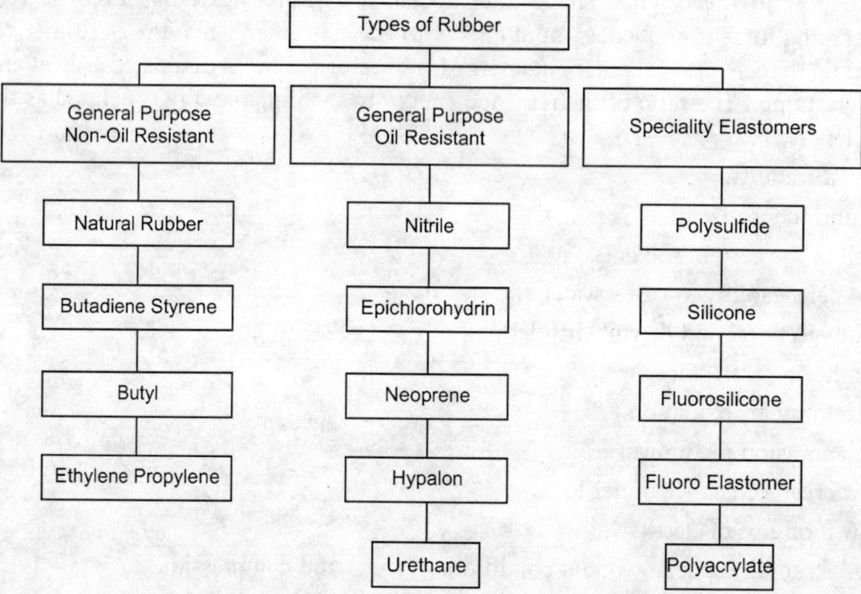

Fig. 19.1. Classification of thermosetting elastomers

Table 19.2. Selected properties of some elastomers

Elastomers	Durometer hardness range	Tensile strength at room temp, psi	Elongation at room temp, %	Temp. range of service, °C	Weather resistance
Natural rubber	20-100	1,000-4,000	100-700	–55-80	Fair
Styrene-butadiene rubber (SBR)	40-100	1,000-3,500	100-700	–55-110	Fair

Elastomers	Durometer hardness range	Tensile strength at room temp, psi	Elongation at room temp, %	Temp. range of service, °C	Weather resistance
Polybutadiene	30-100	1,000-3,000	100-700	–60-100	Fair
Polyisoprene	20-100	1,000-4,000	100-750	–55-80	Fair
Polychloroprene	20-90	1,000-4,000	100-700	–55-100	Very good
Polyurethane	62-95 A 40-80 D	1,000-8,000	100-700	–70-120	Excellent
Polyisobutylene	30-100	1,000-3,000	100-700	–55-100	Very good

Synthetic Rubber

Synthetic high polymers possessing similar physical properties as that of natural rubber are called synthetic rubber. Usually synthetic rubber is an improvement over natural rubber, especially with respect to its resistance to oils, gas, solvents, etc. Synthetic rubbers are made in chemical plants using petrochemicals as their starting point.

Some synthetic rubbers are made by polymerization of only one monomer, for example Neoprene is made by polymerization of chloroprene, while synthetic rubbers like Buna- S, Buna N and Butyl rubber etc. are copolymers as these have more than one monomer.

Synthetic rubbers include elastomers that could be crosslinked such as polybutadiene, polyisoprene, and ethylene-propylene-diene tere-polymer. It also includes thermoplastic elastomers that are not crosslinked and are adapted for special purposes such as automobile bumpers and wire and cable coatings. These materials could be scraped and reused. However, they cannot replace all traditional rubber since they do not have the wide temperature performance range of thermoset rubber. Synthetic rubber supplements natural rubber in tyres for two wheelers,other automobiles and animal drawn vehicles.

Important Elastomer Additives

1. Fillers
2. Accelerators
3. Activator
4. Stabilizers
5. Plasticizers
6. Anti-degradation
7. Flame retardants

Classification of Elastomer Additives

According to the modification that occurs of elastomer when using the additives, the elastomer additives can be classified into five types which include:

- Mechanical property modifier.
- Chemical property modifier.
- Processing modifier.
- Surface property modifier.
- Aesthetic property modifier.

Types of Elastomers

Unsaturated Elastomers: This type of elastomers can be cured by sulfur vulcanization processand can also be cured by non-sulfur vulcanization if desired, for examples:

- Natural rubber
- Synthetic polyisoprene
- Butadiene rubber
- Neoprene rubber
- Nitrile rubber
- Butyl rubber
- Halogenated butyl rubbers (chloro butyl rubber, bromo butyl rubber)
- Hydrogenated Nitrile Rubbers
- Styrene-butadiene Rubber.

Saturated Elastomers: This type of elastomers cannot be cured by sulfur vulcanization process, for examples:

- Ethylene propylene rubber (EPR)
- Ethylene-vinyl acetate (EVA)
- Polyacrylic rubber
- Silicone rubber
- Fluoroelastomers
- Polyether blocks amides
- Chloro sulfonated polyethylene rubber.

Elastomer Adhesives :The examples of elastomers adhesives are polyurethane adhesive, adhesives based on silicones, adhesives based on modified silane. The types of elastomer adhesives include:

1. Chemically Reactive Adhesives.
2. Evaporation Adhesives (Diffusion Adhesives)
3. Hot Melt Adhesives
4. Pressure Sensitive Adhesives
5. Conductive Adhesives.

THERMOPLASTIC ELASTOMERS (TPFs)

Thermoplastic elastomers (TPE's), are plastic polymers with the physical properties of rubbers. They are soft, flexible, and possess the resilience needed of rubbers. However, they are processed like thermoplastics by extrusion and injection molding.

TPE's are more economical to produce than traditional thermoset materials because fewer steps are required to manufacture them than to manufacture and vulcanize thermoset rubber. An important property of these polymers is that they are recyclable.

Thermoplastic elastomers are multiphase composites, in which the phases are intimately depressed. In many cases, the phases are chemically bonded by block or graft copolymerization. At least one of the phases consists of a material that is hard at room temperature.

Important TPE's include blends of semi crystalline thermoplastic polyolefins such as propylene copolymers, with ethylene-propylene terepolymer elastomer. Block copolymers of styrene with other monomers such as butadiene, isoprene, and ethylene or ethylene/propylene are the most widely used TPE's. Styrene-butadiene-styrene (SBS) accounts for 70% of global styrene block copolymers (SBC). Polyurethane thermoplastic elastomers are relatively more expensive than other TPE's. However, they are noted for their flexibility, strength, toughness, and abrasion and chemical resistance. Blends of polyvinyl chloride with elastomers such as butyl are widely used in Japan.

Random block copolymers of polyesters (hard segments) and amorphous glycol soft segments, alloys of ethylene interpolymers, and chlorinated polyolefins are among the evolving thermoplastic elastomers.

Important properties of TPE's are their flexibility, softness, and resilience. However, compared to vulcanizable rubbers, they are inferior in resistance to deformation and solvents.

Important uses for TPE's include shoe soles, pressure sensitive adhesives, insulation, and recyclable bumpers.

Advantages and Disadvantages of Thermoplastic Elastomers(TPEs) are tabulated below.

Table 19.3. Advantages and Disadvantages of Thermoplastic Elastomers (TPEs)

Advantages	Disadvantages
TPEs require little or no compounding since they are fully formulated and are ready for use as received. Offer rapid, efficient, and economical means of fabricating rubber articles. Processing of TPEs is cheaper and simpler requiring fewer steps than does a thermoset rubber. A single step is required to shape TPEs into final article, whereas thermoset rubbers require mixing and vulcanization in addition to the shaping step. Processing times for TPEs are shorter since the long cycle time required for vulcanization is eliminated. Cycle time is typically seconds instead of the minutes required for thermoset rubbers. This enhances productivity and reduces costs. Being thermoplastic, TPE scrap generated during processing can be recycled without any loss in functional properties. In contrast, thermoset elastomers are not recyclable. As a result of the fewer processing steps and shorter cycle time, lower energy is consumed in TPE processing than processing of thermoset rubbers. It is easier to obtain finished parts having closer tolerances with TPEs because of consistent material base and greater ease and control of processing. Elastomers are generally purchased on weight basis but used on volume basis. Consequently, TPEs offer cost advantages since they normally have lower specific gravity than thermoset rubbers.	Most TPEs may require drying before processing TPEs melt at elevated temperatures, consequently, they are unsuitable for applications requiring even brief exposures to temperatures above their metling points. Many thermoset rubbers can withstand such a short exposure.

Elastomer Applications: The natural and synthetic rubbers are widely used for various applications, because of their unusual properties such as high elasticity, extensibility, resiliency, durability, high abrasion resistance, high strength and other properties, which are unmatched by other types of materials. The synthetic rubber is used to manufacturing most types of rubber products, but natural rubber represents the basic constituent of some rubber products. **Rubber materials are used in many fields which includes:**

Industrial Products: Such as rubber bands, fuel resistant parts are, electrical insulation, drive belts, conveyor belts, floor coverings, elevator belts, rafts, foam rubber, hoses, tubes, paints, fertilizer and bridge bearings.

Sporting Field: Such as golf balls, football balls, sporting goods and toys balloons.

Medical Field: such as catheters, gloves, surgical purposes, medical implants and medical equipment.

Consumer Products: Such as rain coats, sponges, pencil erasers, soap, shoes, and adhesives.

Seismic Materials: For example many buildings are fitted with seismic rubber bearings.

Transport Field: The essential uses of rubber is in the transportation such as automotive (truck or bus) tires, bicycles tires and aircraft tires. Most commonly, tires consist of both natural rubber and synthetic rubber. Natural rubber has a greater resistance to heat compared to synthetic rubber, therefore tire products account for more than 50% of natural rubber usage, and the tire manufacturing is the world's largest consumer of natural rubber.

Industrial Applications of Elastomer Materials

- **Gaskets:** Gaskets are mechanical seals that fill the space between imperfect mating surfaces.
- **Seals similar to gasket applications:** Elastomers are heavily utilized as seal in the industrial machinery. Two types of seals are hydraulic seals and pneumatic seals. Hydraulic seal is used for prevention of fluid leakage, this type of seals use elastomers which withstands high pressures, high temperature and flexural forces. While, the pneumatics seals use elastomers which withstand lower pressure.
- **Noise reduction and dampening.** Many elastomer materials are used to reduce both vibration and noises, also uses to dampen natural sound and airborne sound over a broad temperature and frequency. The best example of this application can be found in the automotive industry such as fans to prevent transfer the vibration from the fan to its structure, also less sound radiation and noise generation when fans are in operation.

Table 9.4. Common elastomers and their uses

Common name (chemical composition)	Abbreviation	Uses and properties
Natural rubber (polyisoprene)	NR	General-purpose tires, bushings and couplings, seals, footwear, belting. Good resilience.
Styrene-butadiene-rubber (random co-polymer)	SBR	Tire tread, footwear, wire and cable covering adhesives. High hysteresis.
Butadiene rubber	BR	Tire tread, hose, belts. Very low hysteresis, high rebound
Butyl rubber (from isobutene and 0.5-3% isoprene)	IIR	Inner tubes, cable sheathing, tank liners, roofing, seals, coated fabrics. Very low rebound, high hysteresis
Silicons (typically polydimethyl-siloxane)		Medical applications, flexible molds, gaskets, seals. Extreme use temperature range
Polyurethanes		Sealing and joining, printing rollers, fibers, industrial tires, footwear, wire and cable covering
Nitrile rubber (random copolymer of butadiene) and acrylonitrile	NBR	Seals automotive parts that are in contact with oils and gas, footwear, hose. Outstanding resistance to oils and gas, little swelling in organic liquids
Ethylene-propylene rubbers (random copolymers with 60-80% ethylene)	EP or EPM	Cable insulation, window strips. Outstanding insulative properties
Ethylene-propylene-diene (random terpolymer)	EPDM	Good resistance to weathering, resistant to ozone attack
Chloroprene rubber (polychloroprene)	CR	Wire and cable insulation, hose, footwear, mechanical automotive products. Good resistance to oil and fire, good weatherability
Acrylonitrile-butadiene-styrene (terpolymers)	ABS	Oil hose fuel tanks, gaskets, pipe and fittings, appliance and automotive housings. Resistant to oils and gas
Ionomers (largely copolymers of ethyl-ene and acid-containing monomer reacted with metal ions)		Golf ball covers, shoe soles, weather stripping. Tough, flame-resistant, good clarity, good electrical properties, abrasion-resistant

Fluoroelastomers (fluorine-containing copolymers)		Wire and cable insulation, aerospace applications. Outstanding resistance to continuous exposure to high temperatures, chemicals and fluids
Epichlorohydrin (epoxy copolymers)		Seals, gaskets, wire and cable insulation. Good resistance to chemicals
Polysulfide		Adhesive, sealants, hose binders. Outstanding resistance to oil and organic solvents.

Reclaimed Rubber

Rubber obtained from waste rubber articles such as worn out tyres, tubes, gaskets, hoses, foot wears, etc., is called reclaimed rubber.

The process of reclamation of rubber is carried out as follows:

- The waste is cut into small pieces and powdered by using a 'cracker'.
- Then iron impurities, if any present, is removed by using electromagnetic separator.
- The purified waste is digested with caustic soda solution at 200°C under pressure for 8 to 15 hours in 'steam jacked autoclaves'. This process hydrolyses the fibres present in the waste rubber.
- After the removal of fibres, reclaiming agents like petroleum or coal tar based oils and softeners are added.
- Sulphur gets removed as sodium sulphide and rubber gets devulcanised.
- The rubber is thoroughly washed with water spray and dried in hot air driers.
- Finally, the reclaimed rubber is mixed with small portions of reinforcing agents like clay, carbon block etc.

Properties of Reclaimed Rubber

- Reclaimed rubber has less tensile strength, low elasticity and possesses very low wear resistance when compared to natural rubber.
- However, it is much cheaper and has uniform composition.
- It has better aging property.
- It is quite easy for fabrication.

Uses of Reclaimed Rubber

Reclaimed rubber is used for the manufacture of tyres, tubes, automobile floor mats, belts, hoses, battery containers, mountings, shoes, etc.

20 Polyisoprene (Natural Rubber)

NATURAL RUBBER (cis-1,4-POLYISOPRENE)

Natural Rubber is a high molecular weight hydrocarbon polymer represented by the formula $(C_5H_8)_x$. It is obtained from a milk emulsion called latex by tapping the bark of the tree. "Hevea brasiliensis". It is a polymer of isoprene units. The polymer chain of natural rubber is made of 2000 to 3000 monomer units. Natural rubber is a high polymer of isoprene (2-methyl-1,3-butadiene).

Natural rubber is a coiled linear 1, 4-polymer of isoprene.

$$CH_2{=}\overset{\overset{\displaystyle CH_3}{|}}{C}-CH{=}CH_2$$

Isoprene

In the polymer chain of natural rubber, the residual double bonds are located between C_2 and C_3 of the isoprene unit. All these double bonds have cis configuration and thus natural rubber is cis-1,4-polyisoprene.

In the natural rubber, there is no polar substituent. The only intermolecular forces are van der Waal's type. The cis-configuration gives the polymeric chain of natural rubber a coiled structure. As a result, it can be stretched by the application of a force. When the force is removed, the chain returns back to its original coiled shape.

It is obtained from the latex of rubber trees found in tropical and semi-tropical countries such as India (southern part), Indonesia, Malaysia, Ceylon, South America, etc. The latex contains about 25-40% of rubber hydrocarbons dispersed in water along with stabilizer proteins and some fatty acids. It is a natural polymer and possess remarkable elasticity. It undergoes long range reversible extension under relatively small applied force. This elasticity makes it valuable for a variety of uses.

The desired valuable properties of plasticity, elasticity, toughness, hardness, softness, abrasion, resistance, impermeability, etc. are achieved by the process of compounding and vulcanization of rubber.

Natural rubber is a stereoregular polymer composed of isoprene units attached in a cis configuration. This arrangement gives the rubber high resilience and strength.

Isoprene can be polymerized to form polyisoprene using free radical initiators, but a random polymer is obtained. Stereoregular polyisoprene is obtained when Ziegler-Natta catalysts or anionic initiators are used. The most important coordination catalyst is α-$TiCl_3$ co catalyzed with aluminum alkyls. The polymerization rate and cis content depends upon Al/Ti ratio, which should be greater than one. Lower ratios predominantly produce the trans structure.

Polyisoprene is a synthetic polymer (elastomer) that can be vulcanized by the addition of sulfur. cis-Polyisoprene has properties similar to that of natural rubber. It is characterized by high tensile strength and insensitivity to temperature changes, but it has low abrasion resistance. It is attacked by oxygen and hydrocarbons.

trans-Polyisoprene is similar to Gutta percha, which is produced from the leaves and bark of the sapotacea tree. It has different properties from the cis form and cannot be vulcanized. Few commercial uses are based on trans-polyisoprene.

Important uses of cis-polyisoprene include the production of tires, specialized mechanical products, conveyor belts, footwear, and insulation.

Processing of Natural Rubber

By cutting the bark of rubber tree the milky colloidal rubber milk is obtained. The main constituent of rubber latex is coagulated by using 5% acetic acid and made in to sheets. The rubber sheets are cured under mild heat and then subjected to further processing.

Preparation of Natural Rubber from Latex

- Latex is rubber milk containing about 30 to 45% of rubber.
- The rubber milk is diluted with water and allowed to stand for some time.
- The clear liquid from the top is treated with acetic acid or formic acid to precipitate rubber.
- The precipitated rubber is collected and passed through rollers to get sheets of rubber.
- Rubber sheets are finally dried by smoking. This rubber is called 'Smoked rubber'.
- During the coagulation of rubber milk with acetic or formic acid, retardants like sodium bisulphite ($NaHSO_3$) are added to prevent oxidation of rubber. This is called 'Creep rubber'.

Crepe rubber: To the rubber latex a small amount of sodium bisulphate is added to bleach the colour and feed in to roller which produce 1mm or more thickness sheets which are dried in air at about 40-50°C. The dried thin sheet of rubber are known as "smoked crepe rubber".

Mastication: Rubber becomes soft and gummy mass when subjected to severe mechanical agitation. This process is known as mastication. Mastication is followed by the addition of certain chemical (compounding) which is carried out on roll mills or internal mixers. After mastication is complete, the rubber mix is prepared for vulcanization.

Drawbacks of natural raw rubber: Natural rubber as such has no practical utility value. It is because of many inherent deficiencies which are:

- **Rubber is brittle at low temperature and becomes very soft at high temperatures. Thus, it can be used only in the limited temperature range of 10-60°C**
- **It is too soft to be used for heavy duty operation. Its tensile strength is only 200 Kg/cm^2.**
- **On stretching, it undergoes permanent deformation.**
- **Not resistant to mineral oils, organic solvents and even action of water. It has large water absorption capacity.**
- It is soft and soluble in many hydrocarbon solvents and carbon disulphide (CS_2).
- It has high water absorption power.
- It is a sticky substance, hence is rather difficult to handle.

- It is readily attacked by acids and alkalies.
- It has poor abrasion resistance.
- It is sensitive to oxidative degradation

Compounding of Natural Rubber

Natural rubber is compounded with following substances to get quality rubber:

- **Reinforcing agents or Hardeners:** They are compounded with natural rubber to get hard rubber. Example: Carbon powder, Zinc oxide, etc.
- **Softeners**: They are compounded with natural rubber to get soft spongy rubber. Example: vegetable oils, Stearic acid, paraffin oil etc.
- **Anti-Oxidants:** They are added to prevent aerial oxidation of rubber. Example: Naphthol.
- **Colouring matter:** They are added to give different colour to rubber. Example: Metallic oxides e.g., Zinc oxide – White; Lead chromate – Yellow; Chromium oxide – Green; Carbon black – Black
- **Accelerators:** They are added to speed up the vulcanization reaction of rubber. Example: Lime-magnesia, White lead, etc.
- **Fillers**: Fillers are added to (*i*) reduce the cost, (*ii*) increase the bulk and (*iii*) introduce new characteristics. Example: Textile wastes, Asbestos, Mica, Gypsum, Talc, etc.
- **Vulcanization of rubber:** Vulcanization is compounding of rubber with sulphur. Vulcanization is done by heating rubber with sulphur at 110-140°C in CO_2 atmosphere.

Applications of Natural Rubber after Compounding

(*i*) The major application of natural rubber is in the manufacture of tyres.

(*ii*) In heavy duty tyres, the major portion of the rubber used is natural rubber.

(*iii*) The tank linings in chemical plants where corrosive chemicals are stored are prepared from rubber.

(*iv*) To reduce machine vibrations, rubber is used for sandwiching between two metal surfaces.

(*v*) Foam rubber is used for making cushions, matrices, padding etc. toys and sports items are manufactured from natural rubber.

(*vi*) Gutta percha (trans-1,4- polyisoprene) is used for making submarine cables, golf ball covers, tissue or adhesive etc.

Styrene Butadiene Rubber (SBR-Buna-S)

STYRENE BUTADIENE RUBBER (SBR-BUNA-S)

SBR is produced by the copolymerization of butadiene (75%) and styrene (25%) using free radical initiators. A random copolymer is obtained. Polymerisation is carried out in a stirred tank reactor at around 50°C in the presence of free radical initiators, e.g., potassium sulfate, in an ammoniacal aqueous solution.

Currently, more SBR is produced by copolymerizing the two monomers with anionic or coordination catalysts. The formed copolymer has better mechanical properties and a narrower molecular weight distribution. A random copolymer with ordered sequence can also be made in solution using butyl lithium, provided that the two monomers are charged slowly. Block copolymers of butadiene and styrene may be produced in solution using coordination or anionic catalysts. Butadiene polymerizes first until it is consumed, then styrene starts to polymerize. SBR produced by coordinaton catalysts has better tensile strength than that produced by free radical initiators.

Manufacture of SBR

The production of Styrene Butadiene Rubber (SBR) follows addition polymerization. Its glass transition temperature is approximately –55°C, but may vary with the Styrene content. It can be used extensively within the temperature range of –40 to 100°C. It may be blended with natural rubber or can be used by itself. It has good abrasion resistance properties and the addition of additives may endow it with aging stability. SBR is produced by the copolymerization of Butadiene and Styrene by emulsion polymerization in 3 : 1 weight ratio.

The fresh and recycle monomers are purified by 20 % aqueous caustic purification. Then the stream is continuously passed through 6 to 12 glass – lined or stainless steel reactors. The residence time in these reactors ranges from 5 to 15 hours. Steam heating, water cooling and refrigeration are carried out in these reactors. These reactors are equipped for producing both cold and hot SBR. Cold SBR is produced at 5°C and 1 atm gage by refrigeration techniques ; whereas hot SBR is manufactured at 50°C, and 3-4 atm gage.

The polymerization product is then stripped with chemical inhibitors and the latex resulting from the reactors are collected in blow down tanks. The latex is then sent to a falling film stripper to remove the butadiene; and the bottoms are fed to a perforated plate column where styrene is stripped out.

The monomer – free emulsion is then added with certain compounding ingredients in blend tanks. The latex is coagulated to rubber. The crumbs of rubber float onto shaker screens where the catalyst, emulsifiers and other solubles are removed by the action with wash water. The crumbs are dried in continuous belt dryers with hot air and are subsequently pressed into bales.

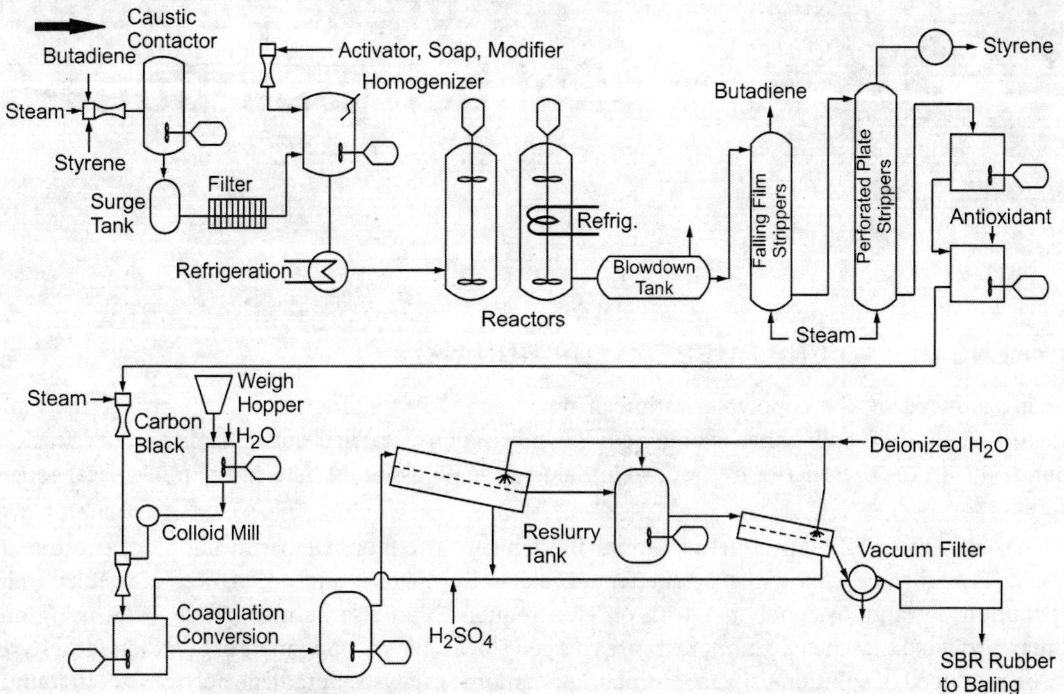

Fig. 21.1. Manufacture of Styrene Butadiene Rubber (SBR)

Properties and Uses of SBR

SBR is a general purpose rubber widely used for tyres either by itself or as a high percentage blend with the natural rubber. It is superior to natural rubber in all properties except tensile strength, resistance and heat build up under heavy loading. It is also used for mechanical goods and electrical insulation.

Styrene butadiene rubber (SBR) is the most widely used synthetic rubber. The main use of SBR is for tire production. Other uses include footwear, coatings, carpet backing, and adhesives.

SBR Consumption pattern in India during 2010-11 is as follows:

Tyres-60%,Conveyor belts-7%, Retreads-5%, Auto components-4%, Other usages-24%

Nitrile Rubber (Buna-N)

NITRILE RUBBER (BUNA-N)

Nitrile rubber (or acrylonitrile rubber) is a copolymer of acrylonitrile and butadiene. The copolymerization occurs in an aqueous emulsion. When free radicals are used, a random copolymer is obtained. Alternating copolymers are produced when a Ziegler-Natta catalyst is employed. Molecular weight can be controlled by adding modifiers and inhibitors. When the polymerization reaches approximately 65%, the reaction mixture is vacuum distilled in presence of steam to recover the monomer.

Nitrile rubbers are available in many grades varying in acrylonitrile content. Increase in acrylonitrile improves resistance to fuels and oil, tensile strength and modulus, processing behavior, heat resistance; increases abrasion resistance and hardness, permeability resistance to gas diffusion; decreases low temperature flexibility, resilience and elasticity and plasticizer compatibility.

A broad range of properties can be obtained from properly compounded nitrile rubber. Nitrile rubber is made by emulsion copolymerisation of butadiene and acrylonitrile at 5°C. The steps involved are polymerisation, coagulation, washing and drying. A basic polymerisation recipe in addition to the monomer contains water, stabilizers, emulsifiers, shortstop catalyst activator and electrolytes. Following polymerisation cycle material is transferred to blow down tank in which short stop catalyst and antioxidant are added and residual monomers are recovered. Then finally, the latex is concentrated, coagulated, washed, dewatered and finally dried.

In the polymerization process the monomer is emulsified in water, a free radical generating catalyst is added and the mixture is agitated. After the polymerization the material from the polymerization reactor is transferred to blow tank in which short stop and antioxidant are added and the residual monomers are removed. The latex formed is concentrated and coagulated into fine crumbs by addition of salt and acids. This followed by washing, dewatering and drying. The dried crumbs are compacted.

Process flow sheet of nitrile rubber manufacture is shown below in figure.

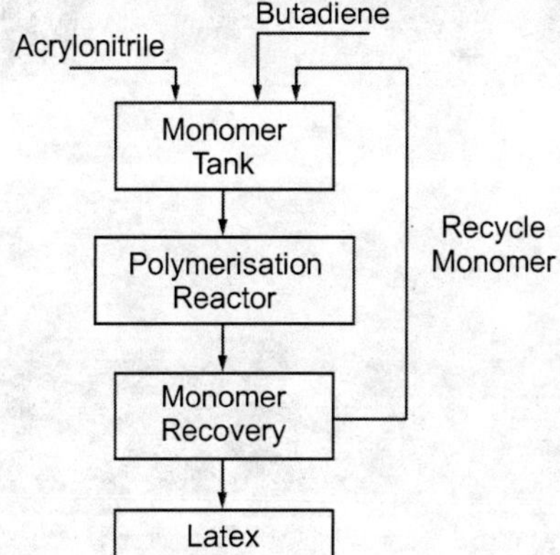

Fig. 22.1. Process flow sheet of (Nitrile rubber (Acrylonitrile rubber) manufacture

Properties and Uses of Nitrile Rubber

The ratio of acrylonitrile/butadiene can be adjusted to obtain a polymer with specific properties. Increasing the acrylonitrile ratio increases oil resistance of the rubber, but decreases its plasticizer compatibility. Nitrile rubber has the special property of being resistant to hydrocarbon liquids.

Within a range of 18-50 % acrylonitrile, with increase in acrylonitrile content:

- Resistance to oil, fuel, abrasion and heat increases and higher tensile strength, hardness and gas impermeability results.
- Low temperature resistance, resilience and plasticizer compatibility decreases
- Requires less sulphur, more accelerators than SBR, highly oil resistant.

Nitrile rubber may be reinforced by phenolic resins and PVC resulting in improved resistance to ozone and weathering, better gloss, bright colors, high resistance to abrasion and oil.

Nitrile rubber is produced in different grades depending on the end use of the polymer. Low acrylonitrile rubber is flexible at low temperatures and is generally used in gaskets, O-rings, and adhesives. The medium type is used in less flexible articles such as kitchen mats and shoe soles. High acrylonitrile polymers are more rigid and highly resistant to hydrocarbons and oils and are used in fuel tanks and hoses, hydraulic equipment, and gaskets. Nitrile rubber is also used in oil field parts, diaphragm, gloves, belts, wire cable insulation, hosepipes, foot wear shoe products, molded rubber goods, fuel hoses, collapsible containers etc.

Nitrile rubber Consumption pattern in India during 2010-11 is as follows:

Auto components-50%, Conveyor belts-22%, Other usages-28%

Butyl Rubber (Polyisobutylene)

BUTYL RUBBER (POLYISOBUTYLENE)

Butyl rubber is a copolymer of isobutylene (97.5%) and isoprene (2.5%).The polymerization is carried out at low temperature (below –95°C) using $AlCl_3CO$ catalyzed with a small amount of water. The product is a linear random copolymer that can be cured to a thermosetting polymer. This is made possible through the presence of some unsaturation from isoprene.

Polyisobutylene is made by polymerization of mainly isobutylene. Isobutylene can be recovered from C_4 stream from steam cracker and FCC. Other route for isobutylene production are: dehydrogenation of isobutene *n*-butane isomerisation in gas phase using platinum catalyst. Butyl rubber is made by slurry polymerisation. The polymerisation is carried out in slurry of monomer in methyl chloride using an aluminium chloride catalyst at –100 to –90°C. The rubber is precipitated by adding water and finally washed and dried.

Properties and Uses of Butyl Rubber

Polyisobutylene is gas impermeable and because of this property it is used for making balloon.

Butyl rubber has unique elastomeric qualities, low rate of gas permeability, thermal stability, good ozone and weathering resistance, vibration damping and higher coefficients of friction, chemical and moisture resistance. Used in tubes, tyre inner liner due to low permeability of air, automotive mechanical parts, adhesives, and sealant.

Butyl rubber vulcanizates have tensile strengths up to 140 kg/cm^2 and are characterized by low permeability to air and a high resistance to many chemicals and to oxidation. These properties make it a suitable rubber for the production of tire inner tubes and inner liners of tubeless tires. The major use of butyl rubber is for inner tubes. Other uses include wire and cable insulation, steam hoses, mechanical goods, and adhesives. Chlorinated butyl is a low molecular weight polymer used as an adhesive and a sealant.

CHLORO BUTYL RUBBER

Chlorobutyl rubber is made from isobutylene and 1-3% isoprene. Introducing a continuous stream of chlorine gas in a hexane solution of butyl, which is prepared by low temperature copolymerisation of isobutylene, and isoprene in methyl chloride makes Chlorobutyl.

Chlorobutyl rubber possesses greater vulcanisation flexibility and is used for making tubeless tyres, tyre side wall components, heat resistant truck inner tubes, hose pipes, gaskets, conveyor belts, adhesive, sealants, tyre curing bays, tank lining etc.

Neoprene Rubber (Polychloroprene)

NEOPRENE RUBBER (POLYCHLOROPRENE)

Neoprene rubber (polychloroprene) is a polymer of chloroprene (($CH_2CH — CCl == CH_2$).

Chloroprene is made either via acetylene route or from butadiene. Butadiene process is commonly used.

Acetylene route involves dimerisation of acetylene to monovinyl acetylene followed by reaction of monovinyl acetylene with HCl. Chloroprene production from butadiene involves following three steps :

- Chlorination of Butadiene to 1,4 dichloro 2-butene
- Isomerisation of 1,4 dichloro-2-butene to 3,4-dichloro 1-butene
- Dehydrochlorination of 3, 4-dichloro1-butene to chloroprene in presence of caustic soda resulting in formation of chloroprene ($CH_2CH — CCl == CH_2$)

Polychloroprene is made by emulsion polymerisation process using resin acid soap emulsifier. Polymerisation is carried out at 40°C in presence of sulphur. Polychloroprene is the oldest synthetic rubber. It is produced by the polymerization of 2-chloro-1,3-butadiene in a water emulsion with potassium sulfate as a catalyst.

Properties and Uses of Neoprene Rubber

The neoprene rubber is a random polymer that is vulcanized with sulfur or with metal oxides (zinc oxide, magnesium oxide etc.). Vulcanization with sulfur is very slow, and an accelerator is usually required. Neoprene vulcanizates have a high tensile strength, excellent oil resistance (better than natural rubber), and heat resistance.

Uses: Adhesives, transportation industry (Automotive – gaskets, V-belts, shock absorber covers, wire jackets, molded seats, aviation-wire cable, gaskets, seats etc.), rail brake hose, track mountings, construction industry, hose and belting.

Neoprene rubber can be used for tyre production, but it is expensive. Major uses include cable coatings, mechanical goods, gaskets, conveyor belts, and cables.

25 Ethylene Propylene Rubber (EPR)

ETHYLENE PROPYLENE RUBBER (EPR)

Copolymerization of ethylene with propylene results in a random, noncrystalline copolymer that is a chemically inert and rubbery material. EPR is a saturated copolymer that can be cross-linked through the combination of the free radicals generated by peroxides or radiation.

Ethylene-propylene rubber (EPR) is a stereoregular copolymer of ethylene and propylene. Elastomers of this type do not possess the double bonds necessary for crosslinking. A third monomer, usually a monoconjugated diene, is used to provide the residual double bonds needed for cross linking. 1,4-Hexadiene and ethylidene norbornene are examples of these dienes. The main polymer chain is completely saturated while the unsaturated part is pending from the main chain. The product elastomer, termed ethylene propylene terepolymer (EPT), can be crosslinked using sulfur. Crosslinking EPR is also possible without using a third component (a diene). This can be done with peroxides.

Ethylene–propylene elastomers are made by solution polymerization of ethylene and propylene in a solvent such as hexane using Ziegler–Natta catalysts. EPDM terpolymers can be similarly made by adding 3 to 9% of any of the above dienes to the monoolefin mixture.

To incorporate sites for vulcanization, an unsaturated terpolymer can be prepared from ethylene, propylene, and a small amount (3 to 9%) of a nonconjugated diene (EPDM). The diene is either dicyclopentadiene, ethylidene norbornene, or 1,4-hexadiene. The resulting unsaturated terpolymer can be vulcanized by traditional techniques. Each of the termonomers confers different characteristics on the final elastomer.

Important properties of vulcanized EPR and EPT include resistance to abrasion, oxidation, and heat and ozone; but they are susceptible to hydrocarbons attack.

The main use of ethylene propylene rubber is to produce automotive parts such as gaskets, mechanical goods, wire, and cable coating. It may also be used to produce tires.

Even though tire use is small, automotive applications are still the largest market for ethylene propylene elastomers. They are used predominantly in radiator and heater hoses, seals, gaskets, grommets, and weather stripping. Blends of polypropylene and EPDM are used as material in the manufacture of car bumpers, fender extensions, and rub strips. Other applications of ethylene propylene elastomers include appliance parts, wire and cable insulation, and modification of polyolefins for improved impact and stress.

❑❑❑

Silicone Rubbers (Polysiloxanes)

Structures and Properties of Silicones

Silicones are synthetic polymers with a silicon-oxygen backbone similar to that in silicon dioxide (silica), but with organic groups attached to the silicon atoms by C-Si bonds. The silicone chain exposes organic groups to the outside.

R R
Si O Si O
R R

Thus, despite having a very polar chain, the physical properties of silicones are similar to those of an alkane. However, the —Si — O — framework of the silicone gives the polymers thermal stability, as in silica, and so they can be used where comparable organic materials would melt or decompose.

To distinguish between different silicones, systematic names are used, based on the monomer. The simplest silicon compound is silane, SiH_4 which belongs to the homologous series of silanes. Silanes correspond to the alkanes whose simplest member is methane, CH_4.

Silane   Methane

The presence of the oxygen atoms in the silicone chain is indicated by using the systematic name, siloxane, so termed as it contains a silicon atom, an oxygen atom and it is saturated as in an alkane.

Thus, if the groups attached to the siloxane chain are phenyl groups, the resulting silicone is called poly (diphenyl siloxane) and has repeating units along the chain.

O—Si—

The most widely used silicones are those which have methyl groups along the backbone. Properties such as solubility in organic solvents, water-repellence and flexibility can be altered by substituting other organic groups for the methyls. For example, silicones with phenyl groups are more flexible polymers than those with methyl groups. They are also better lubricants and are superior solvents for organic compounds.

The structure of the repeating units of silicones can be represented as:

$$-\underset{R}{\overset{R}{|}}\!\!Si-O-\underset{R}{\overset{R}{|}}\!\!Si-O-$$

where R represents organic groups attached to the silicone backbone, for example:

CH_3-	(phenyl ring)	$HC=CH_2$ (attached at HC)	$CH_2CH_2CF_3$ (attached at CH_2)
Methyl	Phenyl	ethenyl	Trifluoropropyl

Manufacture of Silicones

Silicones are manufactured from pure silicon which has been obtained by the reduction of silicon dioxide (silica) in the form of sand with carbon at high temperatures:

$$SIO_2\,(s) + 2C\,(s) \longrightarrow Si\,(s) + 2\,CO\,(g)$$

The production of silicones from silicon takes place in three stages:

(*a*) Synthesis of chlorosilanes

(*b*) Hydrolysis of chlorosilanes

(*c*) Condensation polymerization

(*a*) **Synthesis of chlorosilanes:** Silicon is first converted into chlorosilanes, e.g., $RSiCl_3$, R_2SiCl_2 and R_3SiCl, where R is an organic group.

When chloromethane is passed through heated silicon at about 550 K under slight pressure and in the presence of a copper catalyst (often copper itself but other copper-containing materials can be used, for example, brass or copper(II) chloride), a volatile mixture of chlorosilanes distils over. For example: The mixture of liquids contains these three compounds:

$$SI\,(s) + 2CH_3C\,(g) \longrightarrow (CH_3)_2SiCl_2\,(g)$$

$$Cl-\underset{Cl}{\overset{CH_3}{|}}\!\!Si-Cl \qquad Cl-\underset{CH_3}{\overset{CH_3}{|}}\!\!Si-Cl \qquad Cl-\underset{CH_3}{\overset{CH_3}{|}}\!\!Si-CH_3$$

CH_3SiCl_3 $(CH_3)_2SiCl_2$ $(CH_3)_3SiCl$

Careful distillation of the liquid mixture of chlorosilanes produces pure fractions of each chlorosilane. Dimethyl dichloro silane is the main product (70-90%, the amount depending on the conditions used).

(*b*) **Hydrolysis of chlorosilanes:** A dichlorosilane is hydrolyzed to a molecule with two hydroxyl groups:

$$Cl-\underset{R}{\overset{R}{\underset{|}{\overset{|}{Si}}}}-Cl + 2\,H_2O \longrightarrow HO-\underset{R}{\overset{R}{\underset{|}{\overset{|}{Si}}}}-OH + 2\,HCl$$

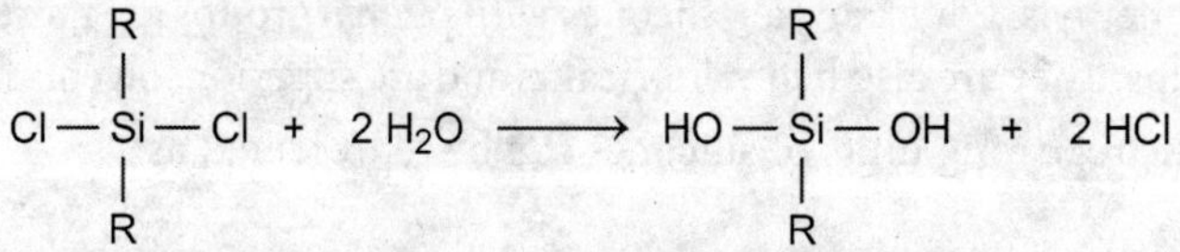

The product is a disilanol. The suffix-*ol* in a silanol is to show that the molecule contains at least one hydroxyl group attached to a silicon atom and the simplest example is dimethyl disilanol:

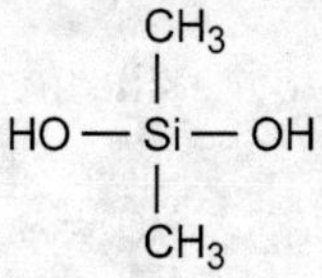

$$HO-\underset{CH_3}{\overset{CH_3}{\underset{|}{\overset{|}{Si}}}}-OH$$

Dimethyldisilanol

This nomenclature is similar to that of the alcohols, the simplest alcohol with two hydroxyl groups being ethane-1,2-diol:

$$HO-CH_2CH_2-OH$$

Ethane-1,2-diol

The hydroxyl groups of silanols react spontaneously to form a siloxane:

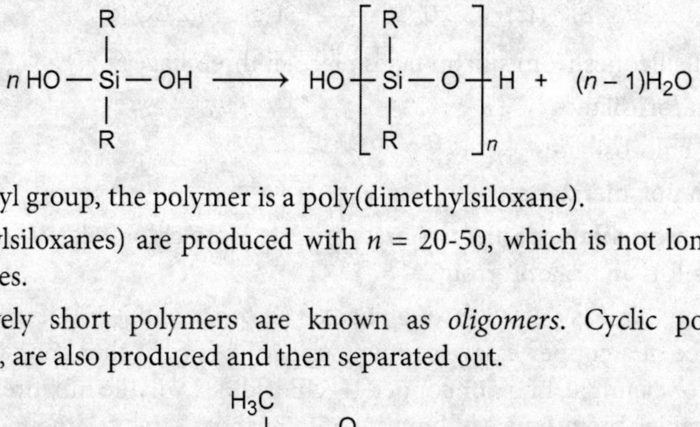

$$n\,HO-\underset{R}{\overset{R}{\underset{|}{\overset{|}{Si}}}}-OH \longrightarrow HO\left[\underset{R}{\overset{R}{\underset{|}{\overset{|}{Si}}}}-O\right]_n H + (n-1)H_2O$$

If R is a methyl group, the polymer is a poly(dimethylsiloxane).

Poly(dimethylsiloxanes) are produced with n = 20-50, which is not long enough to produce useful silicones.

These relatively short polymers are known as *oligomers*. Cyclic polymers, for example $((CH_3)_2SiO)_4$, are also produced and then separated out.

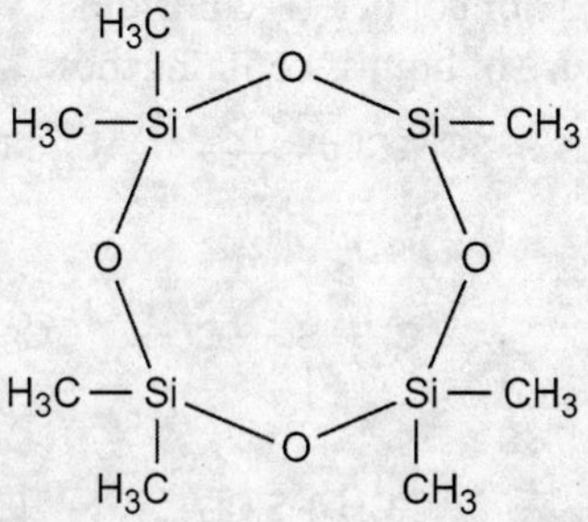

[Structure: cyclic tetramer — eight-membered ring of alternating Si and O atoms, each Si bearing two CH_3 groups (octamethylcyclotetrasiloxane)]

The oligomers are washed and dried. The hydrochloric acid is recycled and reacts with methanol to regenerate chloromethane:

(*c*) **Condensation polymerization:** The oligomers condense rapidly in the presence of an acid catalyst to form long chain polymers:

$$HO\left[\begin{matrix} CH_3 \\ | \\ Si-O \\ | \\ CH_3 \end{matrix}\right]_m H + HO\left[\begin{matrix} CH_3 \\ | \\ Si-O \\ | \\ CH_3 \end{matrix}\right]_n H \longrightarrow HO\left[\begin{matrix} CH_3 \\ | \\ Si-O \\ | \\ CH_3 \end{matrix}\right]_{m+n} H + H_2O$$

These materials are silicone fluids

The value of $(m + n)$ is usually between 2000 and 4000. The production of longer chains is favoured if the water is removed, for example by working under vacuum.

To form silicone gels, elastomers and resins, the long siloxane chains are induced to cross-link. There are four main ways of doing this:

(*i*) Cross-linking is often effected by first synthesizing silanes with a functional group, in place of a methyl group, that will react further. For example, a silane containing an ethenyl (vinyl) group such as ethenyl methyl dichloro silane, can be added to, for example, dimethyldichlorosilane. However with ethenyl groups in the chain, the chains are also able to undergo free radical addition reactions, in a similar way to the free radical polymerization of chloroethene (vinyl chloride). This leads to cross-linking between the polymer chains. As with this polymerization, the addition reactions are initiated by radicals supplied on decomposition of an organic peroxide (for example, dicumyl peroxide):

$$R-O-O-R \longrightarrow 2RO\cdot$$

where, $R = C_6H_5-\overset{\displaystyle CH_3}{\underset{\displaystyle CH_3}{\overset{|}{\underset{|}{C}}}}-$

(*ii*) Cross-linking can also be achieved by using siloxanes with ethenyl (vinyl) groups and other siloxanes containing Si-H groups, with a platinum compound as catalyst:

$$\sim\!\overset{\displaystyle CH_3}{\underset{\displaystyle CH_3}{\overset{|}{\underset{|}{Si}}}}-CH{=}CH_2 + HO-\overset{\wr}{\underset{\wr}{Si}}\sim \longrightarrow \sim\!\overset{\displaystyle CH_3}{\underset{\displaystyle CH_3}{\overset{|}{\underset{|}{Si}}}}-CH_2-CH_2-\overset{\wr}{\underset{\wr}{Si}}\sim$$

Dimethyldisilanol   Dimethyldisilanol

(*iii*) A further way of producing cross-linking is to have an ethanoyl group in the silane. When these silicones are exposed to the air, the moisture reacts with the functional group, yielding a cross-linked silicone. An organometallic tin compound catalyses the reaction. These systems are often used as sealants and can be used in the home. The other product formed is ethanoic acid which can be recognized by its vinegary smell.

(*iv*) If some methyl trichlorosilane is added to the reactant, say dimethyldichlorosilane, the three chlorine atoms are hydrolyzed, thus producing a three dimensional network.

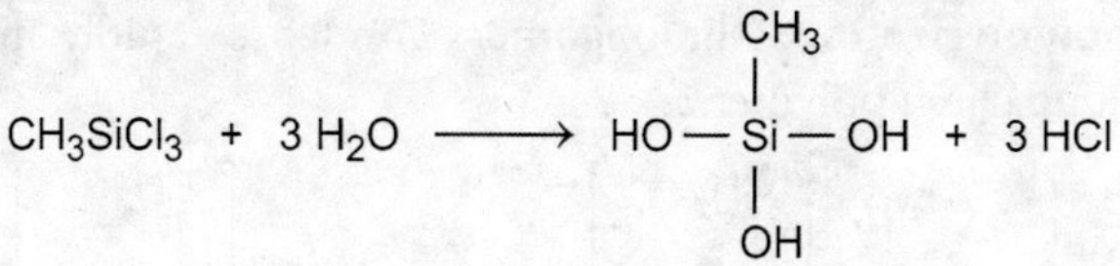

$$CH_3SiCl_3 + 3\,H_2O \longrightarrow HO-\underset{\underset{OH}{|}}{\overset{\overset{CH_3}{|}}{Si}}-OH + 3\,HCl$$

In all four methods, the physical properties of the silicone can be modified by varying the proportions of the reactants, which controls the extent of cross-linking and hence how *rubbery* is the product.

Properties and Uses of Silicone Polymers

Silicon elastomer are made by ring opening reaction caused by action of alkali on monomer acyclic siloxane characterised by exceptional mechanical and electrical performance under extreme temperature condition. Used in aerospace, appliances, electrical industry, construction industry, automotive industry, gaskets sealings, spark plug boots, hose, rubber rolls.

Silicones have unique properties amongst polymers because of the simultaneous presence of organic groups attached to a chain of inorganic atoms. They are used in many industries including those devoted to electronics, paints, construction and food.

Methyl silicone polymers of the linear type produce oils with a low rate of change of viscosity at low temperatures with fluidity down to –85°C. These oils are non volatile, non-toxic and extremely stable with no sludging or oxidation up to 150°C. Their high cost limits usage to special applications such as gage fluids and dashpot liquids. The use of synthetic lubricants of a polymeric composition is synonymous with silicones. Silicone polymers have found use as high temperature rubber and resin coatings. Their unusual physical properties as fluid polymers account for much of the marketing efforts in silicone polymers.

Silicon fluids make excellent liquid dielectrics because of their low power loss, low water absorption and heat stability.

By additional polymerization and compounding with thickening agents, various types of stable greases are produced with a working range from –75 to 260°C, a much wider span than covered by petroleum grade greases.

Silicones can be Sub-divided into Four Classes

(*a*) Silicone fluids
(*b*) Silicone gels
(*c*) Silicone elastomers (rubbers)
(*d*) Silicone resins

Their physical form and uses depend on the structure of the polymer.

(*a*) **Silicone fluids** are typically straight chains of poly(dimethylsiloxane), with the repeating structure:

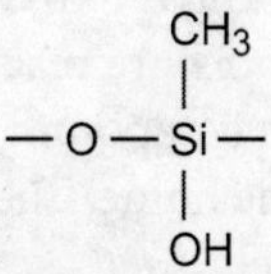

$$-O-\underset{\underset{OH}{|}}{\overset{\overset{CH_3}{|}}{Si}}-$$

They usually have trimethylsilyl groups, $Si(CH_3)_3$ at each end of the chain:

$$H_3C-\underset{CH_3}{\overset{CH_3}{Si}}-\left[O-\underset{CH_3}{\overset{CH_3}{Si}}\right]_n-O-\underset{CH_3}{\overset{CH_3}{Si}}-CH_3$$

Poly(dimethyl)siloxane

The silicones with short chains are *fluids* which, compared to hydrocarbons, have a more or less constant viscosity over a wide temperature range (200 to 450 K). They also have very low vapour pressures.

The low surface tension of silicone fluids gives them unique surface properties. They are, for example, used as lubricants in polishes (a mixture of wax and a silicone fluid dissolved in an organic solvent), in paints and for water-proofing fabrics, paper and leather. They also have anti-foaming properties and have been used, for example, to suppress the foaming of detergents in sewage disposal plants.

They have a low enthalpy of vaporization and a smooth, silky feel and thus are attractive as a basis for personal care products such as perspirants and skin care lotions.

A range of fluids is made by mixing polysiloxanes of low molecular masses with others with higher molecular masses. Some use the cyclic silicones which are formed during the preparation of the linear polysiloxanes.

(*b*) **Silicone gels** are based on the poly(dimethylsiloxane) chains but with a few cross-links between the chains, giving it a very open three-dimensional network. Often the cross-linking is done after a silicone fluid, together with a reactive group, is poured into a mould and then warmed or catalysed so that there is interaction to form cross-linking between the polymer chains. This is a very effective technique for protecting sensitive electronic equipment from damage from vibration and the polymer also acts as an electrical insulator. Pads containing a silicone gel are also used as shock absorbers in shoes, particularly in high-performance trainers and running shoes.

(*c*) **Silicone elastomers (rubbers)** are made by introducing even more cross-linking into the linear chain polymers. The structure is somewhat similar to natural rubber and they behave as elastomers. Their structure is determined by the amount of cross linking and the length of the chains. *These silicone elastomer particles are used in skin creams. Their small and perfectly spherical shape combined with a rubbery texture improves the feel of the cream as it is applied to the skin.* Although their strength at normal temperatures is inferior to that of natural rubber, silicone rubbers are more stable at both low temperatures (200 K) and high temperatures (450-600 K) and are generally more resistant to chemical attack. Silica is added as a filler to make the elastomer stronger.

(*d*) **Silicone resins** have a three-dimensional structure with the atoms arranged tetrahedrally about the silicon atoms. The resins are usually applied as a solution in an organic solvent, and are used as an electrical insulating varnish or for paints where water repellence is desired, for example, to protect masonry. They are also used to give an 'anti-stick' surface to materials coming into contact with 'sticky' materials such as dough and other foodstuffs.

Hydroxyl groups on the resin react with hydroxyl groups that are on the surfaces of various inorganic surfaces such as silica and glass, thus making the surface water-repellent. A large range of silanes, known as *coupling agents*, has been developed to enable chemists to bond an inorganic substrate (such as glass, minerals and metals) to organic materials (for example, organic polymers such as the acrylics, polyamides, urethanes and polyalkenes). The resulting coatings confer the surface properties of a silicone to a very wide range of materials. Similar mechanisms enable some resins to be used as adhesives.

MANUFACTURE, PROPERTIES & APPLICATIONS OF IMPORTANT SYNTHETIC FIBRES

Cellulosic Fibres

CELLULOSIC FIBRES

Fibre filament is a thread like structure of varying length, capable of being woven or spun into larger units. Fibers are solid materials characterized by a high ratio of length to diameter. They can be manufactured from a natural origin such as silk, wool, and cotton, or derived from a natural fiber such as rayon. They may also be synthesized from certain monomers by polymerization (synthetic fibers). In general, polymers with high melting points, high crystallinity, and moderate thermal stability and tensile strengths are suitable for fiber production.

Man-made fibers include, in addition to synthetic fibers, those derived from cellulose (cotton, wood) but modified by chemical treatment such as rayon, cellophane, and cellulose acetate. These are sometimes termed "regenerated cellulose fibers." Rayon and cellophane have shorter chains than the original cellulose due to degradation by alkaline treatment.

Natural fibres such as cotton, wool, silk and linen are being replaced by synthetic fibres. These synthetics are prepared from molten linear polymers or a solution of polymers by extrusion through a die containing 20 to 15000 small holes, solidifying and a group of threads as a filament. This process is called spinning.

All of the synthetic fibres have one or more advantages in terms of high tensile strength, good temperature stability and excellent resistance to chemicals, abrasion, solvents and moisture.

Major fiber-making polymers are those of polyesters (PET, PBT), polyamides (nylons), polyacrylics (Acrylic fibres), and polyolefins(Polypropylene fibres). Polyesters and polyamides are produced by step-growth condensation polymerization reactions, while polyacrylics and polyolefins are synthesized by chain-growth addition polymerization.

Advantages of Synthetic Fibres

Synthetic fibers are more durable than most natural fibers and will readily pick-up different dyes. In addition, many synthetic fibers offer consumer-friendly functions such as stretching, waterproofing and stain resistance. Sunlight, moisture, and oils from human skin cause all fibers to break down and wear away. Natural fibers tend to be much more sensitive than synthetic fibres. This is mainly because natural products are biodegradable. Natural fibers are susceptible to larval insect infestation; synthetic fibers are not a good food source for fabric-damaging insects.

Compared to natural fibers, many synthetic fibers are more water resistant and stain resistant. Some are even specially enhanced to withstand damage from water or stains. Some fabrics are also designed to stretch in specific ways, which makes them more comfortable to wear.

Cotton production is resource intensive: it takes significant amounts of water to farm cotton. Wool producing sheep also need water and grazing land in order to survive. Though, synthetic fiber production does involve some carbon emissions, the environmental footprint of many fibers is much lower.

Disadvantages of Synthetic Fibres

Most of synthetic fibers' disadvantages are related to their low melting temperature:

- Synthetic fibers burn more readily than natural fibres.
- Prone to heat damage.
- Melt relatively easily.
- Prone to damage by hot washing.
- More electrostatic charge is generated by rubbing than with natural fibers.
- Not skin friendly, so it is uncomfortable for long wearing.
- Some people are allergic.
- Non-biodegradable in comparison to natural fibers.

Definition of Cellulose Acetate Fibre:

A manufactured fibre in which the fibre forming substance is cellulose acetate. Where not less than 92% of the hydroxyl groups are acetylated, the term triacetate may be used as a generic description of the fibre.

Basic Principles of Cellulose Acetate Fibre Production

Acetate is derived from cellulose by reacting purified cellulose from wood pulp with acetic acid and acetic anhydride in the presence of sulphuric acid. It is then put through a controlled, partial hydrolysis to remove the sulphate and a sufficient number of acetate groups to give the product the desired properties. The anhydro glucose unit, is the fundamental repeating structure of cellulose, has three hydroxyl groups which can react to form acetate esters. The most common form of cellulose acetate fibre has an acetate group on approximately two of every three hydroxyls. This cellulose diacetate is known as secondary acetate, or simply as "acetate". After it is formed, cellulose acetate is dissolved in acetone for extrusion. As the filaments emerge from the spinneret, the solvent is evaporated in warm air (dry spinning), producing fine filaments of cellulose acetate.

Manufacturing Process for Cellulose Acetate Fibre

Raw Materials used for the production of cellulose acetate fibre are Purified cotton linters, wood pulp, acetic anhydride, acetic acid and sulfuric acid. The process involves steeping of purified cotton in acetic acid which makes Swelling and makes cellulose more reactive. The pretreated cotton with acetic acid is then acetylated with excess acetic acid, acetic anhydride, with sulphuric acid to promote the reaction.

The purified cotton linters or cellulose is fed to the acetylator containing acetic anhydride, acetic acid conc.and Sulphuric acid and acetylation is carried out at 25-30°C. The reaction mixture called acid dope is allowed for ripening for about 10-20 hrs. During reopening, conversion of acetate groups takes place. After reopening, the mixture is diluted with water with continuous stirring. During the process flakes acetate rayon is precipitated which is dried and sent to spinning bath where dry spinning of acetate rayon takes place by dissolving in solvent and passing through spinneret. The solvent is evaporated by hot air. The dope coming from the spinnerate is passed downwards to feed roller and finally to bobbin where spinning is done at higher speed.

Process flow diagram for the manufacture of cellulose acetate fibre is given below in Figure.

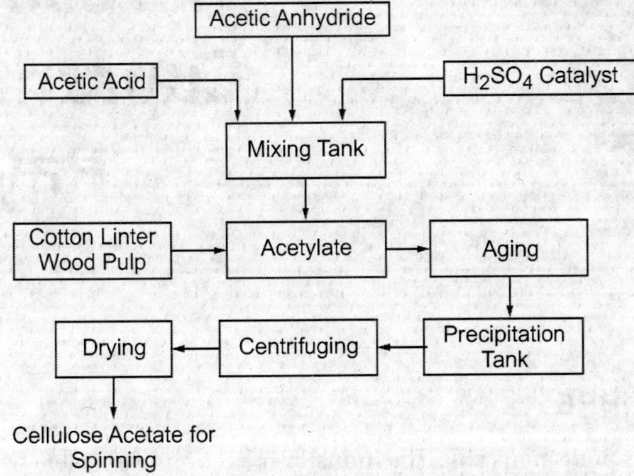

Fig. 27.1. Manufacture of cellulose acetate

The viscose process and the cuprammonium process produce a filament of regenerated cellulose, while the acetate forms a thread that is a definite chemical compound of cellulose, cellulose acetate. Although each of these processes is quite different as far as details of procedure are concerned, they all follow the same general outline: solution of the cellulose through a chemical reaction, aging or ripening of the solution (peculiar to viscose), filtration and removal of air, spinning of the fiber, combining the filaments into yarn, purifying the yarn (not necessary for acetate), and finishing (bleaching, washing, oiling, and drying).

Cellulose Acetate Fibre Characteristics

- Luxurious feel and appearance
- Excellent drapability and softness
- Shrink, moth and mildew resistant
- Wide range of colours and lustres
- Relatively fast drying

Special dyes have been developed for acetate since it does not accept dyes ordinarily used for cotton and rayon. This dye selectivity makes it possible to obtain multi-colour effects in fabrics made from a combination of fibres (cross-dyeing)

In cross-dyeing, yarns of one fibre (e.g., acetate) and those of another fibre (cotton or rayon) are woven into a fabric in a desired pattern. After the fabric has been dyed in one bath, this pattern will appear in different colours or shades according to the distribution of the respective fibres. Solution-dyed or spun-dyed acetate provides excellent colour fastness under the effects of sunlight, perspiration, air contaminants and washing.

Some Major Cellulose Acetate Fibre Uses

- **Apparel:** Blouses, dresses, linings, wedding and party attire, home furnishings, draperies, upholstery
- **Industrial uses:** Cigarette filters

Viscose Rayon & Cuprammonium Rayon Fibre

VISCOSE RAYON FIBRE

Viscose rayon is known as poor man's silk. The industry is becoming obsolete worldwide due to material constraint and adverse environmental effects. Viscose itself suffers from some deficiency - it is a weak fibre, especially when wet. There are few manufacturers of viscose (both VFY-viscose filament yarn and VSF- viscose staple fibre) in India. The industry is surviving today because of the closure of viscose plants abroad due to environmental reasons.

The raw material for the production of Viscose rayon is wood pulp and the production process generates toxic effluents which have adverse environmental effects. Because of its cellulosic base, it has some properties which are similar to those of cotton and any other natural cellulosic fibres. It has moderate resistance to acids and alkalis. However, it may be damaged by relatively weak acids. Moreover, due to the inelasticity of fibers, it has poor abrasion resistance.

The raw materials used in this process are cellulose wood pulp sheets or cotton linters. These are placed in a steeping press with contact in aqueous NaOH Solution for a period of 2-4 hours at normal room temperature. A hydraulic ram presses out the excess alkali and the sheets are shredded to crumbs and aged for 2-3 days. The aging process has its direct consequence on the viscosity of the solution.

Later, CS_2 is added in a rotating drum mixer over a period of 3 hours. The orange cellulose xanthate which forms is transferred to a solubilizer, wherein it is mixed into dilute caustic. The mixing of cellulose xanthate and dilute caustic yields a orange colored viscous solution which contains 7-8 % cellulose and 6.5-7 % NaOH. It is digested at room temperature for 4-5 days. Thereafter, the solution is filtered and fed to spinning machines.

The spinning is carried out in extrusion spinnerettes, which are made up of platinum or gold alloys. The orifices of these spinnerets have a diameter of 0.1-0.2 mm for continuous filament yarns and of diameter 0.05-0.1 mm for short fibre shapes. The solution extruded from the spinnerets is contacted with an acid bath which precipitates the filaments without causing them to break or stick together. The processing treatments of washing, desulphurizing, bleaching and conditioning takes place continuously and in order after the filaments are wound on a series of plastic rolls.

Manufacturing process of viscose rayon is shown below in figure.

Fig. 28.1. Manufacture of viscous rayon

CUPRAAMONIUM RAYON

Cupraammonium rayon is made from reaction of Cellulose with copper salt and ammonia. After bleaching, cellulose is added in ammoniacal solution of copper sulphate resulting in formation of cuprammonium cellulose which is spun into water and the yarn is washed with acid to remove traces of ammonia and dried. Process flow diagram for the manufacture of cuprammoium rayon is given below in figure.

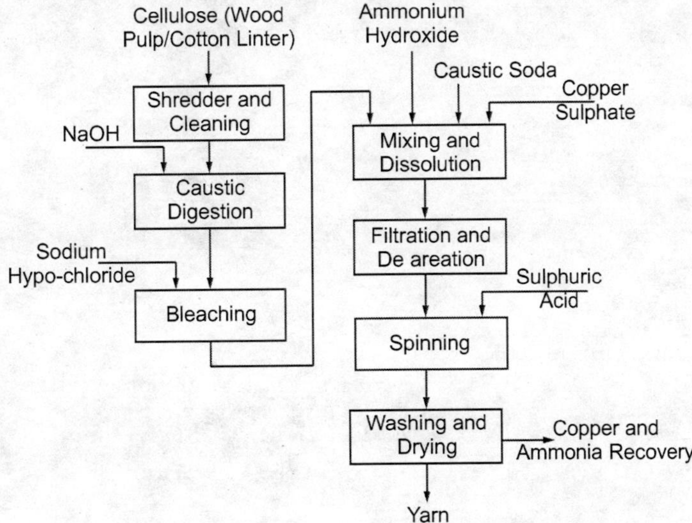

Fig. 28.2. Manufacture of cuprammonium rayon

Cuprammonium or cupro regenerated cellulose is obtained by the cuprammonium process. The name cuprammonium is obtained from the aqueous cuprammonium hydroxide solvent in which the cellulose is dissolved to form the spinning solution.

Polyamides (Nylon-6 & Nylon-66) Fibres

POLYAMIDES (NYLON FIBRES) $\left[-\overset{H}{\overset{|}{N}}-\underset{O}{\underset{\|}{C}}-R- \right]_n$

Polyamides are polymers which contain repeating amide, — CO — NH —, linkages.

Proteins are examples of naturally occurring polyamides. The best known manufactured polyamides are often called nylons (the trade name given by the manufacturer, DuPont) and these are aliphatic polyamides.

However, other manufactured polyamides are also important and these include an aromatic polyamide, Kevlar and plastics produced from carbamide (urea). The nomenclature for describing the linear, aliphatic polyamides (the nylons) is based on the **number of carbon atoms** in the repeating unit.

6	$-\overset{H}{\overset{|}{N}}-(CH_2)_5-\underset{O}{\underset{\|}{C}}-$
6, 6	$-\overset{H}{\overset{|}{N}}-(CH_2)_6-\overset{H}{\overset{|}{N}}-\underset{O}{\underset{\|}{C}}-(CH_2)_4-\underset{O}{\underset{\|}{C}}-$
6, 10	$-\overset{H}{\overset{|}{N}}-(CH_2)_6-\overset{H}{\overset{|}{N}}-\underset{O}{\underset{\|}{C}}-(CH_2)_8-\underset{O}{\underset{\|}{C}}-$
11	$-\overset{H}{\overset{|}{N}}-(CH_2)_{10}-\underset{O}{\underset{\|}{C}}-$

12	H–N–$(CH_2)_{11}$–C(=O)–

Manufacture of Polyamide (Nylon) 6 and 6,6

Both polyamides are manufactured from benzene via cyclohexane. Hydrogen is passed through liquid benzene in the presence of a nickel catalyst under pressure:

$$\text{Benzene} + 3\,H_2 \xrightarrow[\text{450 K, 40 atm}]{\text{Nickel catalyst}} \text{Cyclohexane}$$

Cyclohexane is oxidized by passing air through the liquid under pressure in the presence of a catalyst (often a cobalt salt) to yield two products cyclohexanol and cyclohexanone:

$$\text{Cyclohexane} \xrightarrow[\text{10 atm}]{\text{400 K}} \text{"KA" mixture: } \xrightarrow{1/2\, O_2} \text{Cyclohexanol (OH)};\quad \xrightarrow{O_2} \text{Cyclohexanone (=O)}$$

The mixture of cyclohexanol and cyclohexanone is known as **"mixed oil"** or KA (ketone/alcohol). An alternative route to cyclohexanol is via the hydrogenation of phenol using a nickel catalyst at 400 K and 5 atm:

$$\text{Phenol} + 3\,H_2 \xrightarrow[\text{450 K, 5 atm}]{\text{Nickel catalyst}} \text{Cyclohexanol}$$

A more recent route to cyclohexanol is the Asahi process from benzene via its hydrogenation to cyclohexene and subsequent hydration to alcohol. This is more energy efficient than the other processes.

To make **polyamide 6**, pure cyclohexanone is required. When the mixed oil is heated under pressure with copper (II) and chromium (III) oxides, the cyclohexanol, which is a secondary alcohol, is dehydrogenated to the corresponding ketone, cyclohexanone:

$$2\ \text{Cyclohexanol (OH)} \xrightarrow[\text{500 K, pressure}]{CuO + Cr_2O_3 \text{ catalyst}} 2\ \text{(OH)} + H_2$$

Cyclohexanone is then converted into caprolactam via the oxime (produced by the reaction of the ketone with hydroxylamine - in the form of the salt, hydroxylamine hydrogen sulfate):

$$\text{Cyclohexanone} \xrightarrow[400\text{ K}]{NH_3OH^+ \, HSO_4^-} \text{Cyclohexanone oxime} \xrightarrow[\text{Heat}]{20\% \text{ oleum}} \text{Caprolactam}$$

Cyclohexanone oxime

Caprolactam

The isomerisation of the oxime to caprolactam by sulfuric acid is an example of the Beckmann rearrangement in which an oxime is transformed into an amide in the presence of acid.

A zeolite with acidic sites, is also being used to effect the rearrangement. The zeolite is regenerated and saves the use of sulfuric acid.

To produce the polymer, the caprolactam, water (acting as a catalyst) and a molecular mass regulator, e.g., ethanoic acid, are poured into a reaction vessel and heated under nitrogen at 500 K for about 12 hours:

$$n \text{ Caprolactam} \xrightarrow[500\text{ K}]{12\text{ hours}} \left[-N(H)-(CH_2)_5-C(=O)- \right]_n$$

This is an example of a batch process.

Polyamide 6,6 is produced by reacting 1,6-diaminohexane (hexamethylenediamine) with hexanedioic acid (adipic acid) by condensation polymerization.

One of the monomers, hexanedioic acid is also produced from KA mixed oil (cyclohexanol and cyclohexanone). The mixed oil is oxidized in the liquid phase using moderately concentrated (60%) nitric acid and a copper (II) nitrate and ammonium vanadate (V) catalyst, at 330 K to form hexanedioic acid:

$$HO-C(=O)-(CH_2)_4-C(=O)-OH$$

Hexanedioic acid

This process has a considerable disadvantage. A side-product is nitrogen (I) oxide (nitrous oxide), N_2O, a powerful greenhouse gas but it is carefully removed by thermal or catalytic treatment units. The second monomer, 1,6-diaminohexane, is produced from buta-1,3-diene. To form the polymer, the acid and the diamine are then heated together to form a salt.

The chemical reaction for aliphatic dicarboxylic acids and aliphatic diamines to yield an aliphatic polyamide via a condensation polymerization process can be represented, thus:

$$H_2N-R_1-NH_2 + HO(O)C-R_2-C(O)OH$$

$$\downarrow$$

$$H_2N-R_1-N^+H_3 + {}^-O(O)C-R_2-C(O)OH$$

$$\downarrow$$

$$\left[-NH-R_1-NH-C(=O)-R_2-C(=O)-\right]_n + 2\,H_2O$$

The chain length is regulated by controlling process conditions, such as reaction time, temperature and pressure. An aqueous solution of the salt is heated, in the absence of air, to 500 K. A pressure develops in the vessel. The temperature is then raised to 540 K, and the steam is bled off to keep the pressure constant. Eventually, the pressure is reduced and the polymer is extruded under nitrogen to yield a lace which is then granulated.

Uses of Polyamides

The properties of the polyamides (nylons), which include high strength, abrasion resistance, and resilience, make them very important in the manufacture of clothing and carpets. Although these polyamides account for 95% of the material used in women's hosiery, this still only accounts for about 5% of the total fibres used to make clothing. Nevertheless this is more than either the acrylics or wool but it is substantially less than either cotton or polyesters. The children's clothing is made of polyamide 6, impregnated with nanoparticles of titanium dioxide which gives protection against UV radiation, a very effective way of having a sunscreen. The polyamides (nylons) are also used in engineering plastics, for example, in cars, and for making films for food packaging. They are used in films for their good balance between mechanical strength and barrier properties against oxygen, smells and oils. An important development is the use of polyamides to make safety airbags. Polyamides 6 and 6,6 are the most widely used polyamides for fibres and for engineering materials. The remaining commercial polyamides, for example, polyamides 11 and 12, and 6,10, are most used as engineering plastics. Ropes made from polyamides are used by rock and ice climbers. They are not only very strong but they are also stretchy and thus reduce forces in the event of a fall, by spreading the duration of loading transmitted to anchors and to the body via the harness.

Polyamides/Nylons fibres are the second largest group of synthetic fibers after polyesters. Polymers of amine or polyamides are excellent for making fibres. These are well known as nylons.

Polyamides are produced by the reaction between a dicarboxylic acid and a diamine (e.g., nylon 66), ring openings of a lactam, (e.g., nylon 6) or by the polymerization of *w*-amino acids (e.g., nylon 11).

Condensation polymerisation is carried out with the monomer/monomers in autoclaves or stirred tank batch reactors.

Numbers that follow the word "nylon" denote the number of carbons present within a repeating unit and whether one or two monomers are being used in polymer formation. For nylons using a single monomer such as nylon 6 or nylon 12, the numbers 6 and 12 denote the number of carbons in monomers caprolactam and laurolactam, respectively. For nylons using two monomers such as nylon 610, the first number, 6, indicates the number of carbons in the hexamethylene diamine and the other number, 10, is for the second monomer sebacic acid.

Nylon resins are important engineering thermoplastics. Nylons are produced by a condensation reaction of amino acids, a diacid and a diamine, or by ring opening lactams such as caprolactam. The polymers, however, are more important for producing synthetic fibers.

Properties and Uses of Nylons

Nylon, belonging to a class of polymers called polyamides, has served as a substitute for silk in products such as parachutes, flak vests and women's stockings. Its fibers are useful in making fabrics, rope, carpets and musical strings. Whereas in bulk form, nylon is used for mechanical parts including machine screws, gear wheels and power tool casings. In addition, nylon is used in the manufacture of heat-resistant composite materials.

Important properties of nylons are toughness, abrasion and wear resistance, chemical resistance, and ease of processing. Reinforced nylons have higher tensile and impact strengths and lower expansion coefficients than metals. They are replacing metals in many of their applications. Objects made from nylons vary from extruded films used for pharmaceutical packaging to bearings and bushings, to cable and wire insulation.

Nylons have relatively high melting points due to the presence of the amide linkage. They are also highly crystalline, and the degree of crystallinity depends upon factors such as the polymer structure and the distance between the amide linkages. An increase in polymer crystallinity increases its tensile strength, abrasion resistance, and modulus of elasticity.

Hydrogen bonding in polyamides is fairly strong and has a pronounced effect on the physical properties of the polymer such as the crystallinity, melting point, and water absorption. For example, nylon 6, with six carbons, has a melting point of 223°C, while it is only 190°C for nylon 11. This reflects the higher hydrogen bonding in nylon 6 than in nylon 11.

Moisture absorption of nylons differs according to the distance between the amide groups. For example, nylon 4 has a higher moisture absorption than most other nylons, and it is approximately similar to that of cotton. This is a result of the higher hydrophilic (water loving) character of nylon 4. Nylons, however, are to some extent subject to deterioration by light.

This has been explained on the basis of chain breaking and crosslinking. Nylons are liable to attack by mineral acids but are resistant to alkalies. They are difficult to ignite and are self-extinguishing.

In general, most nylons have remarkably similar properties, and the preference of using one nylon over the other is usually dictated by economic considerations except for specialized uses.

Nylons have a variety of uses ranging from tire cord to carpet to hosiery. The most important application is cord followed by apparel. Nylon staple and filaments are extensively used in the carpet industry. Nylon fiber is also used for a variety of other articles such as seat belts, monofilament finishes, and knitwear. Because of its high tenacity and elasticity, it is a valuable fiber for ropes, parachutes, and underwear.

Table 29.1. Melting points of various nylons type and their monomer formula

Monomer	Nylon Type	Monomer Formula	Approximate M.P. °C
Caprolactam	6	$CH_2-(CH_2)_4-C=O$ NH	223
Adipic acid	66	$HOOC-(CH_2)_4COOH$	265
Hexamethylenediamine		$H_2N-(CH_2)_6-NH_2$	
Sebacic acid	610	$HOOC-(CH_2)_8COOH$	215
Hexamethylenediamine		$H_2N-(CH_2)_6-NH_2$	
11-Aminoundecanoic acid	11	$H_2N-(CH_2)_{10}-COOH$	190
Laurolactam	12	$CH_2-(CH_2)_{10}-C=O$ NH	119
Acrylamide	3	$CH_2=CHCONH_2$	320
2-pyrrolidone	4	$CH_2-(CH_2)_2-C=O$ NH	265
Valerolactam	5	$CH_2-(CH_2)_3-C=O$ NH	260
7-aminoheptanoic acid or Ethyl 7-aminoheptanoate	7	$N_2H-(CH_2)_6-COOH$ $H_2H-(CH_2)_6-COOC_2H_5$	233
Capryllactam	8	$CH_2-(CH_2)_6-C=O$ NH	209
9-aminopelargonic acid	9	$H_2N-(CH_2)_8-COOH$	209
10-aminodecanoic acid	10	$H_2N-(CH_2)_9-COOH$	188

Table 29.2. Typical applications of nylons

Market	Typical Applications
Automotive/truck	Speedometer gears, door lock wedges, distributor point blocks; automotive electrical system such as connectors, fuse blocks, generator parts, spark wire separators, wire insulation; monofilament thread for upholstery, license plate bolts and nuts, fuel vapor canisters, fender extensions, mirrors and grilles; pneumatic tubing and lubrication lines, fuel and fuel-vent lines
Electrical/electronics	Connector, tie straps, wire coil, bobbins, tuner gears
Industrial/machinery	Lawn mower carburetor components, window and furniture guides, pump parts, power tools, fans, housings, gears, pulleys, bearings, bushings, cams, sprockets, conveyor rollers, screws, nuts, bolts, washers
Monofilament/film coatings	Fishing lines, fish nets, brush bristles, food and medical packaging, coatings for wire and cable

Appliances	Refrigerators, dishwashers, ranges, hair dryers and curlers, corn poppers, smoke detectors
Consumer items	Combs, brushes, housewares, buttons, rollers, slides, racquetball racquets

NYLON-66 (POLY HEXA METHYLENE ADIPATE)

Nylon 66 (or 6,6 - Nylon) is a condensation polymer formed by the condensation polymerization of Adipic acid (a dibasic acid) and hexamethylenediamine (a diamine). In the nomenclature, the first number refers to the number of carbon atoms in the diamine and the second number designates the number of carbon atoms in the dibasic acid.

Nylon 6,6 is produced by polymerising adipic acid and hexamethylene diamine in an autoclave in an aqueous solution at about 200°C and a pressure of 16 kg/cm^2 for a reaction time of 1-2 hour. The reaction of hexamethylene diamine and adipic acid produces hexamethylene diammonium adipate salt. The product is a dilute salt solution which is concentrated to approximately 60% and charged with acetic acid to a reactor where water is continuously removed. The presence of a small amount of acetic acid limits the degree of polymerization to the desired level.

The temperature is then increased to 270-300°C and the pressure to 17 atmospheres, which favors the formation of the polymer. The pressure is finally reduced to atmospheric to permit further water removal. After a total of three hours, nylon 66 is extruded under nitrogen pressure.

In the production of Nylon-66, first, the amine and acid is mixed in an aqueous solution, with the pH adjusted at 7.8 to form Nylon salt. The salt is then concentrated under vacuum. It is thereafter charged to an autoclave where the rest of the water is removed. The temperature is gradually increased to 280°C in the autoclave to complete the polymerization and the water of condensation is removed.

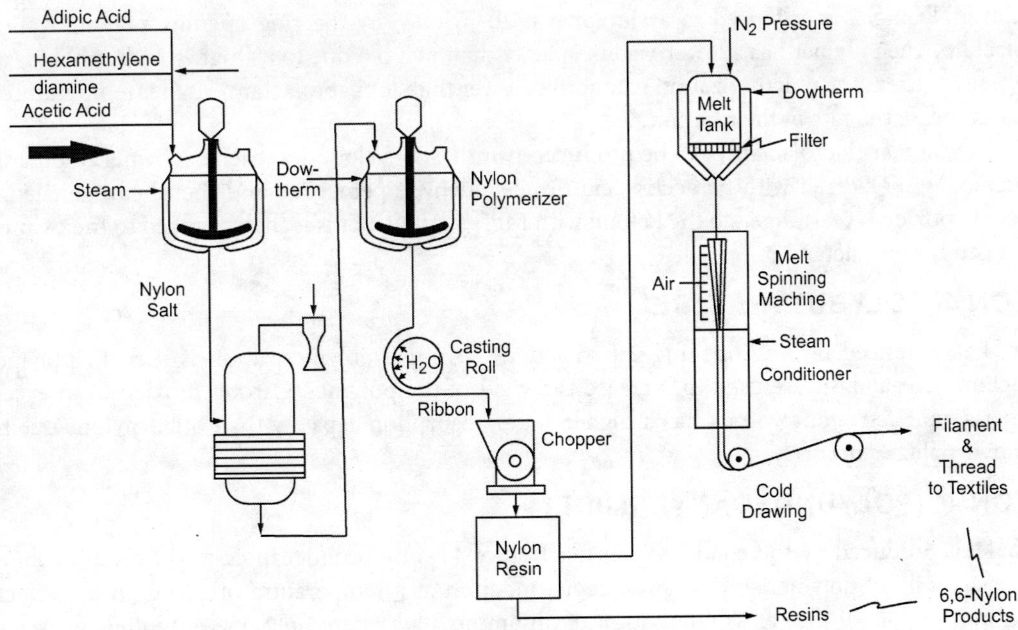

Fig. 29.1. Manufacture of Nylon 66

The final product has a molecular weight of 12000-16000. It is extruded as ribbons onto chilling rolls. Then, it is sent to a chipper which produces small chips, which possesses the convenience of storage and rehandling. These chips are melted, metered through high pressure pumps. After filtration, these are passed via a melt spinnerets to produce nylon fibers. Upon cooling, these filaments harden and are wound on bobbins at a rate of 750 metres per minute or even higher. In order to give the fibre desirable textile properties, the threads are stretched to about 4 times its original length.

The production of Nylon-6 is similar to that of Nylon-66, the only difference being the nature of polymerization. Nylon-6 is manufactured by the step-wise condensation of caprolactum with no net water removal. However, to maintain the thermal equilibrium between monomer and polymer at the melting and spinning temperatures is the only difficult task. Monomer retained on the fibres accounts up to 10%, so the fibres are water washed to remove the soluble caprolactum retained.

NYLON-6 (POLY CAPROLACTUM)

Nylon 6 is produced by the polymerization of caprolactam. The monomer is first mixed with water, which opens the lactam ring and gives *w*-amino acid. The formed amino acid reacts with itself or with at approximately 250-280°C to form the polymer. The reaction time is about 8-12 hours. Temperature control is important, especially for depolymerization, which is directly proportional to reaction temperature and water content. Nylon-6 is a polymer of caprolactum. The number refers to the total carbon atoms in the ring, which can vary from 5 to 12.

Both Nylon-6 and Nylon-66 have similar properties and can be used as thin films and fibres; and for extrusion and injection molding in plastics. The major difference in the physical properties of these two nylons is that the melting point of Nylon-66 is 40-45°C higher than Nylon-6. Moreover, Nylon-66 is harder, rigid and its abrasive – resistance is lower than that of Nylon-6.

NYLON 12

Nylon 12 is a product of the polymerisation of laurolactam at a temperature of about 260°C in an autoclave. Nylon 12 is produced in a similar way to nylon 6 by the ring opening polymerization of laurolactam. The polymer has a lower water capacity than nylon 6 due to its higher hydrophobic (water hating) properties. The polymerization reaction is slower than for caprolactam. Higher temperatures are used to increase the rate of the reaction.

The monomer (laurolactam) can be produced from 1,5,9-cyclododecatriene, a trimer of butadiene. The trimer is epoxidized with peracetic acid or acetaldehyde peracetate and then hydrogenated. The saturated epoxide is rearranged to the ketone with MgI_2 at 100°C. This is then changed to the oxime and rearranged to laurolactam.

NYLON 4 (POLYBUTYRAMIDE)

Nylon 4 is produced by ring opening 2-pyrrolidone. Anionic polymerization is used to polymerize the lactam. Co catalysts are used to increase the yield of the polymer. Carbon dioxide is an excellent polymerization activator. Nylon 4 has a higher water absorption capacity than other nylons due to its lower hydrophobic property.

NYLON 11 (POLYUNDECANYLAMIDE)

Nylon 11 is produced by the condensation reaction of 11-amino undecanoic acid, which is obtained from castor oil, by polymerising in an aqueous medium at a temperature of 220°C in an autoclave. This is an example of the self condensation of an amino acid where only one monomer is used. The monomer is first suspended in water, then heated to melt the monomer and to start the reaction. Water is continuously removed. The polymer is finally withdrawn for storage.

30 Polyesters Fibres

POLYESTERS FIBRES $\left[-O-\overset{\overset{O}{\|}}{C}-R- \right]_n$

Polyesters are polymers formed from a dicarboxylic acid and a diol. They have many uses, depending on how they have been produced and the resulting orientation of the polymer chains.

Polyesters are the most important class of synthetic fibers. In general, polyesters are produced by an esterification reaction of a diol and a diacid. First polyester polymer was prepared by using terephthalic acid and ethylene glycol.

When monomer acids and alcohols are used and some trifunctional groups are present,polyester polymer produced by condensation are termed **alkyd resins** which are used in high grade paints and varnishes. Linear polyesters are formed from bi-functional groups and are suitable for fibres and films. an example of this group are the **Dacron fibre and Mylar film** products formed from condensation polymerisation of dimethyl terephthalate(DMT) and ethylene glycol. This polyester polymer is melt spun to produce Dacron fibre and roll extruded to give Mylar film. Coloring of the polyester finished products must be done in the melting stage because of the chemical inertness of the polyester.

Dacron polyester fibre has high wrinkle resistance compared to other fibres particularly when wet. Dacron textiles retain creases and pleats when washed and dried. It is the main ingredient of "wash and wear " clothing. Blends of 65% Dacron with wool, cotton or rayon is used to prevent excess build up of static electricity and to allow moisture absorption and diffusion required for wearing comfort.

Polyester filament yarn is used in the manufacture of all polyester, polyester-viscose, polyester-cotton, polyester-viscose cotton suitings and shirtings,sarees, bedsheets and other dress materials. Easy washability, wrinkle resistance, no need of ironing, longer life and better look and feel have made polyester filament the most popular of all types of fabrics.

Mylar polyester film is extremely tough when given a two dimensional stretch which makes it ideal for photographic film and recording tape.

Terephthalic acid (Benzene-1,4-dicarboxylic acid) is manufactured by oxidation of 1,4-dimethylbenzene (commonly known as para-xylene). One method is to pass air into the liquid hydrocarbon dissolved in ethanoic acid under pressure, in the presence of cobalt(ll) and manganese(ll) salts as catalysts, at about 500 K:

$$H_3C-C_6H_4-CH_3 + 3\,O_2 \longrightarrow HOOC-C_6H_4-COOH + 2\,H_2O$$

1,4-dimethylbenzene (para-xylene) → Benzene-1,4-dicarboxylic acid (terephthalic acid)

The pure terephthalic acid used in PET production is obtained from the crude product by further processing to remove colour-forming impurities, and then by crystallization, washing and drying stages.

Polyesters can be produced by an esterification of a dicarboxylic acid and a diol, a transesterification of an ester of a dicarboxylic acid and a diol, or by the reaction between an acid dichloride and a diol.

Polyethylene terephthalate (PET) is a polyester produced by esterifying terephthalic acid (TPA) and ethylene glycol or, more commonly, by the trans- esterification of dimethyl terephthalate and ethylene glycol. This route is favored because the free acid is not soluble in many organic solvents. The reaction occurs in two stages. Methanol is released in the first stage at approximately 200°C with the formation of bis(2-hydroxyethyl) terephthalate. In the second stage, polycondensation occurs, and excess ethylene glycol is driven away at approximately 280°C and at lower pressures ($\approx$ 0.01 atm).

Batch polymerization is still used. However, most new processes use continuous polymerization and direct spinning. Products of a continuous process can be fed directly into melt-spinning heads. This removes the casting, chipping, blending and drying stages that are necessary with batch processing. Granules of PET from both batch and continuous processes are used as feedstocks for extrusion and moulding machines for the production of film and some moulded articles.

An alternative route to PET is by the direct reaction of terephthalic acid and ethylene oxide. The product bis(2-hydroxyethyl) terephthalate reacts in a second step with TPA to form a dimer and ethylene glycol, which is released under reduced pressure at approximately 300°C.

This process differs from the direct esterification and the transesterification routes in that only ethylene glycol is released. In the former two routes, water or methanol are coproduced and the excess glycol released.

Manufacture of Polyethylene Terephthalate (PET) Polyester from DMT (Di-Methyl Terephthalate)

PET from DMT is made by transesterification route by reaction of dimethyl terephthalate with ethylene glycol followed by poly-condensation. Polyester through esterification route was more common earlier due to non-availability of purified terephthalic acid. During transesterification, methanol obtained as by product. Fig. 30.1 describes the manufacturing process of polyester from DMT.

Polyester form DMT Routes:

$$CH_3 \cdot COO-C_6H_4-COO \cdot CH_3 + 2\,HO \cdot CH_2 \cdot CH_2 \cdot OH \longrightarrow HO(CH_2)_2COO-C_6H_4-COO(CH_2)_2OH + 2CH_3 \cdot OH$$

$$nHO \cdot (CH_2)_2COO-C_6H_4-COO(CH_2)OH \longrightarrow HO(CH_2)_2COO-C_6H_4-COO(CH_2)_2OH + (n-1)HOCH_2CH_2OH$$

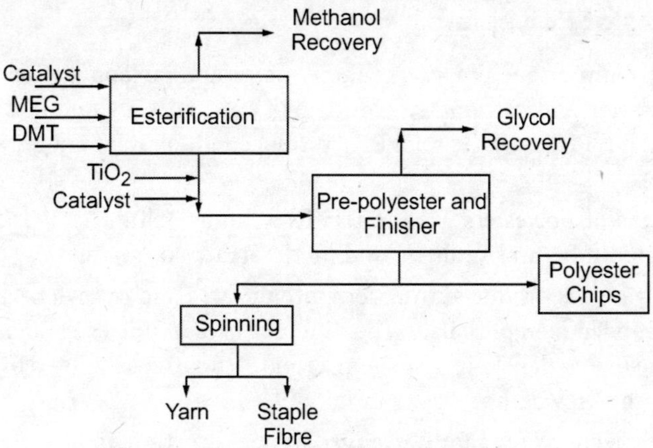

Fig. 30.1. Manufacture of polyester.

Polyethylene terephthalate is an important thermoplastic. However, most PET is consumed in the production of fibers. Polyester fibers contain crystalline as well as nanocrystalline regions. The degree of crystallinity and molecular orientation are important in determining the tensile strength of the fiber (between 18-22 denier) and its shrinkage. The degree of crystallinity and molecular orientation can be determined by X-ray diffraction techniques.

Melt spinning polyesters is preferred to solution spinning because of its lower cost. Due to the hydrophobic nature of the fiber, sulfonated terephthalic acid may be used as a co monomer to provide anionic sites for cationic dyes. Small amounts of aliphatic diacids such as adipic acid may also be used to increase the dyeability of the fibers by disturbing the fiber's crystallinity.

Polyester fibers can be blended with natural fibers such as cotton and wool. The products have better qualities and are used for men's and women's wear, pillow cases, and bedspreads. Fiberfill, made from polyesters, is used in mattresses, pillows, and sleeping bags. High tenacity polymers for tire cord reinforcement are equivalent in strength to nylon tire cords and are superior because they do not "flat spot." V-belts and fire hoses made from industrial filaments are another market for polyesters.

Thermoplastic polyesters are among the large volume engineering thermoplastics produced by condensation polymerization of terephthalic acid with ethylene glycol or 1,4-butanediol. These materials are used to produce film for magnetic tapes due to their abrasion and chemical resistance, low water absorption, and low gas permeability. Polyethylene terephthalate (PET) is also used to make plastic bottles (approximately 25% of plastic bottles are made from PET). Similar to nylons, the most important use of PET is for producing synthetic fibers. Polybutylene terephthalate (PBT) is another thermoplastic polyester produced by the condensation reaction of terephthalic acid and 1,4-butanediol.

The polymer is either produced in a bulk or a solution process. It is among the fastest growing engineering thermoplastics, and leads the market of reinforced plastics.

Unsaturated polyesters are a group of polymers and resins used in coatings or for castings with styrene. These polymers normally have maleic anhydride moiety or an unsaturated fatty acid to impart the required unsaturation. A typical example is the reaction between maleic anhydride and ethylene glycol.

Phthalic anhydride, a polyol, and an unsaturated fatty acid are usually copolymerized to unsaturated polyesters for coating purposes. Many other combinations in variable ratios are possible for preparing these resins.

Properties and Uses of Polyesters

Polyesters are extremely important polymers. Their most familiar applications are in clothing, food packaging and plastic water and carbonated soft drinks bottles. It is often known by its trivial name, polyethylene terephthalate (PET). Of the total PET produced globally, about 65% is used to make fibres, 5% for film and 30% for packaging.

Important properties of polyesters are the relatively high melting temperatures (≈ 265°C), high resistance to weather conditions and sunlight, and moderate tensile strength.

Another useful polyester is produced from terephthalic acid and propylene glycol. It is known by its trivial name, polytrimethylene terephthalate. The different uses of polyesters depend on their structure. The benzene rings in the molecular chain give them a rigid structure, leading to high melting points (over 500 K) and great strength. They do not discolour in light.

In PET *fibres*, the molecules are mainly arranged in one direction, in *film*, they are in two directions and for packaging, they are in three directions.

As fibres: The polyester is produced as small granules. These are melted and squeezed through fine holes and the resulting filaments spun to form a fibre. This fibre, commonly known as **Terylene or Dacron**, is widely used in clothing (for example, in suits, shirts and skirts) either alone or in blends with other natural or manufactured fibres, principally cotton. It is also used for filling anoraks and bedding duvets to give good heat insulation. Other uses include car tyre cords, conveyor belts and hoses, where its strength and resistance to wear are paramount. Polyesters are often used to make the suits and parachutes for sky divers.

As films : The polyester can also be made into thin films which can be used in food packaging, audio and video tapes, electrical insulation, and X-ray films.

As packaging : A relatively newer use is for packaging, for example for bottles. The small granules of the polyester are heated to about 500 K and further polymerization takes place. This heating is sometimes called solid-state polymerizing. The polymer is melted, moulded and then stretched. The molecules are now orientated in three directions giving the plastic great strength. For beverage packaging, the barrier properties of the PET to certain gases, the excellent mass to strength ratio and the recyclability of the final product are very important features influencing the choice of PET over other plastics or more traditional packaging materials.

Polypropylene Fibres

POLYPROPYLENE FIBRES

Polypropylene fibers represent a small percent of the total polypropylene production. (Most polypropylene is used as a thermoplastic.) The fibers are usually manufactured from isotactic polypropylene.

Properties and Uses of Propylene Fibre

Important characteristics of polypropylene are high abrasion resistance, strength, low static buildup, and resistance to chemicals. Crystallinity of fiber grade polypropylene is moderate and when heated, it starts to soften at 145°C and then melts at 170°C. Melt spinning is normally used to produce the fibers.

Polypropylene fibers are used for face pile of needle felt, tufted carpets, upholstery fabrics, etc. Polypropylene multi filament yarn is used in various applications like manufacture of fish nets, ropes, sewing threads, synthetic furnishing, upholstery, tapestry, carpets etc. It is cheaper than nylon and other synthetic and natural fibre yarns. Yarn produced from isotactic propylene has high strength and durability, low density, rich woolen texture and excellent resistance to all types of chemicals, acids and alkalies. Its low density yields higher length of yarn as compared to other textile yarns of same weight and thickness.

Acrylic Fibres (Polyacrylonitrile)

ACRYLIC FIBRES (POLYACRYLONITRILE)

Polyacrylonitrile, is manufactured from propylene via acrylonitrile.

$$\left[CH_2 - \underset{\substack{| \\ H}}{\overset{\substack{N \\ |||}}{C}} \right]_n$$

Acrylic fibre belongs to one of the four major classes of synthetic fibres-polyester, polyamide, acrylic and polyolefin(polypropylene) fibres. Polyacrylonitrile $(CH_2CHCN)_n$ is the major component of acrylic fibre.First acrylic fibre was Du pont's orlon. It is made by polymerising acrylonitrile in a similar manner as vinyl polymers. The ivory white polymer is then dissolved in an organic solvent (e.g., dimethyl formamide) or an inorganic solvent (e.g., conc. $ZnCl_2$ or sodium thiosulphate solution). The solution is filtered and then dry spun(in warm air) or wet spun (in a coagulating solution). The fibre is then stretched to several times its original length to orient the polymer molecules into long parallel chains for strength.

A comonomer is usually incorporated to improve fibre property such as dyeability. When large amounts (>15%) of such comonomers are used, then the fibres are referred to as **modified acrylics or modacrylics**. Union Carbide's **Dynel** contains 40% acrylonitrile units and 60% vinyl chloride units in the polymer chain.Acrylic fibres consists of copolymers with at least 85% by weight of acrylonitrile. Modacrylic fibres contain at least 30-80% acrylonitrile.

It is very widely used in co-polymers, particularly in fabrics and when materials need to be made hard and shock proof.

Manufacture of Polyacrylonitrile (Acrylic) Fibres

(*a*) **The monomer:** Acrylonitrile, the monomer, is manufactured from propylene. The alkene is mixed with ammonia and oxygen (from air) (1:1:2 volume ratio) and passed over a mixture of bismuth (III) and molybdenum (VI) oxides:

As it is a very exothermic reaction, and the temperature must be controlled at 600 K, a fluidized bed reactor is used.

$$H_2C=CH-CH_3 + NH_3 + 1\frac{1}{2}O_2 \longrightarrow H_2C=CH-C\equiv N + 3H_2O$$

A small amount of hydrogen cyanide (3-6%) is also formed, which can be used in the manufacture of methyl 2-methylpropenoate. The process has been modified, in Japan, to use propane as the feedstock. It will become particularly important if propane becomes much cheaper than propylene. The catalyst used is based on vanadium(V) and antimony (III) oxides.

(*b*) **The polymer:** The polymer is manufactured by radical polymerization initiated by either a peroxide or by a mixture of potassium peroxy disulfate, $K_2S_2O_8$ and a reducing agent such as potassium hydrogen sulfite, $KHSO_3$.

About equal amounts of the stereoregular polymer, isotactic and syndiotactic, are produced. The polymerization is either in solution or as a slurry. Acrylic fibres are third largest class of synthetic fibre after polyester and nylons.

Process Technology of Acrylic Fibre Production

Most of the acrylic polymers manufactured for fiber grade are made through suspension polymerisation that gives high percentage of conversion, better product whiteness, shorter residence time and easy control of polymerisation. Emulsion polymerization is used in case of modified acrylic fibre. Inorganic compounds such as persulphate, chlorates or hydrogen peroxide are used as radical generators. Redox initiation is normally used in production of acrylic fibre. The most common redox system consist of ammonium or potassium persulphate (oxidizer), sodium bisulphate (reducing agent) and ferric or ferrous ion (catalyst).

Spinning: Spinning includes solution/dope preparation, spinning techniques and finishing operation including after treatment, cutting and bailing.

Manufacturing process of acrylic fibre is shown below in Figure.

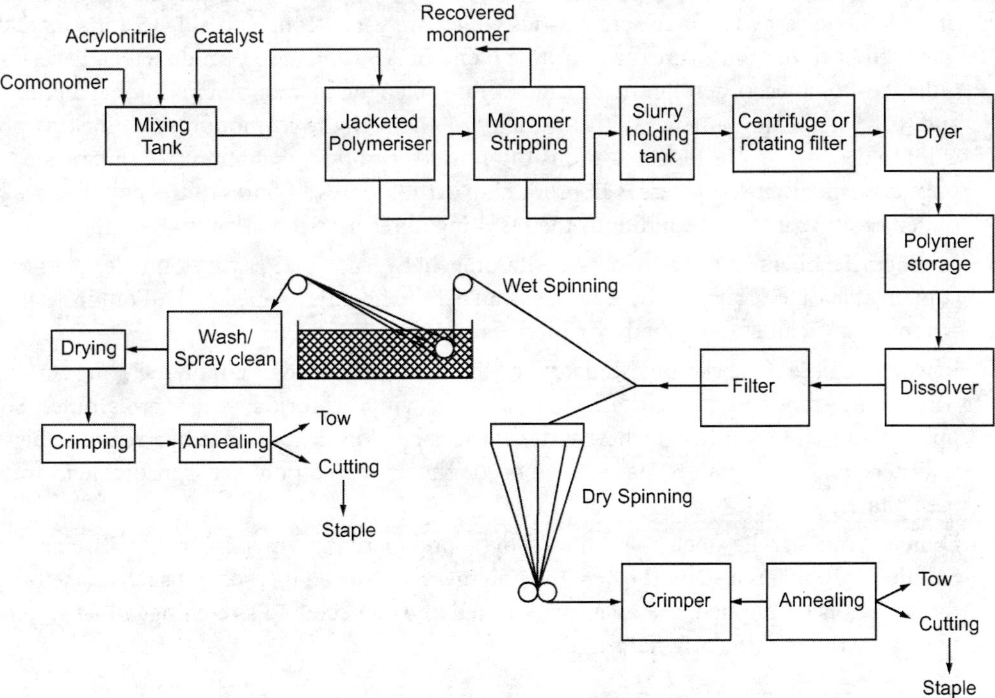

Fig. 32.1. Manufacture of acrylic fibre

In dry spinning of acrylic fibre dimethyl formamide (DMF) is used. The DMF spin dope contains the polymer in the DMF, thermal stabilizers, delustrant. It is passed through spinnerate placed at top of the solvent removal tower. The DMF evaporated by circulating inert gas through tower at about 300-350°C to remove the solvent.

In wet spinning, sodium thiocyanate are commonly used as solvent. Wet spinning fiber is spun into a liquid bath containing a solvent non-solvent mixture called coagulant. Nonsolvent is usually water. The fibre emerging from spin bath are washed and dried followed by cutting and bailing.

(*c*) **Co-polymers of acrylonitrile:** Polymerization takes place as for the homopolymer by radical polymerization. The two monomers are mixed prior to addition of the initiator. When, ethenyl ethanoate is used as the co-polymer, polymerization is initiated with small amounts of potassium hydrogen sulfite and potassium peroxodisulfate which get incorporated into the co- polymer, giving it sites which can bind to colorants and make them fast. Alternatively, a small amount of a third monomer containing, for example a sulfonic acid group, serves the same purpose.

The co-polymer contains a more or less regular alternation of the individual monomers.

Two of the most used acrylic fibres are formed from the co-polymerization of acrylonitrile with vinyl acetate and acrylonitrile with methyl acrylate. The former is often mixed with cotton fibres to produce a light fabric, used in women's clothes. The latter is often used with wool. The co-polymer of acrylonitrile and methyl acrylate is a wool substitute and is often mixed with wool itself for heavier fabrics, used in pullovers, jumpers and in suits. The soft tops for high quality cars are produced from almost pure homopolymer.

The co-polymers of acrylonitrile with styrene known as SAN and with butadiene and phenylethene, known as ABS, are plastics which are very strong and able to withstand shocks. The high heels of sandles are made from a blend of ABS and a polyamide which is very strong. Other co-polymers of acrylonitrile include those when the co-monomer is methyl methacrylate) and with 1,1-dichloroethene. With 1,1-dichloroethene as the co-monomer, a block co-polymer is formed which is fire-resistant and is often used in children's clothing. An increasing use of polyacrylonitrile co-polymers is in producing carbon fibres. If fibres of the polymer are heated under strictly controlled conditions the resulting fibres have remarkable strength.

Modacrylic fibers are copolymers containing between 35–85% acrylonitrile. Acrylic fibers contain at least 85% acrylonitrile. **Orlon** is an acrylic fiber developed by DuPont in 1949; Dynel is a modacrylic fiber developed by Union Carbide in 1951.

Polyacrylic fibres (Acrylic and Modacrylic Fibers) are produced by copolymerizing acrylonitrile with other monomers such as vinyl acetate, vinyl chloride, and acrylamide. Solution polymerization may be used where water is the solvent in the presence of a redox catalyst. Free radical or anionic initiators may also be used. The produced polymer is insoluble in water and precipitates.

Dynel, a modacrylic fiber, is produced by copolymerizing vinyl chloride with acrylonitrile. Solution spinning is also used where the polymer is dissolved in a solvent such as acetone. After the solvent is evaporated, the fibers are washed and subjected to stretching, which extends the fiber 4–10 times of the original length.

Properties and Uses of Polyacrylonitrile

Important properties of acrylics are resistance to solvents and sunlight, resistance to creasing, and quick drying. Acrylic fiber breaking strength ranges between 1500 and 2650 kg/cm^2 and they have a water absorption of approximately 5%. Dynel, due to the presence of chlorine, is less flammable than many other synthetic fibers.

Acrylic fibres have good weather resistance and resistance to chemical attack. The main characteristics of acrylic fibres are warmth, light weight, ease in washing and low flammability which makes it suitable for use in the manufacture of woolen and worsted fabrics, sweaters, pullovers, blankets, furnishings, carpets, rugs, curtains, draperies, knitted garments etc.

Copolymers of acrylonitrile are sensitive to heat, and melt spinning is not used. Solution spinning (wet or dry) is the preferred process where a polar solvent such as dimethyl formamide is used. In dry spinning the solvent is evaporated and recovered.

Major uses of acrylic fibers are woven and knitted clothing fabrics (for apparel), carpets, and upholstery. Acrylic fibers are characterized by having properties similar to wool and have replaced wool in many uses such as blankets, carpets, and sweaters.

Polyacrylonitrile itself is a very harsh fibre, like horse hair. An almost pure homopolymer is used when a very tough fabric is needed, for example for awnings, a soft top of a car or in brake linings. It is even used to reinforce concrete and in road construction. However, the vast majority of the polymer is co-polymerized. Although these co-polymers often contain more than 85% of acrylonitrile units, they are much softer. The fibres formed from them are known as 'acrylic' fibres.

APPENDICES

33 Composite Materials

COMPOSITE MATERIALS

A composite is structural materials that consist of two or more combined constituents that are combined at a macroscopic level and are not soluble in each other. One constituent is called the reinforcing phase and the one in which it is embedded is called the matrix. The reinforcing phase material may be in the form of fibres, particles, or flakes. The matrix phase materials are generally continuous. Examples of composite systems include concrete reinforced with steel and epoxy reinforced with graphite fibres, etc.

Composites consist of:

1. Combination of two or more materials – Composite = matrix + fiber (filler)

   Matrix:

   (*i*) Material component that surrounds the fiber.

   (*ii*) Usually a ductile, or tough, material with low density

   (*iii*) Strength usually = 1/10 (or less) than that of fiber

   (*iv*) Examples include: thermoplastic or thermoset

   (*a*) Thermoset most common (epoxy, phenolic)

   (*v*) Serves to hold the fiber (filler) in a favorable orientation. Fiber reinforcing material

   Filler:

   (*vi*) Materials that are strong with low densities

   (*vii*) Examples include glass, carbon or particles.

2. Designed to display a combination of the best characteristics of each material i.e. fiberglass acquires strength from glass and flexibility from the polymer.
3. Matrix and filler bonded together (adhesive) or mechanically locked together.

   Where are composites used?

CFRP – carbon fiber reinforced composite.

GFRP – glass fiber reinforced composite

Features of Composite Materials

- A material which is composed of two or more materials at a microscopic scale and have chemically distinct phases.
- Heterogeneous solid consisting of 2 or more different materials that are mechanically or metallurgically bonded together.
- Constituent materials have significantly different properties and retains its identity in the composite and maintains its characteristic structure and properties.

Need for Composite Materials

To enhance certain desired properties of materials like:

- Strength
- Toughness
- Wear resistance
- Fatigue life
- Thermal/Electrical conductivity
- Energy dissipation
- Stiffness
- Corrosion resistance
- Reduced weight
- Thermal/Electrical insulation
- Acoustic insulation

Major Constituents of a Composite Material

1. Matrix-base material
2. Reinforcement-added material to give some desired properties

Functions of Reinforcement is to

- Contribute to desired properties
- Enhance load carrying capacity
- Transfer the strength to matrix.

Functions of Matrix

- Holds the fibres together
- Protects the fibres from environment
- Protects the fibres from abrasion (with each other)
- Helps to maintain the distribution of fibres
- Distributes the loads evenly between fibres
- Enhances some of the properties of the resulting material and structural component (that fibre alone is not able to impart). These properties are component (that fibre alone is not able to impart). These properties are such as: transverse strength of a lamina Impact resistance.
- Provides better finish to final product

Advantages of Composite Materials

(*a*) High strength to weight ratio (low density high tensile strength) or high specific strength ratio.
  (*i*) 1020 HRS spec strength = 1 (E6 in)
  (*ii*) Graphite/Epoxy, spec strength = 5 (E6 in)
(*b*) High creep resistance
(*c*) High tensile strength at elevated temperatures
(*d*) High toughness
(*e*) Generally perform better than steel or aluminum in applications where cyclic loads are encountered leading to potential fatigue failure (i.e. helicopter blades).
(*f*) Impact loads or vibration – composites can be specially formulated with high toughness and high damping to reduce these load inputs.
(*g*) Some composites can have much higher wear resistance than metals.
(*h*) Corrosion resistance
(*i*) Dimensional changes due to temp changes can be much less.
(*j*) Anisotropic – bi-directional properties can be design advantage (i.e., helicopter blades)

Disadvantages (or Limitations) of Composite Materials

(*a*) Material costs
(*b*) Fabrication/ manufacturing difficulties
(*c*) Repair can be difficult
(*d*) Wider range of variability (statistical spread)
(*e*) Operating temperature can be an issue for polymeric matrix (i.e. 500 F). Less an issue for metal matrix (2,700 F).
(*f*) Properties non-isotropic makes design difficult.
(*i*) Example – video test in line w/fibers 10X stronger vs fibers oriented at an angle.
(*g*) Inspection and testing typically more complex.

Factors Influencing the Properties of the Composite

- Properties of the individual components of the composite
- Relative amounts of the components in the composite
- Size, shape and distribution of discontinuous components in the composite
- Degree of bonding between components
- Orientation of various components

CLASSIFICATION OF COMPOSITES

Classification of Composite Materials by Matrix

1. Metal matrix
2. Ceramic matrix
3. Polymer (Resin) matrix

Most common – also called fiber reinforced polymer

Classification of Composite Materials by Matrix

1. Polymer matrix composites or Fiber Reinforced Polymer (FRP)
   (*a*) Fibers generally glass, carbon or kevlar
   (*b*) Matrix can be:
      (*i*) Thermoplastics: PE, Nylon, PS, PP, PC, PVC
      (*ii*) Thermosets: Epoxy, polyester, phenolics
   (*c*) Have high strength and stiffness to weight ratio
   (*d*) Aerospace, sporting goods marine
   (*e*) Examples: GFRP aka fiberglass (polyester or epoxy and glass), CFRP (polyester or epoxy and carbon), KFRP (polyester or epoxy and Keflar)

Layered Composites

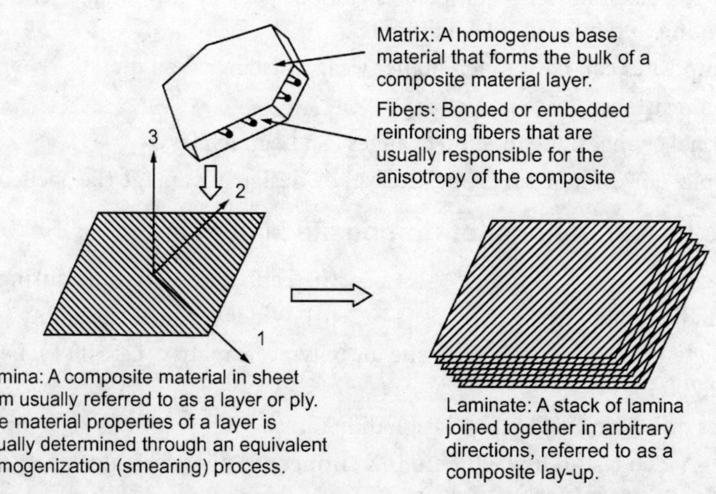

Ranking of Most Common Fibers for FRP

Property	Glass	Carbon	Kevlar
Strength	Worst	In - between	Best
Stiffness	Worst	Best	In – between
Cost	Best	Worst	In – between
Weight	Worst	Best	In-between

Classification of Composite Materials by Matrix:

Metal matrix composites (MMC):

(*a*) Metal matrix: Al, Ti, Mg, Fe, Cu, Ni

(*b*) Example: Al-SiC (silicon carbide)

(*c*) Example: Al-Al_2O_3 (aluminum oxide)

High strength, high stiffness, abrasion resistance, dimensional stability, high temperature and toughness.

Ceramic Matrix Composites (CMC):

(*a*) Silicon carbide-silicon carbide (SiC-SiC)

(*b*) Same material both matrix and filler but filler different form such as whiskers, chopped fibers or strands to achieve preferred properties.

The strength of the composite depends primarily on the amount, arrangement and type of fiber (or particle) reinforcement in the resin.

Typically, the higher the reinforcement content, the greater the strength. In some cases, glass fibers are combined with other fibers, such as carbon or aramid (Kevlar29 and Kevlar49), to create a "hybrid" composite that combines the properties of more than one reinforcing material.

Classification of Composite by Filler Type

– Particle-reinforced composites
– Fiber-reinforced composites
– Structural composites

Particle Reinforced Composites:

1. Particles used for reinforcing include:
   (*a*) ceramics and glasses such as small mineral particles,
   (*b*) metal particles such as aluminum,
   (*c*) and amorphous materials, including polymers and carbon black.
2. Particles are used to increase the modulus of the matrix, to decrease the permeability of the matrix, or to decrease the ductility of the matrix.
3. Particle reinforced composites support higher tensile, compressive and shear stresses
4. Particles are also used to produce inexpensive composites.

5. Examples:
   (*a*) automobile tire which has carbon black particles in a matrix of elastomeric polymer.
   (*b*) spheroidized steel where cementite is transformed into a spherical shape which improves the machinability of the material.
   (*c*) concrete where the aggregtes (sand and gravel) are the particles and cement is the matrix.

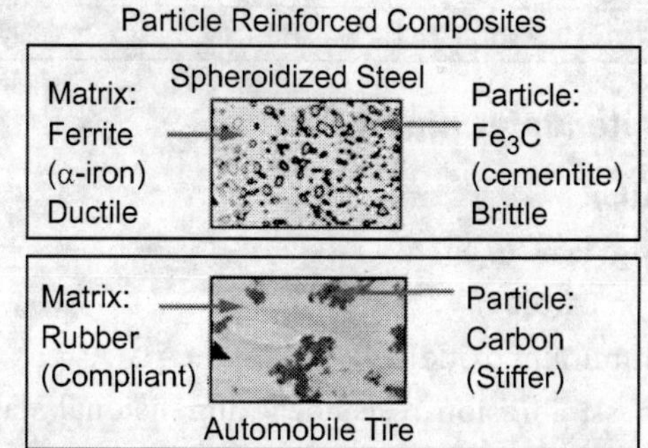

Fig. 33.1. Examples for particle-reinforced composites. (Spheroidized steel and automobile tire)

Fiber-reinforced Composites

1. Reinforcing fibers can be made of metals, ceramics, glasses, or polymers that have been turned into graphite and known as carbon fibers. Fibers increase the modulus of the matrix material. The strong covalent bonds along the fiber's length gives them a very high modulus in this direction because to break or extend the fiber the bonds must also be broken or moved. Fibers are difficult to process into composites which makes fiber-reinforced composites relatively expensive. Fiber-reinforced composites are used in some of the most advanced, and therefore most expensive, sports equipment, such as a time-trial racing bicycle frame which consists of carbon fibers in a thermoset polymer matrix. Body parts of race cars and some automobiles are composites made of glass fibers (or fiberglass) in a thermoset matrix.
2. The arrangement or orientation of the fibers relative to one another, the fiber concentration, and the distribution all have a significant influence on the strength and other properties of fiber-reinforced composites. Applications involving totally multidirectional applied stresses normally use discontinuous fibers, which are randomly oriented in the matrix material. Consideration of orientation and fiber length for a particular composites depends on the level and nature of the applied stress as well as fabrication cost. Production rates for short-fiber composites (both aligned and randomly oriented) are rapid, and intricate shapes can be formed which are not possible with continuous fiber reinforcement.

Structural Composites

1. The properties of structural composites depend on:
   (*a*) Constituents (*b*) Geometrical design

Common Structural Composite Types are

Sandwich panels: Consist of two strong outer sheets which are called face sheets and may be made of aluminum alloys, fiber reinforced plastics, titanium alloys, steel. Face sheets carry most of the loading and stresses. Core may be a honeycomb structure which has less density than the face sheets and resists perpendicular stresses and provides shear rigidity. Sandwich panels can be used in variety of applications which include roofs, floors, walls of buildings and in aircraft, for wings, fuselage and tailplane skins.

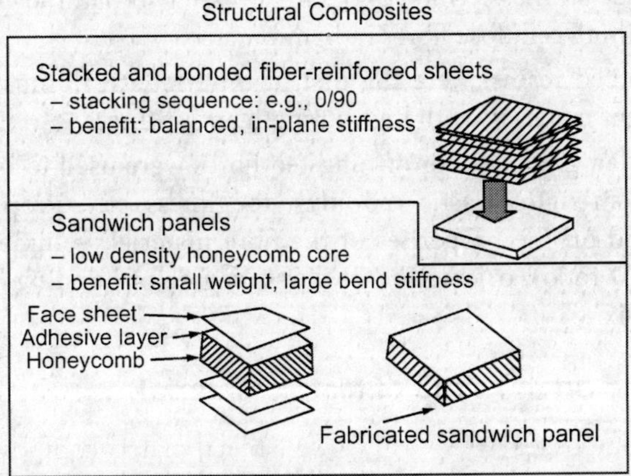

I. Classification Based on Geometry

1. Laminar or layered composites
2. Particulate composites
3. Fiber reinforced composites

1. **Laminar composites:** Several layers stacked up together form laminar composites. Each layer is called as ply. When all the layers or plies are stacked in the same orientation(degrees), it is called lamina. When plies are stacked at various angles, it is called laminate.

   Layers of wood with different orientations are adhesively bonded together to form plywood. Swelling and shrinkage tendencies are reduced, strength and fracture resistance increased due to this bonding. Safety glass has a layer of polymeric adhesive placed between two pieces of glass, which serves to retain the glass when broken.

   Aramid-Aluminium Laminates (Arall) consist of thin sheets of aluminium bonded with adhesive impregnated aramid fibres.

   Bimetallic strips consists of two metals with different coefficients of thermal expansion bonded together in a laminate. Changes in temperature produce flexing or curvature in the product which may be employed in thermostat and other heat sensing applications.

2. **Particulate composites:** Particulate composites are dispersion strengthened materials where a small amount of hard, brittle, small particles are dispersed throughout a softer and more ductile matrix.

   e.g., Sintered aluminium powder (SAP)- Aluminium matrix strengthened by 14% of Aluminium oxide and a nickel alloy containing 1-2% of Thoria. Cemented carbides like Tungsten carbide, tantalum carbide, tungsten carbide are embedded in metal matrix like Cobalt. Hard, stiff carbide can withstand high temperatures and pressure of cutting, whereas toughness is imparted by the cobalt in the composite.

   Particulate toughened ceramics using zirconia and alumina matrices are used in bearings, bushings, valve seats, die inserts and cutting tool inserts. One of the major advantages of particulate composites is that its properties are often isotropic- uniform in all directions, which is a very important aspect in engineering applications.

3. **Fibre reinforced composites:** Continuous or discontinuous thin fibres of a material are embedded in the matrix. The matrix supports and transmits force to the fibres, protects them from environment and handling, provides ductility and toughness, while fibres carry most of the load and impart stiffness. Wood and bamboo are two naturally occurring fibre reinforced composites containing cellulose fibres in lignin matrix.

   Glass fibre reinforced resins, were the first of the modern fibrous composites attempted to produce lightweight materials with high strength and stiffness.

   Later carbon, Kevlar, graphite, boron-tungsten fibres were used to improve various properties like high tensile strength, elastic modulus, toughness etc., Properties of fibre reinforced composites depend on the properties of the fibre material, volume fraction of fibres, aspect ratio of fibres (L/D ratio), orientation of fibres, degree of bonding between fibre and matrix, properties of matrix.

II. Classification Based on the Type of Matrix Material

1. Polymer matrix composites
2. Metal matrix composites
3. Ceramic matrix composites
4. Carbon and graphite matrix composites

1. **Polymer matrix composites:** Polymers include thermoplastic or thermosetting plastics. Thermoplastics such as thermoplastic polyimides, polyphenylene sulphide, polyether ketone are usually used as matrix materials due to their high temperature properties. Thermosets include epoxies, polyesters, bismaleimides. The advantage of these composites is the easiness in fabricating the composite, where the disadvantage is that it can be used for temperatures below 300°C only.
2. **Metal matrix composites:** Aluminium, copper, magnesium titanium, nickel, superalloy are the common metals used as matrix. Compared to organic matrix materials, higher heat resistance, improved thermal and electrical conductivity can be achieved.
3. **Ceramic matrix composites:** They offer light weight, high temperature strength, good dimensional and environmental stability. The crystalline ceramics, based on alumina, silicon carbide, silicon nitride, boron nitride, zirconia are used as matrix materials.
4. **Carbon and graphite matrix composites:** Heat resistant matrix that could operate at temperatures up to 3300°C. Racing car disc brakes, aerospace turbines, jet engine components, rocket nozzles, surgical implants are some of its applications.

III. Classification Based on the Form of Reinforced Fibres

1. Continuous fibre
   A. Unidirectional fibres
   B. Woven fibres
2. Discontinuous fibre
   A. Random orientation
   B. Preferred orientation

Advantages of Composites

- Excellent strength to weight and stiffness to weight ratio is the hallmark of composite material
- Laminate pattern and ply build up in a part can be changed or adjusted to acquire the necessary mechanical properties in various directions.

- In the field of aerodynamics, its easy to achieve smooth aerodynamic profile for drag reduction. Complex double curvature parts with a smooth surface finish can be made in one manufacturing operation.
- Production cost is reduced as composites have a wide range of manufacturing processes.
- Composites offer excellent resistance to corrosion, chemical attack and outdoor weathering.

Drawbacks of Composites

- Cost of the material is high when compared to traditional materials
- Defects like micro cracking, delamination, voids, missing layers, fibre breakage, improper curing are common problems. Care should be taken to have controlled fabrication conditions to avoid these problems.
- Repairing is difficult and secondary machining operations during finishing and assembly weakens the parent composite material.

Types of Fiber Reinforcement

The fiber reinforcement provides the structural performance required of the final part. The fibers or filaments come in many chemical types and forms and are the primary contributor to the stiffness, strength and other properties of the composite. The dominant chemical types of commercially available fibers are: fiberglass, aramid, carbon, polyester and vectran. Other fiber types may be suitable for special applications. The dominant forms that fibers are sold include: Strands (or roving or tow or yarn. Many fibers or filaments stranded together in a bundle, wound in a spool or reel,) woven fabrics (flattened strands of filaments woven in a variety of weaves to a type of fabric or cloth,) unidirectional (strands laid side by side and stitched or held together by other means, forming a kind of fabric that bares reinforcement only in the fill direction,) multiaxial (unidirectional woven fabrics stitched together in a combination of orientations,) and chopped strand mat (chopped strands held together with some kind of glue or "binder" in the form of a non-woven fabric.) All fibers designated for use in composites undergo chemical treatments and are coated with some kind of "sizing". Sizing is a chemical that binds the filaments together, reduces abrasion, facilitates impregnation and acts like a coupling agent that enhances compatibility with one or more types of resin.

Selection of reinforcement type greatly influences costs. This is not only because of the quality and cost of the material itself, but also because higher cost material usually requires highly skilled personnel, more sophisticated production facilities and often more labour hours.

There are many different types of fibers that can be used to reinforce polymer matrix composites. The most common are carbon fibers (AS4, IM7, etc.) and fiberglass (S-glass, E-glass, etc.). As with the matrix, the fiber chosen will be determined by the end application.

Carbon (Graphite) Fibers

Carbon fibers are conductive, have an excellent combination of high modulus and high tensile strength, have a very low (slightly negative) CTE and offer good resistance to high temperatures.

Carbon fibers are frequently categorized using tensile modulus. There are five categories of carbon fibers generally used in composites; low modulus, standard modulus, intermediate modulus, high modulus and ultra-high modulus. The exact cut-off for these categories will vary depending on the reference consulted, but in general, low modulus fibers have a tensile modulus of less than 30Msi and ultra-high modulus fibers have tensile modulus greater than 75Msi. As a point of comparison, steel has a tensile modulus of 29Msi.

As the modulus increases, the fibers tend to get more brittle, more expensive and harder to handle. Further, the tensile strength of the fibers generally increases as the modulus increases from low to intermediate, but then tends to fall off in the high and ultra-high modulus fibers. i.e., the tensile strength of carbon fibers tends to be the greatest for the intermediate modulus fibers. For these reasons, standard and intermediate modulus fibers tend to give the best overall performance, unless the application is very stiffness oriented. This is illustrated even more clearly when fiber price and availability is also taken into consideration.

Carbon has the highest strength and highest price of all reinforcement fibers used in composites today. It was invented in the UK in early 1960s at the Royal Aircraft Establishment at Farnborough, Hampshire. The most common method of making carbon fiber is the oxidation and thermal pyrolysis of an organic precursor, polyacrylonitrile (a polymer fiber based on acrylonitrile.) When heated in the correct conditions, the non-carbon constituents evaporate away. The resulting fiber is 93-95% carbon. Instead of PAN as a precursor, carbon fiber can also be manufactured from pitch or rayon. The size, or thickness of Carbon tows is measured in "k" or thousands of filaments. A 3k tow has 3,000 filaments and a 12k has 12,000. Carbon fibers exhibit: substantially better strength and stiffness values than all the others, outstanding temperature performance, high electrical and low thermal conductivity. Impact or damage tolerance of pure carbon composite products can be from relatively low to very poor, and greatly depends on processing method. Despite that, when weight on a composite product is important, carbon fiber is the reinforcement to use.

Carbon fibre is the reinforcement material of choice for "advanced" composites, Carbon fibre exhibits excellent fatigue resistance which do not suffer from stress rupture compared with glass or aramid fibres. Carbon fibres are supplied in tows and may vary from 1000 fibres per tow to hundreds of thousands per tow.

Untreated carbon fibres do not wet easily, so adhesion to the matrix must be achieved by mechanical interference coupled with surface treatment and chemical bonding between the fibre and the matrix.

Carbon reinforced composites are often used for low strength applications requiring good electrical properties due to the high conductivity of carbon fibre. Most carbon fibres are derived from polyacrylonitrile, but for even higher conductivity, fibres derived from pitch can have three times the conductivity of copper.

Fiberglass

Fiberglass is, as its name implies, glass that has been spun into the form of fibers. Fiberglass is not as strong or stiff as carbon fibers, but it has characteristics that make it desirable in many applications. Fiberglass is non- conductive (i.e., an insulator) and it is generally invisible to most types of transmissions. This makes it a good choice when dealing with electrical or broadcast applications.

There are five major types of fiberglass. A-glass (alkali glass) which has good chemical resistance, but lower electrical properties. C-glass (chemical glass) which has very high chemical resistance. E-glass (electrical glass) which is an excellent insulator and resists attacks from water. S-Glass (structural glass) which is optimized for mechanical properties. D-glass (dielectric glass) which has the best electrical properties but lacks in mechanical properties when compared to E and S glass.

E-glass and S-glass are, by far, the most common types found in composites. These types have good combinations of chemical resistance, mechanical properties and insulating properties. Of the two, E-glass offers the more attractive economics, and S-glass offers better mechanical performance.

Fiberglass was first discovered in 1893 and was firstly made commercially available in 1936 as insulation material. It became popular in the 1950's, when some of the health hazards associated with asbestos were becoming apparent. Due to the similarity in shape between the fiberglass and the asbestos fibers, fiberglass was able to effectively replace asbestos in many applications such as electrical, thermal, and acoustic insulation and structural reinforcement. Today it is the dominant reinforcement fiber in composite construction, accounting for over 90% of worldwide consumption. This is simply because it has good strength to weight characteristics, can be processed easily and sells at a low price. Glass filaments are made relatively easily by extruding molten glass. The diameters of the fibers produced range from 5 to 25 microns. Many different types and qualities of glass sell for significantly different cost. The most widespread quality is "E-glass", "E" from the word Electric implies that it is an electrical insulator. It is low cost product, used mostly in the marine industry. Others are "S-glass" and "S2-glass", the letter "S" comes from the word Strength implying that it has improved mechanical properties. These types are much more expensive and used mostly in armor applications. "S" is certified as for the production parameters. "AR glass" is resistant to alkali chemical attack. "C" or "T-glass" is resistant to acid and corrosion. "A-glass" is glass with more alkali content similar to window glass, and costs a bit less. Generally, when cost is a major driving force in the selection of a reinforcing material, fiberglass is usually preferred.

Glass fibre is the commonest reinforcing material used in polymer matrix composites. These have high tensile strength but low modulus compared with other fibres.

Typical Variants are:

E-glass

E-glass accounts for 90% of the glass fibre market and is used mainly in a polyester matrix.

The 'E' in E-glass stands for 'electrical' and was intended to indicate that the material had low electrical conductivity. The major driver, however, for its command of the market is due to the fact that it is the cheapest glass fibre and is therefore the preferred material in general purpose products.

Continuous and chopped E-glass are widely used in product manufacture. Its advantages are relatively low cost combined with high tensile strength and modulus, with individual filament strengths around 3500 MPa and modulus around 80GPa. Elongation-to-break is nearly 5%. Two types are available one contains boron, the other boron free.

The ultimate use temperature of E-glass is around 500°C. The maximum service temperature for the composite will, however, be dictated by the matrix material.

The corrosion resistance of E-glass without boron is approximately seven times the corrosion resistance of the boron-containing E-glasses. Boron free E-glasses have approximately a 10% higher dielectric constant than boron containing E- glasses when measured at room temperature making them less suitable for electronic circuit boards and aerospace applications.

ECR-glass

ECR-glass was manufactured specifically to resist acid and alkali exposure. These fibres are boron-free with a modified structure to enhance short and long term acid alkali resistance. Their mechanical properties are similar to E- glass but they have less weight loss based on sulphuric acid exposure. The enhanced corrosion resistance results in an increase in cost.

S-glass, R-glass and Te-glass

S-glass is 10-15% stronger than E-glass, but is otherwise similar in its properties to E-glass and R-glass. It is boron free and heat resistance is better than that of E-glass. S-glass fibres have modified silicate network giving an increase in mechanical properties. S-glass is typically selected for many structural applications because due to its enhanced mechanical and temperature performance. It should be noted, that higher melt temperatures are required for S-glass, requiring more process energy making these fibres more expensive than E-glass.

Silica/quartz

Silica/quartz glass fibres have increasing silicon oxide with enhanced high temperature performance. Applications include communication and satellite equipment for protection against electrical discharges.

D-glass

D-glass fibres are low volume specialist fibres with low dielectric constant which can be 40% lower than E-glass, They are used in circuit boards because of their low dielectric constant.

These variations of glass can be obtained / supplied in several different forms including:

- Rovings
- Chopped strand mat
- Woven rovings
- Sheet moulding compound

Aramid Fibre

Aramid was invented by DuPont in 1960's ("Kevlar" is the registered trade name of Dupont aramid) as a result of research on nylon (polyamide) fibers. It was firstly introduced in the market in the 1970's as tire reinforcement and like fiberglass, as an asbestos substitute. The chemical structure of aramid shows the aromatic benzene rings along the polymeric backbone. The word aramid is a contraction of its chemical description–aromatic polyamides. Strength and modulus of aramid are very good, density very low, UV resistance low and compression and shear strength are similar to E glass. Its value comes from excellent toughness and resistance to impact, damage, abrasion and heat (up to 500°C.)

The superior toughness of aramid is an outcome of the energy consuming failure mechanism of its fibers. This energy absorbing failure mechanism makes it ideal for use in armor, military and ballistic applications, like helmets and bullet-proof vests. Among many other very important uses, it is used for firefighting protection, on the underside of airplanes (protection against stone hits during takeoff and landing) and the underside of race cars. It is used generally in important structures whenever impact, abrasion and/or heat is anticipated. It also blends very well and works together with other fibers, like glass and carbon. Carbon-aramid "hybrid" constructions bare the high strength and stiffness of carbon and the impact protection of aramid. Processing aramid in composite fabrication is somewhat more difficult than fiberglass and carbon. Toughness makes fabrics difficult to cut with conventional methods. Fiber wet out is more difficult than fiberglass and carbon. Orthophthalic polyester resin will not adhere well to aramid (isophthalic polyester is much better). Post fabricating aramid components, e.g. trimming off the edges is also difficult. Quality cutting tools are recommended.

Aramid fibres have the highest strength to weight ratio compared to other commercially available fibres. Kevlar manufactured by DuPont is one familiar brand name. Aramid fibre exhibits similar tensile strength to glass fibre, but can have modulus at least two times as great. Aramid is very tough allowing significant energy absorption but, compared to carbon, it is lower in compressive strength and has poorer adhesion to the matrix. It is also susceptible to moisture absorption.

Boron Fibre

Boron fibre actually predates carbon fibre as a high-modulus reinforcement material. The cost of boron, however, has seen its demise, with its replacement with carbon fibre. They do not differ greatly from glass fibre in tensile strength, but can have modulus five times that of glass. Since the objective of reinforcement is to stiffen, this is a significant advantage. Their use is confined to niche markets, where the modulus advantage over carbon fibre is critical.

Other Fibers

While carbon fibers and fiberglass are the most common reinforcements in thermoplastic composites, there are other options. Aramid fibers (such as Kevlar ® and Twaron ®) and boron fibers have been used in composites and offer some beneficial properties (excellent toughness and compressive strength, respectively). However, they have characteristics that have limited their use (susceptibility to light/ difficulty machining and brittleness, respectively). Still others include ceramic fibers like SiC or aluminum oxide. These may be attractive for their compression, insulating, or high- temperature properties.

High modulus polyester (HMP) – Diolen

HMP and Diolen are high strength polyester fibers. Such fibers have been used in industrial applications for many decades, in a wide field of applications like tire reinforcement, safety belts, ropes and nets. Recognition of their value in composite applications for aerospace, defense, marine and transportation has come in the late years. They are used both as primary reinforcement and in hybrid arrangements with fiberglass and aramid. Toughness makes them ideal for use on the outside layers of composite construction for impact protection. Placed on the outside layers they also protect aramid and glass from moisture (and blistering) and being smooth, they reduce print-through. At the same time, their very low density also makes them ideal for use in the middle layers, where working as a core material, they add stiffness to the structure. Other attractive features include good impact and fatigue resistance, and potential for vibration damping. Generally, properties of Diolen resemble the properties of aramid at lower level, but its reasonable price and reliability make it attractive. Diolen performance at elevated temperature falls off. It should not be used in above 90°C applications. Processing is easier than that of aramid, but not as easy as fiberglass and carbon.

Vectran

Vectran is the outcome of research on liquid crystal polymer (LCP) fibers. This fiber is formed by melt extrusion of LCP through fine diameter capillaries. It is used in aerospace composites and many other high value industry applications. It exhibits high strength and modulus, excellent creep resistance, high abrasion resistance, excellent flex/fold characteristics, minimal moisture absorption, excellent chemical resistance, high dielectric strength, outstanding cut resistance, high impact resistance, outstanding vibration damping characteristics, very good property retention at high/low temperatures (melting point of 330°C, with progressive strength loss from 220°C.) They have gold color, similar to aramid, but are usually painted in other colors for cosmetic purposes. Vectran also adheres to resins better than aramid. Among others, it should be used in applications where weight and impact resistance are important.

RESINS

Polyester Resin

Polyester resins are the most widely used resin systems, particularly in the marine industry. Polyester resins such as these are of the 'unsaturated' type. Unsaturated polyester resin is a thermoset, capable of being cured from a liquid or solid state when subject to the right conditions. There is a whole range of polyesters made from different acids, glycols and monomers, all having varying properties.

There are two principle types of polyester resin used as standard laminating systems in the composites industry. Orthophthalic polyester resin is the standard economic resin used. Isophthalic polyester resin is now becoming the preferred material in industries such as marine where its superior water resistance is desirable. Most polyester resins are viscous, pale coloured liquids consisting of a solution of a polyester in a monomer which is usually styrene. The addition of styrene in amounts of up to 50% helps to make the resin easier to handle by reducing its viscosity. The styrene also performs the vital function of enabling the resin to cure from a liquid to a solid by 'cross- linking' the molecular chains of the polyester, without the evolution of any by- products. These resins can therefore be moulded without the use of pressure and are called 'contact' or 'low pressure' resins. Polyester resins have a limited storage life as they will set or 'gel' on their own over a long period of time. Often small quantities of inhibitor are added during the resin manufacture to slow this gelling action. Colouring of the resin mix can be carried out with pigments. The choice of a suitable pigment material, even though only added at about 3% resin weight, must be carefully considered as it is easy to affect the curing reaction and degrade the final laminate by use of unsuitable pigments.

Filler materials are used extensively with polyester resins for a variety of reasons including:

- To reduce the cost of the moulding
- To facilitate the moulding process
- To impart specific properties to the moulding

Fillers are added in quantities up to 50% of the resin weight although such addition levels will affect the flexural and tensile strength of the laminate. The use of fillers can be beneficial in the laminating or casting of thick components where otherwise considerable exothermic heating can occur. Addition of certain fillers can also contribute to increasing the fire-resistance of the laminate.

Epoxy Resin

Epoxies are thermosetting polymer resins where the resin molecule contains one or more epoxide groups. Epoxy resins are formed from a long chain molecular structure similar to vinyl ester with reactive sites at either end. In the epoxy resin, however, these reactive sites are formed by epoxy groups instead of ester groups. The absence of ester groups means that the epoxy resin has particularly good water resistance. The epoxy molecule also contains two ring groups at its centre which are able to absorb both mechanical and thermal stresses better than linear groups and therefore give the epoxy resin very good stiffness, toughness and heat resistant properties.

The chemistry can be adjusted to perfect the molecular weight or viscosity as required by the end use. There are two primary types of epoxies, glycidyl epoxy and non-glycidyl. Glycidyl epoxy resins can be further defined as either glycidyl-amine, glycidyl-ester, or glycidyl-ether. Non-gylcidyl epoxy resins are either aliphatic or cycloaliphatic resins.

One of the most common glycidyl epoxy resins is created using Bisphenol-A, and is synthesized in a reaction with epichlorohydrin. The other frequently used type of epoxy is known as novolac based epoxy resin. Epoxy resins are cured with the addition of a curing agent, which is commonly called a hardener. The most common type of curing agent is amine based. Unlike in polyester or vinyl ester resins where the resin is catalyzed with a small (1-3%) addition of a catalyst, epoxy resins usually require the addition of the curing agent at a much higher ratio of resin to hardener, often 1:1 or 2:1. As mentioned, the properties of epoxy can be altered and tweaked to fit the desired need. Epoxy resin can be "toughened" with the addition of thermoplastic polymers.

Vinyl Ester Resin

Vinyl ester resins are similar in their molecular structure to polyesters, but differ primarily in the location of their reactive sites, these being positioned only at the ends of the molecular chains. As the whole length of the molecular chain is available to absorb shock loadings this makes vinyl ester resins tougher and more resilient than polyesters. The vinyl ester molecule also features fewer ester groups. These ester groups are susceptible to water degradation by hydrolysis which means that vinyl esters exhibit better resistance to water and many other chemicals than their polyester counterparts, and are frequently found in applications such as pipelines and chemical storage tanks. With the reduced number of ester groups in a vinyl ester when compared to a polyester, the resin is less prone to damage by hydrolysis. The material is therefore sometimes used as a barrier or 'skin' coat for a polyester laminate that is to be immersed in water, such as in a boat hull. The cured molecular structure of the vinyl ester also means that it tends to be tougher than a polyester, although to really achieve these properties the resin usually needs to have an elevated temperature post cure.

Vinyl esters are halfway between polyesters and epoxies as far as typical properties and toughness, and they are a step up on a cost basis. Their longer double bonded vinyl groups that link the ester groups together contribute a flexible nature to the resin. That flexible nature produces parts that withstand impacts and repeated flexing without developing cracks.

Vinyl esters provide excellent resistance to water, organic solvents and alkalis, but less resistance to acids than polyesters. Vinyl esters are stronger than polyesters and more resilient than epoxies. Molding conditions for vinyl esters are similar to those for polyesters.

METAL MATRIX COMPOSITES (MMC) AND TESTING OF COMPOSITES

A metal matrix composite (MMC) is composite material with at least two constituent parts, one being a metal. The other material may be a different metal or another material, such as a ceramic or organic compound. When at least three materials are present, it is called a hybrid composite. An MMC is complementary to a cermet. MMCs are made by dispersing a reinforcing material into a metal matrix. The reinforcement surface can be coated to prevent a chemical reaction with the matrix. For example, carbon fibers are commonly used in aluminum matrix to synthesize composites showing low density and high strength. However, carbon reacts with aluminum to generate a brittle and water-soluble compound Al_4C_3 on the surface of the fiber. To prevent this reaction, the carbon fibers are coated with nickel or titanium boride.

DIFFUSION BONDING

Diffusion bonding is a solid-state joining process capable of joining a wide range of metal and ceramic combinations to produce both small and large components. The process is dependent on a number of parameters, in particular, time, applied pressure, bonding temperature and method of heat application. Other examples of solid-state joining include cold pressure welding, friction welding, magnetically impelled arc butt (MIAB) welding and explosive welding.

Diffusion bonding itself can be categorised into a number of variants, dependent on the form of pressurisation, the use of interlayers and the formation of a transient liquid phase. Each finds specific application for the range of materials and geometries that need to be joined.

Process Mechanism of Diffusion Bonding

In its simplest form, diffusion bonding involves holding pre-machined components under load at an elevated temperature usually in a protective atmosphere or vacuum. The loads used are usually below those which would cause macro deformation of the parent material(s) and temperatures of 0.5-0.8 T_m (where T_m = melting point in K) are employed. Times at temperature can range from 1 to 60 + minutes, but this depends upon the materials being bonded, the joint properties required and the remaining bonding parameters. Although the majority of bonding operations are performed in vacuum or an inert gas atmosphere, certain bonds can be produced in air.

Solid Phase Diffusion Bonding

Bonding in the solid phase is mainly carried out in vacuum or a protective atmosphere, with heat being applied by radiant, induction, direct or indirect resistance heating. Pressure can be applied uniaxially or isostatically. In the former case, a low pressure (3-10MPa) is used to prevent macro deformation of the parts (i.e. no more than a few percent). This form of the process therefore requires a good surface finish on the mating surfaces as the contribution to bonding provided by plastic yielding is restricted. Typically surface finishes of better than 0.4 μm RA are recommended and in addition the surfaces should be as clean as practical to minimise surface contamination.

In hot isostatic pressing, much higher pressures are possible (100-200MPa) and therefore surface finishes are not so critical, finishes of 0.8 μm RA and greater can be used. A further advantage of this process is that the use of uniform gas pressurisation allows complex geometries to be bonded, as against the generally simple butt or lap joints possible with uniaxial pressurisation.

Where dissimilar materials need to be joined in the solid phase (and in particular metal to ceramic joints), it is possible to introduce single or multiple interlayers of other materials to aid the bonding process and to modify post- bond stress distribution.

Liquid Phase Diffusion Bonding/Diffusion Brazing

This technique is applicable only to dissimilar material combinations or to 'like' materials where a dissimilar metal insert is used. Solid state diffusional processes lead to a change of composition at the bond interface and the bonding temperature is selected as the temperature at which this phase melts.

Alternatively, with the dissimilar metal insert, it melts at a lower temperature than the parent material. Thus a thin layer of liquid spreads along the interface to form a joint at a lower temperature than the melting point of either of the parent materials. A reduction in bonding temperature leads to solidification of the melt, and this phase can subsequently be diffused away into the parent materials by holding at temperature.

The technique has been used particularly for the bonding of aluminium alloys where eutectics can be formed with copper, silver or zinc to assist in the break-up of the stable aluminium surface oxide.

Superplastic Forming/Diffusion Bonding

This technique has been developed specifically within the aerospace industry. The process is used commercially for titanium and its alloys, this material being one that exhibits superplastic properties at elevated temperatures within defined strain rate conditions. These conditions of temperature and pressure coincide with the conditions required for bonding, and therefore the two processes have been combined into one manufacturing operation either in sequence or together. The process (known as SPF/DB or more correctly DB/SPF) is used to produce stiff sandwich structures for airframe parts, or the wide chord, hollow fan blades for aeroengines. Both these involve skins with internally bonded structures as reinforcing elements.

HOT ISOSTATIC PRESSING

Hot isostatic pressing (HIP) is a manufacturing process used to reduce the porosity of metals and influence the density of many ceramic materials. This improves the material's mechanical properties and workability.

The HIP process subjects a component to both elevated temperature and isostatic gas pressure in a high pressure containment vessel. The pressurizing gas most widely used is argon. An inert gas is used, so that the material does not chemically react. The chamber is heated, causing the pressure inside the vessel to increase. Many systems use associated gas pumping to achieve the necessary pressure level. Pressure is applied to the material from all directions (hence the term "isostatic").

For processing castings, the inert gas is applied between 50.7 MPa and 310 MPa), with 100 MPa being most common. Process soak temperatures range from 482 °C for aluminum castings to 1,320°C for nickel-based superalloys. When castings are treated with HIP, the simultaneous application of heat and pressure eliminates internalvoids and micro porosity through a combination of plastic deformation, creep, and diffusion bonding; this process improves fatigue resistance of component.

Primary applications are the reduction of micro shrinkage, the consolidation of powder metals, ceramic composites andmetal cladding. Hot isostatic pressing is also used as part of a sintering (powder metallurgy) process and for fabrication of metal matrix composites.

HIP is a process that subjects a material simultaneously to both high temperature and high gas pressure, usually Argon, in vessels equipped with sophisticated control systems and telemetry.

Typically, the temperature is selected to permit limited plastic deformation of the material being processed in the solid state at an argon gas pressure of 1,000 to 3,000 atmospheres is isostatically exerted on the heated parts for a period of time. The chamber is then slowly cooled, depressurized and the parts removed.

HIP can close internal porosity in a material without distorting external geometry, consolidate powder materials to 100% of theoretical density or form perfect diffusion bonds between similar or dissimilar materials.

All raw materials contain microscopic voids and "bubbles" of gas, from that standpoint, all raw materials can be considered porous. The advantages of HIP for such porous materials include the elimination of all internal porosity in simple or complex shapes with resulting improvement of mechanical properties such as ductility and fatigue life.

The HIP process falls into three categories: Densification, Powder Metallurgy and Composites, diffusing two like, or unlike metals together.

The benefits for powder materials include the ability to produce 100% dense billets and powdered metal near-net shapes at relatively low cost because of the reduction in machining costs. Near net shapes can be pressed to 100% density, Shape control is obtained by cans and mandrels designed using both CAD and FEA programs. Small variations in packing density lead to small distortions in the final shape. Mating surfaces, for instance, will almost always have to be machined. HIP can also achieve 100% diffusion bonds in clad composite materials.

Typical HIP products now include automotive parts, pump bodies, valves, vacuum chambers, bearings, sterile enclosures, etc. anywhere residual porosity causes high rejection rates, unacceptable property levels and surface finishing problems after machining. Commercial alloy applications include steel, stainless steel and aluminum castings. Not limited to metals, the process is very versatile, having been used to densify ceramics, plastics, glasses and many other materials.

Material densification via the HIP process, densifies to the material's theoretical limit. This has several advantages including, improvement of mechanical properties, enhancement of physical assets and ease of manufacturing. Typical applications have been for fatigue-critical applications like nickel and titanium alloy structural castings for jet engines, turbine blades and vanes, and cast cobalt-chromium or titanium alloy orthopedic implants, such as hip joints.

The HIP process allows for the manufacture of composite or dissimilar materials, thus avoiding welds or costly, gasketed assemblies. The same high temperature and pressures may be applied to achieve the diffusion bonding of encapsulated metal powder to a solid material or the diffusion bonding of different solid materials. Thereby providing for the combination of properties such as a stainless steel body diffusion bonded to a titanium nozzle that is nearly free from residual stresses

INFILTRATION

Liquid state fabrication of Metal Matrix Composites involves incorporation of dispersed phase into a molten matrix metal, followed by its Solidification.

In order to provide high level of mechanical properties of the composite, good interfacial bonding(wetting) between the dispersed phase and the liquid matrix should be obtained.

Wetting improvement may be achieved by coating the dispersed phase particles (fibers). Proper coating not only reduces interfacial energy, but also prevents chemical interaction between the dispersed phase and the matrix.

Infiltration is a liquid state method of composite materials fabrication, in which a preformed dispersed phase (ceramic particles, fibers, woven) is soaked in a molten matrix metal, which fills the space between the dispersed phase inclusions.

The motive force of an infiltration process may be either capillary force of the dispersed phase (spontaneous infiltration) or an external pressure (gaseous, mechanical, electromagnetic, centrifugal or ultrasonic) applied to the liquid matrix phase (forced infiltration).

Infiltration is one of the methods of preparation of tungsten-copper composites. The principal steps of the technology are as follows:

- Tungsten Powder preparation with average particle size of about 1-5 mkm.
- Mixing the tungsten powder with a polymer binder.
- Compacting the powder by a molding method (Metal injection molding, die pressing, isostatic pressing). Compaction should provide the pre determined porosity level (apparent density) of the tungsten structure.

Solvent Debinding

Sintering the green compact at 1204-1315°C in Hydrogen atmosphere for 2 hrs. Placing the sintered part on a copper plate (powder) in the infiltration/sintering furnace. Infiltration of the sintered tungsten skeleton porous structure with copper at 1110-1260°C in either hydrogen atmosphere or vacuum for 1 hour.

Gas Pressure Infiltration

Gas Pressure Infiltration is a forced infiltration method of liquid phase fabrication of Metal Matrix Composites, using a pressurized gas for applying pressure on the molten metal and forcing it to penetrate into a preformed dispersed phase.

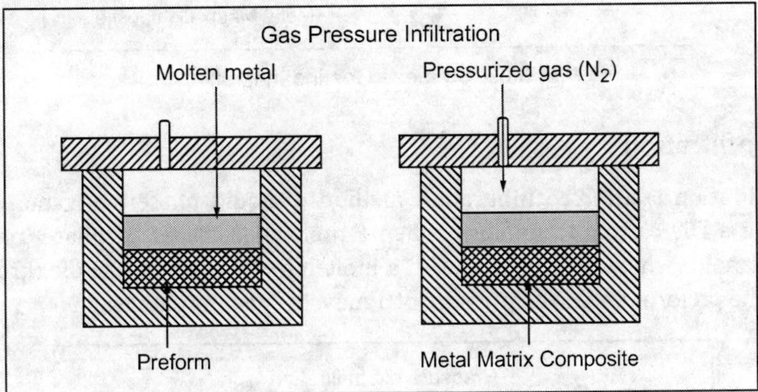

Fig. 33.2. Gas Pressure Infiltration

Gas Pressure Infiltration method is used for manufacturing large composite parts. The method allows using non-coated fibers due to short contact time of the fibers with the hot metal. In contrast to the methods using mechanical force. Gas Pressure Infiltration results in low damage of the fibers.

Squeeze Casting Infiltration

Squeeze Casting Infiltration is a forced infiltration method of liquid phase fabrication of Metal Matrix Composites, using a movable mold part (ram) for applying pressure on the molten metal and forcing it to penetrate into a performed dispersed phase, placed into the lower fixed mold part. Squeeze Casting Infiltration method is similar to the Squeeze casting technique used for metal alloys casting.

Squeeze Casting Infiltration process involves the following steps:

A preform of dispersed phase (particles, fibers) is placed into the lower fixed mold half.

A molten metal in a predetermined amount is poured into the lower mold half. The upper movable mold half (ram) moves downwards and forces the liquid metal to infiltrate the preform.

The infiltrated material solidifies under the pressure.

The part is removed from the mold by means of the ejector pin.

The method is used for manufacturing simple small parts (automotive engine pistons from aluminum alloy reinforced by alumina short fibers)

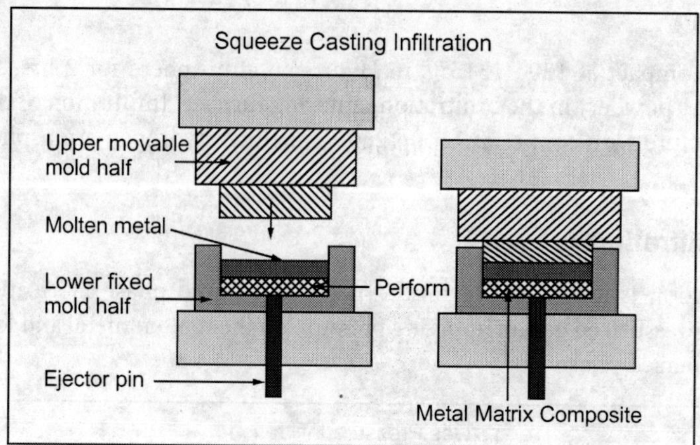

Fig. 33.3. Squeeze Casting Infiltration

Pressure Die Infiltration

Pressure Die Infiltration is a forced infiltration method of liquid phase fabrication of Metal Matrix Composites, using a Die casting technology, when a preformed dispersed phase (particles, fibers) is placed into a die (mold) which is then filled with a molten metal entering the die through a sprue and penetrating into the preform under the pressure of a movable piston (plunger).

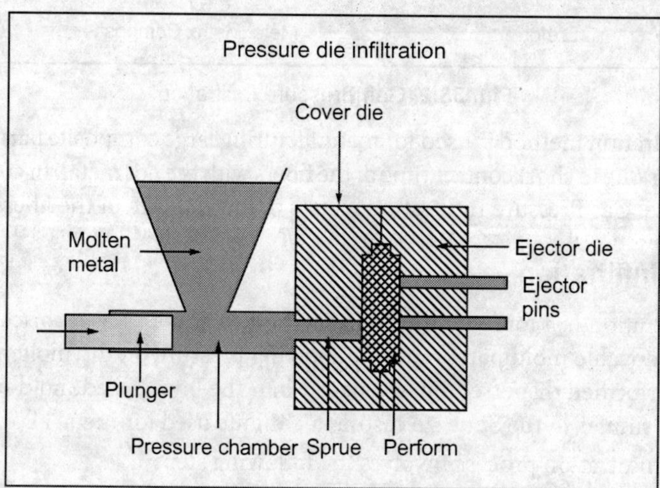

Fig. 33.4. Pressure die infiltration

Testing of Composites

Testing consists of measurement of fundamental properties or measurement of responses to influences such as load, temperature etc., Destructive testing of composites consists of tensile test, impact test, bend test, fatigue test, torsion test, creep test etc.,

TENSILE TEST

Tensile test is performed on ductile materials to determine tensile properties such as limit of proportionality, Yield point or yield strength, Maximum tensile strength, breaking strength, percentage elongation, percentage reduction in area, modulus of elasticity etc., Tensile test is usually carried out in Universal Testing Machine.

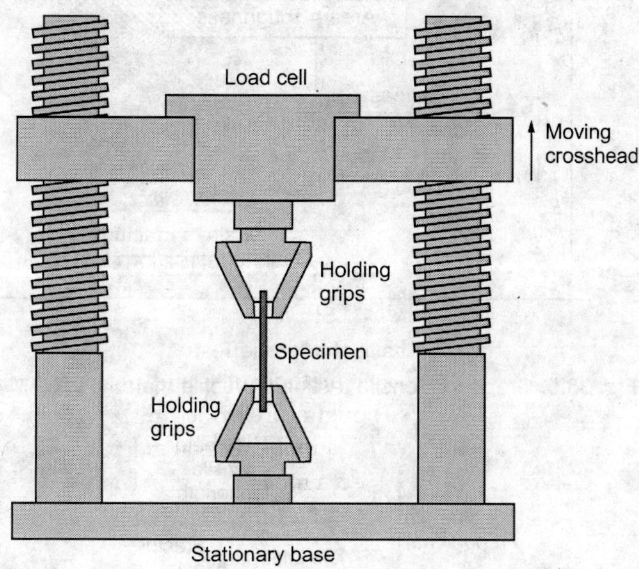

As the name implies, tensile test involves pulling of the test specimen along its length as shown in the figure.

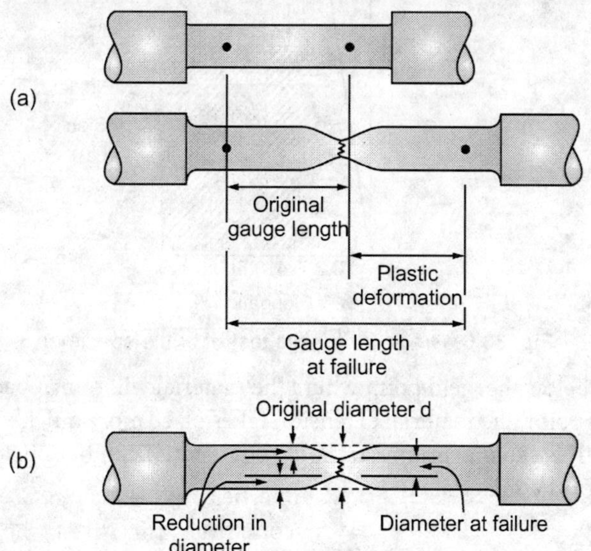

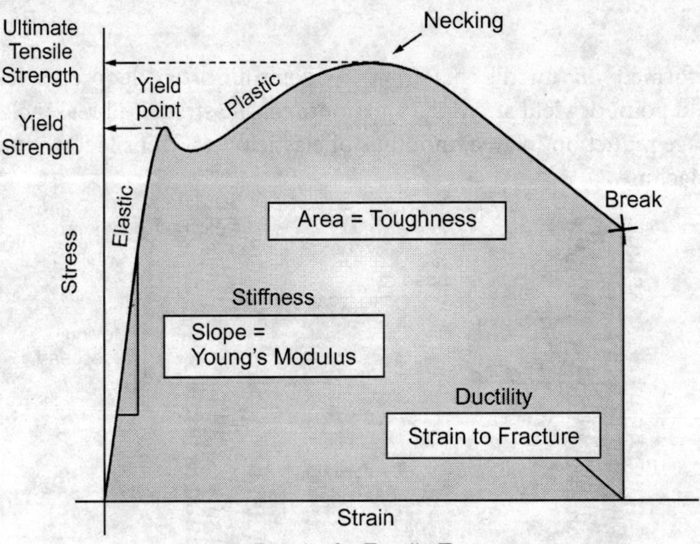

Fig. 33.5. Graph for Tensile testing of ductile material in UTM

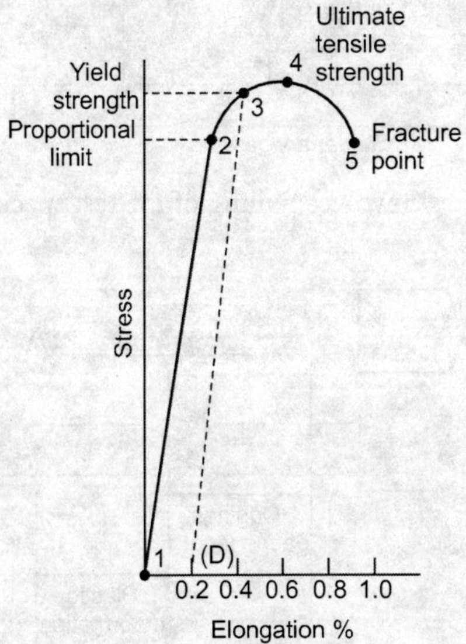

Fig. 33.6. Graph for tensile test of brittle specimen

Elastic region exists below the yield point where the material will return back to its original position when unloaded. At yield point, the material extension takes place more quickly with no or little increase in load. Ultimate strength is the maximum strength the material can bear before breaking. The plastic region does not exist in brittle materials.

COMPRESSIVE TEST

Compression test is conducted in a manner similar to tensile test except that the force is compressive. Since brittle materials are not suited for tension test, they are tested for compression. The compression test specimens are made in cubical shape or cylindrical shape to avoid eccentric loading. The compression load is applied and the corresponding reduction in lengths of the specimen are recorded.

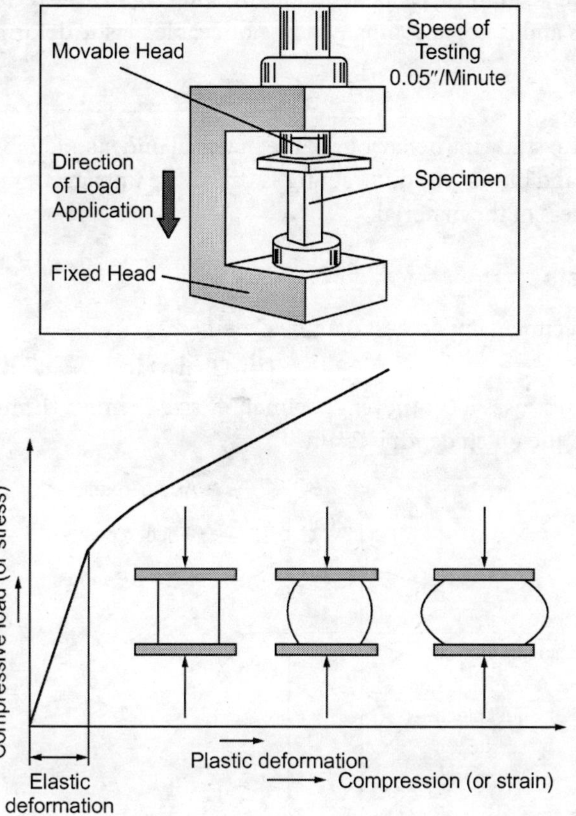

Fig. 33.7. Model graph for compression strength for ductile materials

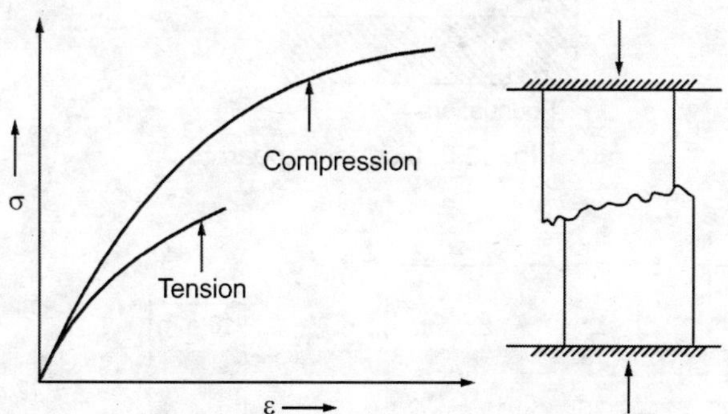

Fig. 33.8. Model graph for compression strength for brittle materials

The ductile materials such as steel, aluminium, copper have stress-strain diagrams similar to the ones we have for tensile test where there would be an elastic region followed by the plastic region. When a ductile material is compressed, it begins to bulge on sides to give a barrel shape, as shown in the above figure. Brittle materials in compression typically have an initial linear region followed by a region in which the shortening increases at a higher rate than does the load. Thus, the compression stress – strain diagram has a shape that is similar to the shape of the tensile diagram. However, brittle materials usually reach much higher ultimate stresses in compression than in tension. Brittle materials in compression behave elastically up to certain load, and then fail suddenly by splitting or by cracking. The brittle fracture is performed by separation and is not accompanied by noticeable plastic deformation.

IMPACT TEST

Impact test is performed to study the behaviour of the material under suddenly applied load. The capacity of the material to withstand blows without fracture is called the impact strength of the material. Impact test indicates the toughness of the material.

Types of Impact Tests

Based on the types of specimen, impact tests can be classified as

(*i*) Izod test (*ii*) Charpy test

(*i*) **Izod test:** Izod test uses a cantilever specimen of size 75mm × 10mm x 10mm with a v notch angle of 45° and the notch depth is 2mm.

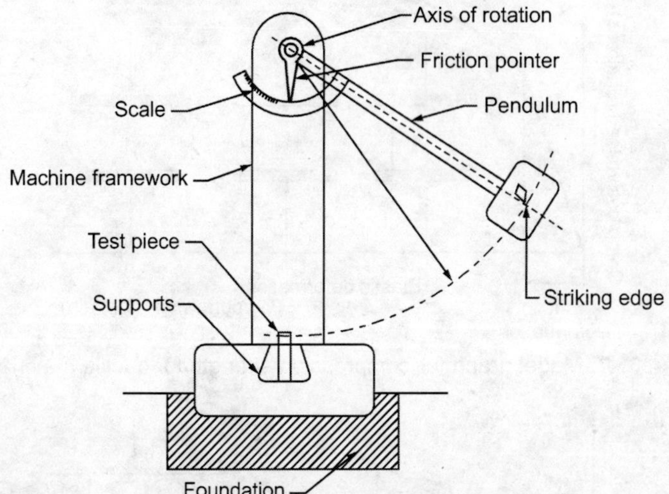

Fig. 33.9. Izod test apparatus

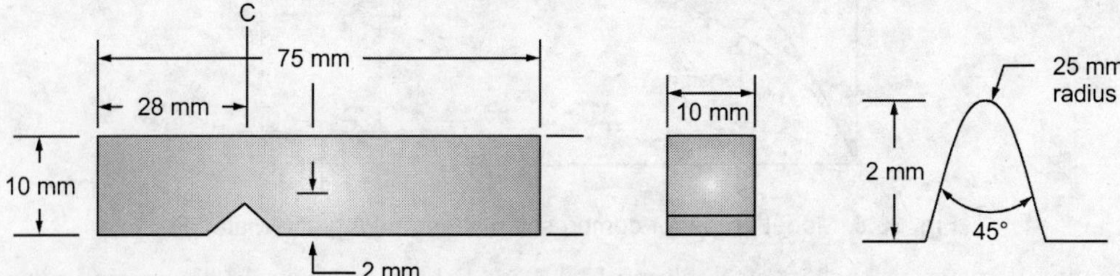

Fig. 33.10. Izod test specimen

(*ii*) **Charpy test:** Charpy test uses a simply supported test specimen of size 55mm × 10mm × 10mm with a v notch angle of 45° and depth of notch is 2mm.

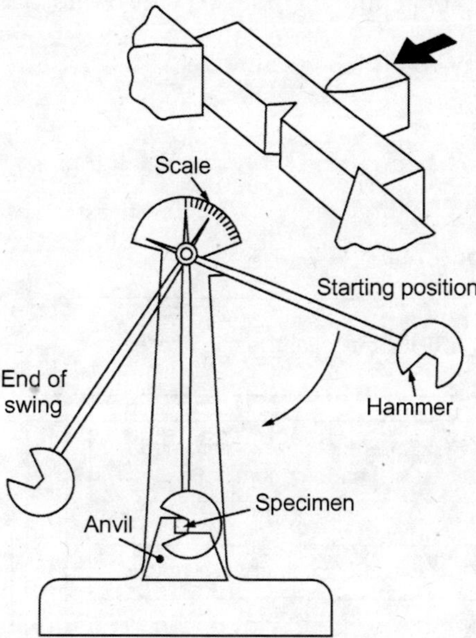

Fig. 33.11. Charpy test apparatus

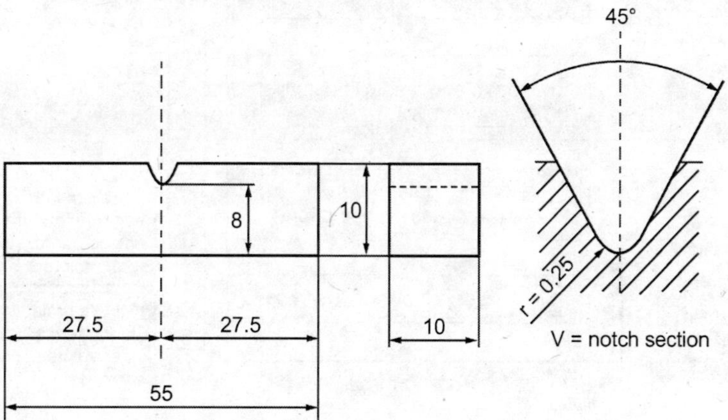

Fig. 33.12. Charpy Test specimen

Impact Testing Procedure

- The specimen is placed in the vice of the anvil.
- The pendulum hammer is raised to known standard height depending on the type of specimen to be tested.
- When the pendulum is released, its potential energy is converted into kinetic energy just before it strikes the specimen.
- The Izod specimen is hit above the V notch and the Charpy specimen will be hit behind the V notch.

- The pendulum after rupturing the specimen rises on the other side of the machine.
- The energy absorbed by the specimen during breaking is the weight of the pendulum times the difference in two heights of pendulum on either side of the machine.

Factors affecting Impact Strength

Factors affecting impact strength are:

- Angle of Notch
- Impact velocity
- Dimensions of notch specimen
- Shape of Notch
- Temperature of the specimen

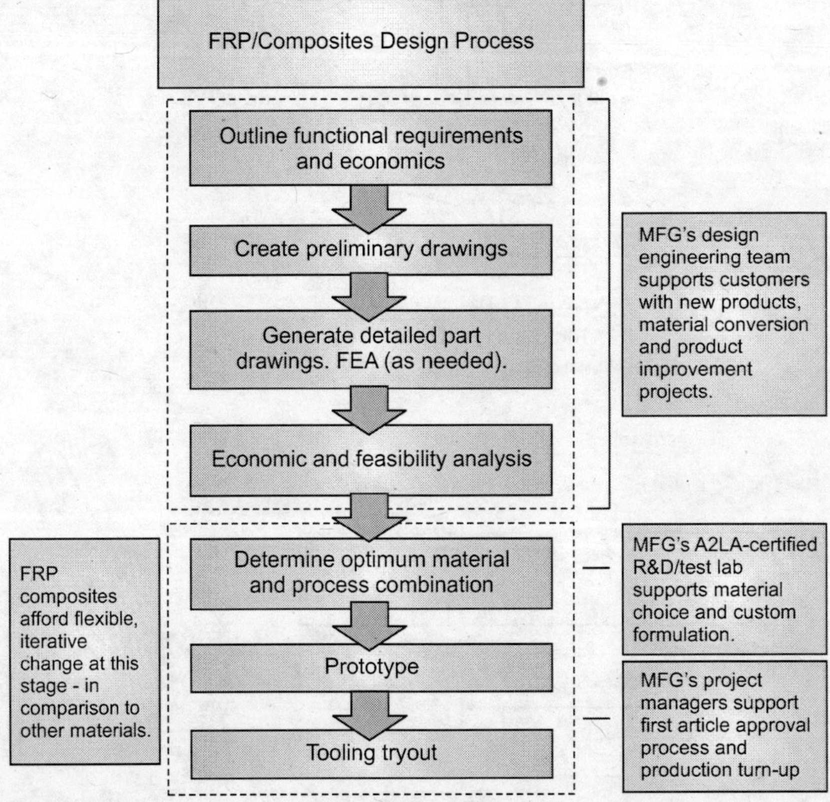

Table 33.1. Markets and typical applications for composites

Market	Typical Applications
Aircraft/aerospace/military	Helicopter blades and shell; control surfaces and floor beams (Boeing 767); F-16s horizontal stabilizer, skin, dorsal access panels and leading edge fairing, and vertical tail fin skins and louver fin leading edge and rudder
Appliances/business equipment	Refrigerator compressor bases, room air conditioner parts, interior dishwasher parts, business machine bases and housings

Construction	Reinforced plastic bathtubs and shower fixtures, dimensionally stable reinforced plastic window lineals and doors, traffic signs, pedestrian shelters, telephone booths, and traffic kiosks
Consumer products	Recreational and sporting goods and recreational-vehicle equipment; microwave dinnerware and office furniture
Corrosion-resistant equipment	Underground gasoline tanks and reinforced plastic pipes and fittings; flooring for chemical processing plants and pulp and paper mills; tank linings for the storage of crude oil; piping and fittings for petroleum refineries; caustic and sulfur dioxide scrubbers; mixing tanks for phosphoric acid fertilizer solutions
Electrical/electronic	Fiber-optic cable; satellite dish and antenna, protective coverboards, light poles, electrical conduit, bus bar insulation, and third-rail insulators and covers
Marine	Sailboat, powerboat, mine sweeper, military craft clad with reinforced plastics
Land transportation	Exterior body parts, suspensions, chassis, load floors, bumpers, drive shafts, lead springs, under- hood components, and automotive frames

Short Questions & Answers

Q.1. What is a polymer?

Ans. A polymer is a compound that consists of long-chain molecules. The molecules consist of repeating units, called mers, connected end to end.

Q.2. What are the three basic categories of polymers?

Ans. The categories are: (1) thermoplastics, (2) thermosetting polymers, and (3) elastomers.

Q.3. How do the properties of polymers compare with those of metals?

Ans. In general, polymers have lower strength, hardness, stiffness, density, and temperature resistance compared to metals. In addition, polymers are low in electrical and thermal conductivity.

Q.4. What are the two methods by which polymerization occurs?

Ans. The two types of polymerization are: (1) addition or chain polymerization and (2) step polymerization, also known as condensation polymerization.

Q.5. What does the *degree of polymerization* indicate?

Ans. The degree of polymerization indicates the average number of mers or repeating units in the polymer molecule.

Q.6. Define the term *tacticity* as it applies to polymers.

Ans. Tacticity refers to the way the atoms or atom groups replacing H atoms in the molecule are arranged.

Q.7. What is *cross-linking* in a polymer and what is its significance?

Ans. Cross-linking is the formation of connections between the long-chain molecules in a polymer. It causes the polymer structure to be permanently altered. If the amount of cross-linking is low, the polymer is transformed into an elastomer; if cross-linking is significant, the polymer is transformed into a thermosetting polymer.

Q.8. What is a *copolymer*?

Ans. A copolymer is a polymer made up of two different types of mers, such as ethylene and propylene.

Q.9. The arrangement of repeating units in a copolymer can vary. What are some of the possible arrangements?

Ans. There are four possible arrangements of the mers along the chain: (1) alternating, (2) random, (3) block, and (4) graft.

Q.10. What is a *terpolymer*?

Ans. A terpolymer is a polymer with three different mer types. An example is ABS (acrylonitrile -butadiene-styrene) plastic.

Q.11. How are a polymer's properties affected when it takes on a crystalline structure?

Ans. Density, stiffness, and melting temperature increase.

Q.12. Does any polymer ever become 100% crystalline?

Ans. No.

Q.13. What are some of the factors that influence a polymer's tendency to crystallize?

Ans. Factors are: (1) only linear polymers can form crystals; (2) copolymers do not form crystals; (3) stereoregularity - isotactic polymers always form crystals, atactic polymers never form crystals, and syndiotactic polymers sometimes form crystals; (4) slow cooling from the molten states promotes crystal formation; (5) plasticizers inhibit crystal formation; and (6) stretching the polymer tends to promote crystallization.

Q.14. Why are fillers added to a polymer?

Ans. Fillers are added to increase strength or simply to reduce the cost of the polymer.

Q.15. What is a *plasticizer*?

Ans. A plasticizer is a chemical added to the polymer to make it softer and more flexible. It is often added to improve the polymer's flow characteristics for shaping.

Q.16. In addition to fillers and plasticizers, what are some other additives used with polymers?

Ans. Other additives include: lubricants - to reduce friction and improve flow; flame retardants; colorants; cross-linking agents, antioxidants, and ultraviolet light absorbers.

Q.17. Describe the difference in mechanical properties as a function of temperature between a highly crystalline thermoplastic and an amorphous thermoplastic (TP).

Ans. A highly crystalline TP retains rigidity during heating until just before its T_m is reached. An amorphous TP shows a significant drop in deformation resistance at its T_g as temperature is raised; it becomes increasingly like a liquid as temperature continues to increase.

Q.18. What is unique about the polymer *cellulose*?

Ans. Cellulose is a polymer that grows in nature. Wood fiber contains about 50% cellulose and cotton fiber is about 95% cellulose.

Q.19. The nylons are members of which polymer group?

Ans. Polyamides.

Q.20. What is the chemical formula of ethylene, the monomer for polyethylene?

Ans. C_2H_4

Q.21. What is the basic difference between low density and high density polyethylene?

Ans. LDPE has a branched structure and is amorphous. HDPE is linear and highly crystalline. These differences account for HDPE higher density, stiffness, and melting point.

Q.22. How do the properties of thermosetting polymers differ from those of thermoplastics?

Ans. Thermosets are more rigid, brittle, capable of higher service temperatures, and cannot be remelted.

Q.23. Cross-linking (curing) of thermosetting plastics is accomplished by one of three ways. Name the three ways.

Ans. The three ways are: (1) temperature-activated systems, in which elevated temperatures accomplish curing; (2) catalyst-activated systems, in which small amounts of a catalyst cause cross-linking; and (3) mixing-activated systems, in which two reactive components are mixed and curing occurs by their chemical reaction.

Q.24. Elastomers and thermosetting polymers are both cross- linked. Why are their properties so different?

Ans. Elastomers are lightly cross-linked, whereas thermosets are highly cross-linked. Light cross-linking allows extensibility; a highly cross-linked structure makes the polymer rigid.

Q.25. What happens to an elastomer when it is below its glass transition temperature?

Ans. An elastomer is hard and brittle below its T_g.

Q.26. What is the primary polymer ingredient in natural rubber?

Ans. Polyisoprene.

Q.27. How are thermoplastic elastomers different from conventional rubbers?

Ans. TPEs are different in two basic ways: (1) they exhibit thermoplastic properties, and (2) their extensibility derives from physical connections between different phases in the polymer.

Q.28. What are some of the reasons why the plastic shaping processes are important?

Ans. Some of the reasons are: (1) many of the processes are net shape processes; (2) in general, less energy employed than in metalworking processes; (3) lower temperatures are required to process plastics than metals or ceramics; (4) great flexibility in geometry; and (5) painting and other finishing processes generally not required.

Q.29. Identify the main categories of plastics shaping processes, as classified by the resulting product geometry.

Ans. The categories are: (1) extrusion, (2) molding, (3) forming of continuous sheets and films, (4) fibers, (5) foamed products, and (6) discrete formed sheets and films.

Q.30. Viscosity is an important property of a polymer melt in plastics shaping processes. Upon what parameters does viscosity depend?

Ans. Viscosity of a polymer melt depends on: (1) temperature, and (2) shear rate. Of course, (3) the molecular weight of the polymer also affects viscosity.

Q.31. How does the viscosity of a polymer melt differ from most fluids that are Newtonian?

Ans. A polymer melt exhibits pseudoplasticity, which means that its value decreases with increasing shear rate.

Q.32. Besides viscosity, what other properties of a polymer melt are important in plastics processing? Briefly define each of the properties identified.

Ans. Other properties include: viscoelasticity, a combination of viscous and elastic properties which cause the melt to exhibit memory - the tendency to return to its previous shape, as exhibited by die swell in extrusion.

Q.33. Define *die swell* in extrusion.

Ans. Die swell is the tendency of the extrudate to expand in cross-section dimensions immediately on exiting the die orifice. It results from the viscoelastic properties of the polymer melt.

Q.34. Briefly describe the plastic extrusion process.

Ans. In plastic extrusion, a polymer melt is compressed to flow through a die orifice and thus the continuous length of the plastic assumes a cross-sectional shape that is approximately the same as the profile of the orifice.

Q.35. The barrel and screw of an extruder are generally divided into three sections; identify the sections.

Ans. (1) feed section, in which the feed stock is fed from the hopper and heated; (2) compression section, in which the polymer changes to a viscous fluid; and (3) metering section, in which pressure is developed to pump the plastic through the die orifice.

Q.36. What are the functions of the screen pack and breaker plate at the die end of the extruder barrel?

Ans. The functions are: (1) filter dirt and lumps, (2) build pressure, (3) straighten the flow and remove memory.

Q.37. What are the various forms of extruded shapes and corresponding dies?

Ans. (1) solid profiles, such as rounds and L-shapes; (2) hollow profiles, such as tubes; (3) wire and cable coating; (4) sheet and film; and (5) filaments (continuous fibers).

Q.38. What is the distinction between plastic sheet and film?

Ans. Thickness. Sheet stock has a thickness greater than 0.020 in (0.5 mm), while film stock is less than 0.020 in (0.5 mm) thick.

Q.39. What is the *blown-film process* for producing film stock?

Ans. The ***blown-film process*** is a widely used process for making thin polyethylene film for packaging. It combines extrusion and blowing to produce a tube of thin film. The process begins with the extrusion of a tube which is immediately drawn upward while still molten and simultaneously expanded in size by air inflated into it through the die mandrel.

Q.40. Describe the *calendaring process*.

Ans. Calendaring is a process for producing sheet and film stock out of rubber or rubbery thermoplastics such as plasticized PVC. In the process, the initial feedstock is passed through a series of rolls to work the material and reduce its thickness to the desired gage.

Q.41. Polymer fibers and filaments are used in several applications; what is the most important application?

Ans. Textiles.

Q.42. Technically, what is the difference between a *fiber* and a *filament*?

Ans. A fiber is a long, thin strand of material whose length is at least 100 times its diameter; a filament is a fiber of continuous length.

Q.43. Among the synthetic fiber materials, which are the most important?

Ans. Polyester is the most important commercially, followed by nylon, acrylics, and rayon.

Q.44. An injection molding machine is divided into two principal components; identify them.

Ans. The components of an injection molding machine are: (1) the injection unit and (2) the clamping unit.

Q.45. What are the two basic types of clamping units?

Ans. The clamping units are: (1) mechanical toggle clamp and (2) hydraulic. In addition, there are hydro mechanical units which combine hydraulic and mechanical actuations.

Q.46. Gates in injection molds have several functions; name them.

Ans. The functions of gates in an injection mold are: (1) to increase shear rate to increase viscosity and temperature of the polymer melt, (2) the thinner cross- section of the gate freezes more rapidly to seal off the cavity; and (3) parts can be more easily broken off the runner at the gate.

Q.47. What are the advantages of a three-plate mold over a two- plate mold in injection molding?

Ans. As the mold opens, the three-plate mold automatically separates the molded part(s) from the runner system.

Q.48. Discuss some of the defects that can occur in plastic injection molding.

Ans. The defects include: short shots, in which the polymer melt solidifies before filling the cavity; flashing, in which the polymer melt is squeezed into the parting surfaces between the mold halves and around ejection pins; sink marks, in which the surface is drawn into the molding by contraction of internal material; and weld lines where the melt has flowed around a core or

Q.49. Describe *structural foam molding.*

Ans. Structural foam molding is an injection molding process in which a gas or gas-producing ingredient is mixed with the polymer melt prior to injection into the mold cavity; this results in the part having a tough outer skin surrounded by a foam core.

Q.50. What are the significant differences in the equipment and operating procedures between injection molding of thermoplastics and injection molding of thermosets?

Ans. The differences in injection molding of thermosets are: (1) shorter barrel length, (2) lower temperatures in the barrel, these first two reasons to prevent premature curing; and (3) use of a heated mold to cause cross-linking of the TS polymer.

Q.51. What is *reaction injection molding*?

Ans. Reaction injection molding involves the mixing of two highly reactive liquid ingredients and immediately injecting the mixture into a mold cavity where chemical reactions leading to solidification occur. The two ingredients form the components used in catalyst-activated or mixing-activated thermoset systems.

Q.52. What kinds of products are produced by blow molding?

Ans. Blow molding is used to produce hollow, seamless containers, such as bottles.

Q.53. What is the starting material form in thermoforming?

Ans. Thermoforming starts with a thermoplastic sheet or film.

Q.54. What is the difference between a positive mold and a negative mold in thermoforming?

Ans. A positive mold has a convex shape; a negative mold has a concave cavity.

Q.55. Why are the molds generally more costly in mechanical thermoforming than in pressure or vacuum thermoforming?

Ans. In mechanical thermoforming, matching mold halves are required; while in other thermoforming processes, only one mold form is required.

Q.56. What are the processes by which polymer foams are produced?

Ans. There are several foaming processes: (1) mechanical agitation - mixing a liquid resin with air by, then hardening the polymer by means of heat or chemical reaction; (2) mixing a physical blowing agent with the polymer - a gas such as nitrogen (N_2) or pentane (C_5H_{12}) which can be dissolved in the polymer melt under pressure, so that the gas comes out of solution and expands when the pressure is subsequently reduced; and (3) mixing the polymer with chemical compounds, called chemical blowing agents, that decompose at elevated temperatures to liberate gases such as CO_2 or N_2 within the melt.

Q.57. What are some of the general considerations that product designers must keep in mind when designing components out of plastics?

Ans. Some of the general considerations are: (1) Plastics are not as strong or stiff as metals and should not be used in applications where high stresses will be encountered. (2) Impact resistance of plastics is general good, better than many ceramics. (3) Service temperatures of plastics are limited relative to engineering metals and ceramics. (4) Thermal expansion is greater for plastics than metals; so dimensional changes due to temperature variations are much more significant than for metals. (5) Many types of plastics degrade from sunlight and certain other forms of radiation. Also, some plastics degrade in oxygen and ozone atmospheres. Finally, plastics are soluble in many common solvents.

Q.58. How is the rubber industry organized?

Ans. The rubber industry is organized into three parts: (1) rubber growing plantations produce natural rubber, (2) the petrochemical industry produces synthetic rubber, and (3) fabricators take the NR and SR and produce finished rubber goods.

Q.59. How is raw rubber recovered from the latex that is tapped from a rubber tree?

Ans. The rubber is usually recovered as follows: (1) the latex is collected into tanks and diluted to half natural concentration; (2) formic or acetic or other acid is added to the solution which causes the rubber to coagulate; (3) the coagulum is then squeezed through rolls to drive off water; and (4) the resulting sheets are dried in smokehouses for several days. The resulting raw rubber is called ***ribbed smoked sheet***.

Q.60. What is the sequence of processing steps required to produce finished rubber goods?

Ans. The typical sequence is: (1) production of the raw rubber, (2) compounding, (3) mixing, (4) shaping, and (5) vulcanization.

Q.61. What are some of the functions of the additives that are combined with rubber during compounding?

Ans. The additives and functions are: vulcanizing chemicals, reinforcing fillers, extenders to reduce cost, antioxidants, coloring pigments, plasticizers to soften the rubber, and blowing agents to make foam rubber.

Q.62. Name the four basic categories of processes used to shape rubber.

Ans. The categories are: (1) extrusion, (2) calendering, (3) coating, and (4) molding.

Q.63. What are some of the operations used to coat rubber onto a fabric to produce reinforced rubber?

Ans. Calendering, skimming, dipping, and spraying.

Q.64. What does vulcanization do to the rubber?

Ans. Vulcanization causes cross-linking of the rubber molecules; this strengthens and stiffens the rubber while extensibility is retained.

Q.65. Name the three basic tire constructions and briefly identify the differences in their construction.

Ans. (*a*) diagonal ply, (*b*) belted bias, and (*c*) radial ply. Diagonal ply and belted bias both have their carcass plys running in a diagonal direction relative to the tire circumference; radial ply has its carcass plies running in a radial direction; belted bias and radial ply tires use belts, which are additional plies around the outside circumference of the tire; whereas diagonal ply tires do not have these belts.

Q.66. What are the three basic steps in the manufacture of a pneumatic tire?

Ans. The three steps are: (1) preform the components, (2) building the carcass and adding the rubber for the sidewall and treads, and (3) molding and curing.

Q.67. What is the purpose of the bead?

Ans. The bead coil provides a rigid support for the tire when it is mounted onto the wheel rim.

Q.68. What is a TPE?

Ans. TPE stands for thermoplastic elastomer; it is a thermoplastic polymer that behaves like a rubber.

Q.69. Many of the design guidelines that are applicable to plastics are also applicable to rubber. However, the extreme flexibility of rubber results in certain differences. What are some examples of these differences?

Ans. Three examples: (1) no draft is needed on the part for mold removal; (2) holes should be molded into rubber parts rather than machined, whereas holes can be machined or molded in a plastic part; and (3) screw threads are not normally used on rubber parts.

Q.70. What is a *composite material*?

Ans. A composite material is a materials system consisting of two or more distinct phases whose combination results in properties that differ from those of its constituents.

Q.71. Identify some of the characteristic properties of composite materials.

Ans. Typical properties include: (1) high strength-to- weight and stiffness-to-weight ratios; (2) good fatigue properties and toughness; (3) anisotropic properties in many cases; and (4) other properties and features that are difficult or impossible to obtain with metals, ceramics, or polymers alone.

Q.72. What does the term *anisotropic* mean?

Ans. Anisotropic means that the properties of a material vary depending on the direction in which they are measured.

Q.73. How are traditional composites distinguished from synthetic composites?

Ans. Traditional composites have been used for decades or centuries; some of them are obtained from sources in nature, such as wood. Synthetic composites are manufactured.

Q.74. Name the three basic categories of composite materials.

Ans. Metal matrix composites (MMCs), ceramic matrix composites (CMCs), and polymer matrix composites (PMCs).

Q.75. What are the common forms of the reinforcing phase in composite materials?

Ans. The forms are: (1) fibers, (2) particles and flakes, and (3) an infiltrated phase in skeletal structures.

Q.76. What is a *whisker*?

Ans. A whisker is a thin, hair like crystal of very high strength.

Q.77. What are the two forms of sandwich structure among laminar composite structures? Briefly describe each.

Ans. The two forms are: (1) foamed-core sandwich, in which the core is polymer foam between two solid skins; and (2) honeycomb, in which the core is a honeycomb structure sandwiched between two solid skins.

Q.78. Give some examples of commercial products which are laminar composite structures.

Ans. Examples given in Table 9.2 are: automotive tires, honeycomb sandwich structures, fiber reinforced polymer structures such as boat hulls, plywood, printed circuit boards, snow skis made from fiber reinforced polymers, and windshield glass.

Q.79. What are the three general factors that determine the properties of a composite material?

Ans. Three factors are : (1) the component materials; (2) the geometric shapes of the constituents - the reinforcing phase in particular - and the resulting structure of the material; and (3) the interaction of the phases.

Q.80. What is the *rule of mixtures*?

Ans. The rule of mixtures applies to certain properties of composite materials; it states that the property value is a weighted average of the property values of the components, the weighting being by proportions of the components in the composite.

Q.81. What is a *cermet*?

Ans. A cermet is a composite material consisting of a ceramic and a metal. In the text, it is defined as a composite consisting of ceramic grains imbedded in a metallic matrix.

Q.82. Cemented carbides are what class of composites?

Ans. Yes; although the cemented carbide industry does not generally think of cemented carbides as cermets, they fit within the definition.

Q.83. What are some of the weaknesses of ceramics that might be corrected in fiber-reinforced ceramic matrix composites?

Ans. Weaknesses of ceramics include: low tensile strength, poor toughness, and susceptibility to thermal cracking.

Q.84. What is the most common fiber material in fiber-reinforced plastics?

Ans. E-glass.

Q.85. What does the term *advanced composites* mean?

Ans. An advanced composite is a PMC in which carbon, Kevlar, or boron fibers are used as the reinforcing material.

Q.86. What is a *hybrid composite*?

Ans. A hybrid composite is a fiber-reinforced PMC in which two or more fibers materials are combined in the FRP.

Q.87. Identify some of the important properties of fiber- reinforced plastic composite materials.

Ans. Properties include: high strength-to-weight ratio, high modulus-to-weight ratio, low density, good fatigue strength, good corrosion resistance, and low thermal expansion for many FRPs.

Q.88. Name some of the important applications of FRPs.

Ans. FRPs are used in modern aircraft as skin parts, automobile body panels, printed circuit boards, tennis rackets, boat hulls, and a variety of other items.

Q.89. What is meant by the term interface in the context of composite materials?

Ans. The interface is the boundary between the component phases in a composite material.

Q.90. What are the principal polymers used in fiber-reinforced polymers?

Ans. Principal polymer matrices in FRPs are unsaturated polyesters and epoxies.

Q.91. What is the difference between a roving and a yarn?

Ans. A roving consists of untwisted filaments, while a yarn consists of twisted fibers.

Q.92. In the context of fiber reinforcement, what is a *mat*?

Ans. A mat is a felt consisting of randomly oriented fibers held loosely together in a binder.

Q.93. Why do we say that particles and flakes are members of the same basic class of reinforcing material?

Ans. Flakes are simply particles that possess very low width-to-thickness ratios.

Q.94. What is *sheet molding compound* (SMC)?

Ans. SMC consists of TS polymer resin, fillers, and chopped glass fibers, all rolled into a sheet of typical thickness = 0.250 inch.

Q.95. How is a prepreg different from a molding compound?

Ans. Prepregs have continuous fibers rather than chopped fibers as in molding compounds.

Q.96. Why are laminated FRP products made by the spray-up method not as strong as similar products made by hand lay-up?

Ans. Because in hand lay-up, orientation of the fibers is controlled; whereas in spray-up, the fibers in each layer are randomly oriented.

Q.97. What is the difference between the wet lay-up approach and the prepreg approach in hand lay-up?

Ans. In wet lay-up, the layer of fiber reinforcement is placed into the mold dry, and the uncured resin is then applied to it to form the composite laminate. In the prepreg approach, layers of fiber preimpregnated with resin are laid into the mold.

Q.98. What is an *autoclave*?

Ans. An autoclave is an enclosed chamber which can supply heat and/or pressure at controlled levels.

Q.99. What are some of the distinguishing characteristics of the closed mold processes for PMCs?

Ans. Characteristics include: (1) they use molds consisting of two halves that open and close during the molding cycle; (2) high pressure is generally associated with the process; and (3) the geometric shapes of the moldings are more complex in three dimensions.

Q.100. Identify some of the different forms of PMC molding compounds.

Ans. PMC molding compounds include SMCs, TMCs, and BMCs.

Q.101. What is *preform molding*?

Ans. Preform molding is a compression molding process in which a precut mat is placed into the lower half of a mold together with a charge of thermosetting resin; the materials are then pressed between heated molds to cure the resin and produce a fiber-reinforced molding.

Q.102. Describe *reinforced reaction injection molding* (RRIM).

Ans. RRIM involves the injection of resins that cure by chemical reaction together with reinforcing fibers into a closed mold. The resulting part is a fiber-reinforced (usually glass fiber) plastic molding.

Q.103. What is *filament winding*?

Ans. Filament winding is a process in which resin- impregnated continuous fibers are wrapped around a rotating mandrel with the internal shape of the FRP product; the resin is cured and the mandrel is removed.

Q.104. What is the advantage of computer numerical control over mechanical control in filament winding?

Ans. CNC allows independent control over mandrel rotation and carriage speed for greater flexibility in relative motions.

Q.105. Describe the *pultrusion* process.

Ans. Pultrusion is a process in which continuous fibers are dipped into a resin and pulled through a shaping die (somewhat like an extrusion die) where the resin cures. The resulting sections are similar to extruded parts except that they are reinforced with continuous fibers.

Q.1067. How does *pulforming* differ from pultrusion?

Ans. Pulforming is pultrusion with the added operation of a shape change in the length (straight length becomes curved) and cross-section (different cross-sections throughout the length).

Q.107. With what kinds of products is *tube rolling* associated?

Ans. Typical products include: bicycle frames and space trusses.

Q.108. How are FRPs cut?

Ans. Uncured FRPs are cut by methods which include: knives, scissors, power shears, Steel-rule blanking dies, laser beam cutting, and water jet cutting. Cured FRPs are cut by WC and HSS cutting tools, diamond cutting tools, and water jet cutting.

Q.109. What is the catalyst used in the Low Pressure Zeigler Process for the manufacture of polyethylene?

Ans. Aluminium triethyl combined with titanium tetrachloride.

Q.110. What does the Ziegler-Natta catalyst consist of?

Ans. The Ziegler Natta catalyst consists of a combination of HF and BF_3.

Q.111. What is the main criteria based upon which Polyethylene is classified?

Ans. Polyethylene is classified based upon its density and molecular bonding.

Q.112. Describe the classification and properties and applications of various types of polyethylene.

Ans. Ethylene is the monomer for manufacturing Polyethylene. Polyethylene is broadly classified into High Density Polyethylene (HDPE), Linear Low Density Polyethylene (LLDPE) and Low Density Polyethylene (LDPE).

High Density Polyethylene(HDPE) is predominantly linear structure, closely packed with density of 0.960 - 0.980 gram/cm^3, crystalline as high as 95% and a melting point as high as 138.5 ^{0}C. It finds its major application in Raffia, Blow Moulding, Film, Pipes and Injection Moulding.

HDPE because of its unique properties and increased availability offers abundant opportunities for growth. There is a huge untapped potential for the sector like loose to packed packaging in Blow moulding containers,automotive sector, infrastructure, and in installing drinking water and sewerage system etc.

Linear Low Density Polyethylene(LLDPE) is linear in structure. The branches are formed due to presence of co monomer. The co-monomer used is Butene (C_4), Hexene (C_6) and Octene (C_8). The low pressure used in manufacture of LLDPE does not allow the formation of large branches. LLDPE finds its application in mono and multi-layered films, rotomoulding,wire and cables and high flow LL is used in master batch compounds etc.

Currently the fast growing sector in LL is Metallocene, followed by wire and cable and high flow grades. Metallocene LL grades are fast moving due to its properties like better strength, high dart, higher line speed, better sealing, downsizing of the film etc.

Low Density Polyethylene(**LDPE**) is predominantly a branched polymer. The properties depend upon the frequency and the length of branches. LDPE helps in processibility, has excellent transparency and good sealing properties and low glass transition temperature. Due to its inherent strength, LD is extensively used as a blend with other polymers. LDPE finds its application in general purpose film, heavy duty film,liquid packaging, injection moulding, extrusion coating, wire and cables, adhesive lamination.

The major application wise share is in the order of general purpose film, extrusion coating, milk packaging, heavy duty, injection moulding, wire and cables and others.Flexible Packaging constitutes 84% of the LD consumption. The growing sectors are general purpose film, heavy duty, milk packaging, injection moulding sectors, adhesive laminates etc.

Important properties of polyethylenes

Polyethylenes	Melting point range °C	Density g/cm^3	Degree of crystallinity %	Stiffness modules psi × 10^3
Branched, Low density	107-121	0.92	60-65	25-30
Medium density	—	0.935	75	60-65
Linear, High density:				
Ziegler type	125-132	0.95	85	90-110
Phillips type	—	0.96	91	130-150

Q.113. Discuss the salient features of Polyethylene (PE) Manufacturing Processes.

Ans. Three classes of PE, namely, LDPE, HDPE, and LLDPE, are produced in different processes. LDPE is produced in a very high-pressure process, and HDPE and LLDPE are obtained in moderately low-pressure processes.

Salient features of PE manufacturing processes is given below in Table.

Table. 34.1. Salient features of polyethylene manufacturing processes

Reactant Polymer		Type of Polymerisation	Catalyst Temperature, °C	Pressure, atm
Ethylene LDPE reactor, liquid phase	Solution or bulk, tubular	H_2O_2	200-240.	3500
Ethylene HDPE reactor, liquid phase	Suspension, stirred tank	$TiCl_4$	60-80.	20-35
Ethylene LLDPE and 1-butene reactor, vapour phase	Suspension, fluidised bed	$TiCl_4$	100	7–20

Q.114. Describe the process of manufacture of Low Density Polyethylene (LDPE).

Ans. In a tubular reactor, pure liquid ethylene is mixed with hydrogen peroxide and forced through the tubular reactor at very high pressure (3500 atm), surrounded by a cooling medium to extract heat of polymerization reaction. Oxygen is used with hydrogen peroxide as the initiator.

The reaction takes place in solution. The heat of the reaction is given as 3650 kJ/kg.

$$nC_2H_4 = (-C_2H_4-)_n - 3650 \text{ kJ/kg.}$$

Figure below is a schematic flow diagram of an LDPE manufacturing process.

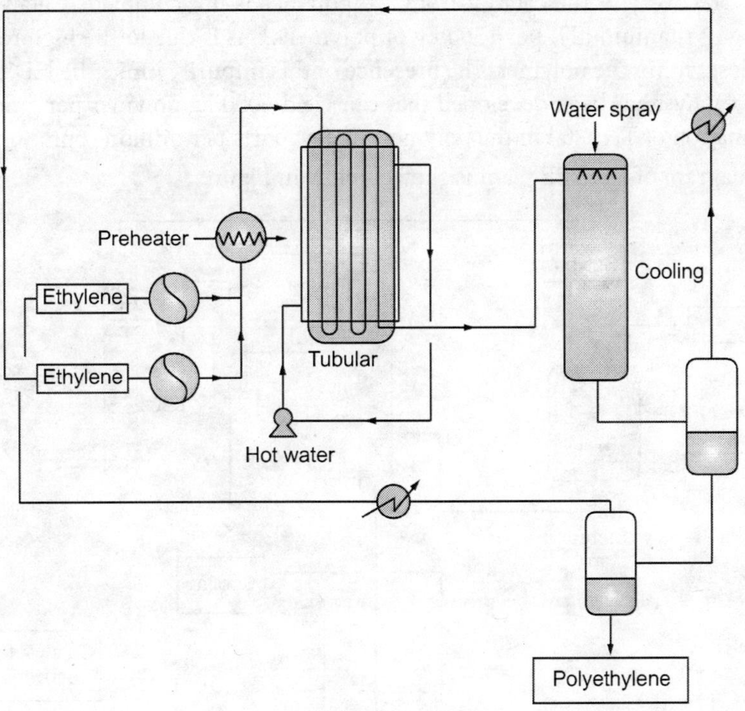

The temperature in the reactor is controlled above 200°C to avoid crystallisation of LDPE, which will damage the reactor tube. Conversion per pass is about 20%. Effluent from the reactor, consisting of the product and the unconverted monomer, is separated by high and low pressure separator vessels. Ethylene is recycled to the reactor. The overall conversion achieved by recycling is about 95-97%. Finally, the molten LDPE is withdrawn from the low pressure separator and extruded, followed by cooling, drying, and pilling. As tubular reactors are prone to plugging and poor heat transfer problems, a thick-walled, stirred tank vessel reactor is used. High speed agitation helps in good heat transfer. If the agitator stops, the temperature will rise to such an extent that explosion may takes place because of the presence of oxygen in the reaction mixture.

LDPE is used for making films, sheets, tubes, blocks, insulation, hoses etc. But, it cannot be used at temperatures above 80°C because deformation will occur. It has excellent dielectric properties, good elasticity up to –60°C, and it is anti-corrosive. It is soft and waxy and is used to make films.

Q.115. Describe the manufacturing process of High Density Polyethylene (HDPE).

Ans. HDPE is manufactured by the suspension polymerisation method. In this method, high purity ethylene is introduced into the reactor vessel in which a catalyst (Ziegler-Nutta catalyst, $TiCl_4$ in alkyl aluminium) is suspended in benzene at a pressure of 20-35 atm and at a temperature of 60-80°C. The ratio of alkyl aluminium and titanium chloride determines the size of the polymer. The greater the ratio, the greater the molecular weight of the polymer.

After the reaction, the polymer mixture is separated from ethylene and inerts in a flash drum. The polymer is water washed and filtered to recover the catalyst (water soluble) and reused. However, recovery of the catalyst from the polymer is not complete. Catalyst consumption is about 1 g of titanium (Ti) per 1500 kg of polymer. This is due to the high residual presence of the catalyst within the polymer. The presence of this impurity limits the HDPE's applications. A modern catalyst has been developed that can yield 6000 kg polymer per gram of Ti, where the contamination of Ti metal in the polymer is a few parts per million (ppm) only.

A flow diagram of a HDPE plant is shown below in Figure.

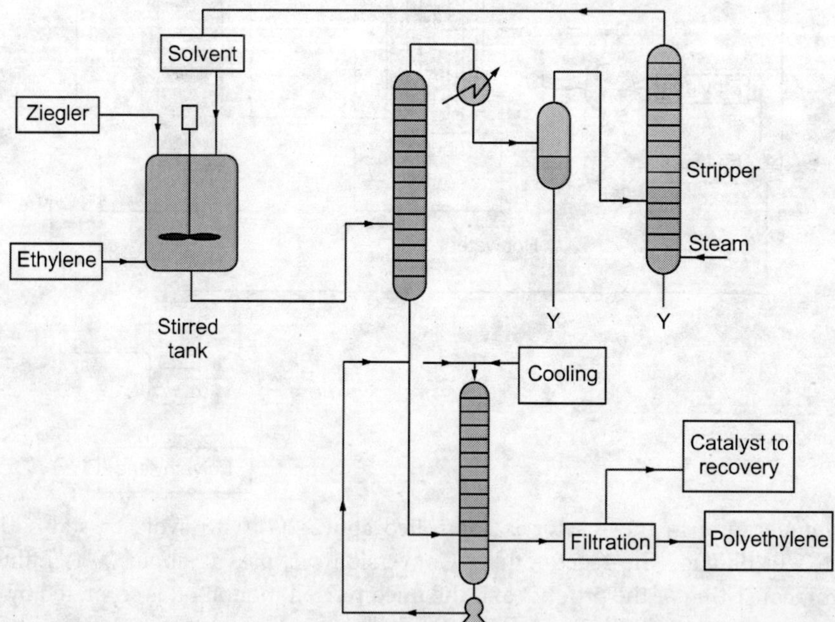

Fig. 34.1. Low pressure Ziegler process for high density polyethylene manufacture.

HDPE has a rigid and translucent property and is suitable for making electrical goods, bottles, ropes, etc. The main economic advantage of HDPE is that it can be manufactured at much lower pressure as compared to LDPE. HDPE has a higher melting temperature than LDPE and has almost the same glass transition temperature.

Q.116. Describe the production process of linear low density polyethylene(LLDPE).

Ans. LLDPE is a copolymer of ethylene and 1-butene having linear structure. It is produced by a low pressure fluidised bed process where the temperature and pressure are 100°C and 7-20 atm. respectively. Gaseous monomers are used to fluidise the polymer granules previously prepared. During polymerisation, additional polymers are produced of specific sizes depending on the seed polymers present. Unreacted monomers are separated from the effluent and recycled to the reactor. A long residence time in the range of 3–5 hours is required for the reaction. Such a process is shown below in Figure. HDPE is also produced by a similar fluidised method in modern plants.

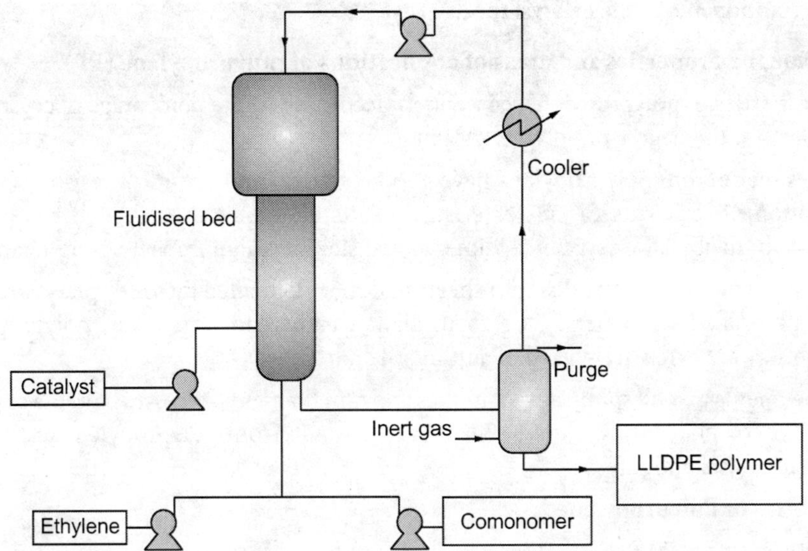

Fig. 34.2. A fluidised bed LLDPE manufacturing unit.

Q.117. Describe the manufacture, properties and applications of Ethylene Vinyl Acetate (EVA).

Ans. Ethylene Vinyl Acetate is manufactured from polymerisation of Ethylene and Co-monomer (Vinyl Acetate/Ethyl Acrylate/Acrylic Acid/Methyl Acrylate). EVA is the most popular ethylene copolymer.

Vinyl Acetate (VA) content in EVA varies from 9 to 40%.The increase in VA content improves the properties like flexibility, ESCR, compatibility, filler acceptability, X-likability adhesion. etc.

EVA due to its flexibility for various processing techniques like Foaming, Moulding, Coating, blending, extrusion makes it useful for broader application range. Resin properties which makes it unique and useful are: very high low temperature impact strength, good elasticity and resilience, high resistance to flex cracking and high ESCR, excellent clarity and gloss, good resistance to weathering etc..

Significant growth of EVA has been possible due to technological developments in shoe manufacturing technology. Apart from footwear (shoes, slippers, beach sandals, etc.) market other applications like film; HMA are also witnessing a healthy growth. Emerging sector "Solar Film" is highly technical and quality sensitive film produced from high VA and growing at a very healthy pace.

Q.118. How Polypropylene (PP) is produced?

Ans. After polyethylene, polypropylene is a valuable polymer and is used as plastic for making pipes, ropes, fibres, etc. It is manufactured by catalytic reaction in a stirred tank reactor, where Ti and aluminium halides are used as catalysts at a temperature of 60-70°C and a pressure of 1-2 MPa. An unreacted monomer is recycled after it is separated from the catalyst and polymer mixture in a flash chamber under vigorous stirring conditions. The mixture of polymer and catalyst is then passed to a centrifugal separator where a catalyst and polypropylene polymer is recovered. Further processing of the spent catalyst in the presence of alcohol is carried out to recover the active components of the catalyst for its reuse.

Q.119. Describe the properties and areas of applications of polypropylene (PP).

Ans. The properties of polypropylene vary widely according to the percentage of crystalline isotactic polymer and the degree of polymerization.

Articles made from polypropylene have good electrical and chemical resistance and low water absorption. Its other useful characteristics are its light weight (lowest thermoplastic polymer density), high abrasion resistance, dimensional stability, high impact strength, and no toxicity.

Polypropylene can be extruded into sheets and thermoformed by solid- phase pressure forming into thin walled containers. Due to its light weight and toughness, polypropylene and its copolymers are extensively used in automobile parts.

Improvements in melt spinning techniques and film filament processes have made polypropylene accessible for fiber production. Low cost fibers made from polypropylene are replacing those made from sisal and jute.

Properties of Polypropylene

Density, g/cm^3: 0.90-0.91

Fill temperature, max. °C: 130

Tensile strength, psi: 3,200–5,000

Water absorption, 24 hr., %: 0.01

Elongation, %: 3-700

Melting point, T_m °C: 176

Thermal expansion, 10-5 in./in. °C: 5.8-10

Specific volume, cm^3/lb: 30.4-30.8

Polypropylene today is one of the youngest members of the polymer which became one of the most invented inventions touching our lives in every possible way.

Main areas of applications of PP are in packaging, furniture, automobiles and domestic appliances sectors.

Q.120. Which are commonly used two polymerization methods for the manufacture of polyvinyl chloride? Mention some of the disadvantages of using PVC as a polymer.

Ans. Two methods for manufacture of PVC are Emulsion Polymerization and Suspension Polymerization. Disadvantage of PVC is that it has low solvent solubility. In addition to this, the heat stability of PVC is poor.

Q.121. Describe the manufacturing process, properties and uses of polyvinyl chloride.

Ans. PVC is a polymer of vinyl chloride, $nCH_2 = CHCl = (-CH_2 - CHCl - CH_2 -)_n$.

It is a thermoplastic, widely used for many applications, and is commonly produced by the emulsion polymerisation method. An aqueous emulsion is produced in a mixer with ammonium salts of fatty acids and a vinyl chloride monomer and the reaction is allowed to continue in a stirred tank reactor. A very low pH is maintained during the reaction. The product containing the polymer latex and unreacted monomer are separated in a gas–liquid separator. The separated vapour of the monomer is recycled after scrubbing with caustic solution to neutralise the acidic vapours. The polymer from the separator is then dried for making powder of polymer. A PVC manufacturing plant is shown below in Figure.

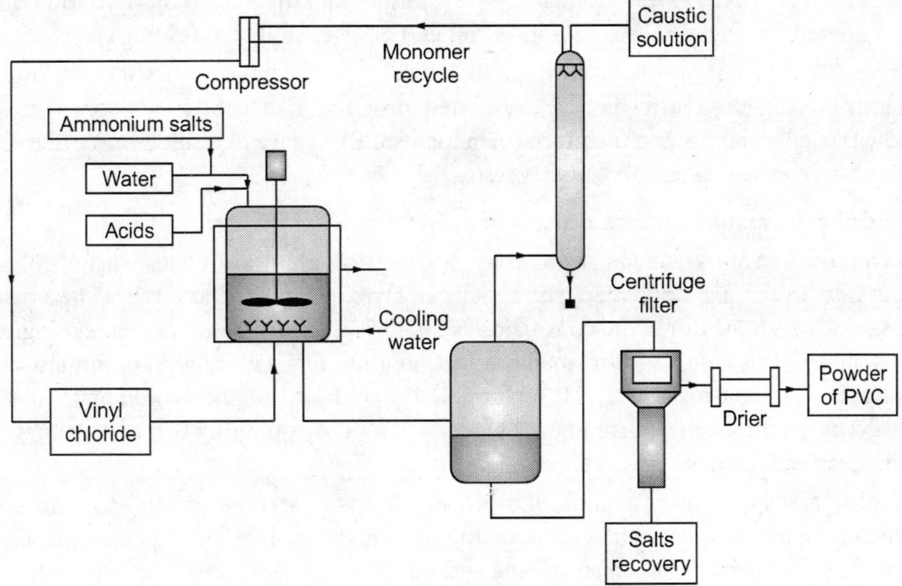

Fig. 34.3. A PVC manufacturing plant.

Two types of the homopolymer are available, the flexible and the rigid. Both types have excellent chemical and abrasion resistance. The flexible types are produced with high porosity to permit plasticizer absorption. Articles made from the rigid type are hard and cannot be stretched more than 40% of their original length. An important property of PVC is that it is self-extinguishing due to presence of the chlorine atom.

Flexible PVC grades account for approximately 50% of PVC production. They go into such items as table cloths, shower curtains, furniture, automobile upholstery, and wire and cable insulation.

Many additives are used with PVC polymers such as plasticizers, antioxidants, and impact modifiers. Heat stabilizers, which are particularly important with PVC resins, extend the useful life of the finished product.

Rigid PVC is used in many items such as pipes, fittings roofing, automobile parts, siding, and bottles.

Q.122. Describe the areas of application of PVC in India.

Ans. The world's most versatile plastic, PVC had a humble beginning by making golf balls and shoe heels. Soon after this discovery, PVC-based products such as raincoats, shower curtains and insulated wire hit the market. PVC is recognized as 'an infrastructure plastic' finding applications in pipes, conduits, ducts, wire and cables, flooring, windows and doors and roofing etc. It is also used in many other sectors including automobiles, medical and healthcare, packaging, sports and leisure.

Pipes are the biggest end use sector of PVC with a share of 70%, other sectors being calendered products, wire and cables, films and sheets, profiles, footwear and medical and healthcare.

Sector wise PVC Industry Consumption in India in decreasing order are: Pipe(for water supply), Pipe (for Irrigation), Pipe (Sewerage), Pipe (Plumbing), Pipe (Flexible), Wire and Cables, Films, Calendered Products, Sheets, Fittings, Profiles, Foot wear, Blow moulding etc.

Green Building concept is increasingly being accepted in the country. PVC is known to save energy during the entire life cycle. Besides; products like windows, doors, partition, wall, paneling, false ceiling, and furniture when made in PVC, save natural resource like wood which would help in saving fast depleting forests.

Q.123. Describe the manufacturing process of polystyrene.

Ans. Polymerisation of styrene can be done by mass, solution, emulsion, or suspension polymerisation methods. In the mass polymerisation method, styrene is heated at 80-85°C in a batch reactor under nitrogen pressure. The conversion is 30–40% in a long residence time, ranging from 40 to 50 hours. This is suitable for small production units. Since no catalyst or initiator is used, this polymer is suitable for electrical insulation. In the solution polymerisation method, styrene and a solvent are mixed in a stirred tank blender followed by pumping through a series of reactors with heating facilities.

In a batch suspension polymerisation process, drops of styrene are dispersed in water in the presence of a benzoyl peroxide initiator. In the emulsion polymerisation method, a stirred tank autoclave is employed, in which styrene and water are emulsified with an initiator. The product is then separated from unreacted monomers, which are then recycled.

Q.124. What are the areas of applications of different types of polystyrene.

Ans. There are three types of polystyrenes with different applications - General purpose Polystyrene (GPPS), High Impact Polystyrene (HIPS) and Expandable Polystyrene (EPS).

General Purpose polystyrene (GPPS) has excellent transparency, processability and its grades are used to cover a wide variety of applications like disposable containers, cutlery, CD storage cases, co-extrusion to obtain gloss layer, toys, stationery and refrigeration components.

High Impact polystyrenes (HIPS) resins are for a wide variety of applications both in injection and extrusion/thermoforming processes. Typical applications include Electronics, household appliances, disposable containers, toys and refrigeration components.

Expandable Polystyrene (EPS) also known as Polystyrene foam is prepared by suspension polymerization of styrene, and then adding foaming agent and white bead-like particles, with relative density of 1.05 are obtained. Thermal conductivity is low. EPS is polystyrene in a rigid foam honeycomb material with good insulation and vibration absorption, high compressive strength, very light weight and resistance to moisture. Its uses include building insulation and sound insulation, sidewall, and within the walls of the Cover, packaging materials and disposable containers.

Polystyrene (PS) end use is in India is mainly into the "Electronics/Appliances" sector, which accounts for almost 60 percent of total demand. PS demand in the India faces competition from other polymers, mainly polypropylene in both packaging and durables, primarily due to cost reasons.

Q.125. How are General purpose Polystyrene (GPPS) different from High impact Polystyrene (HIPS)?

Ans. GPPS is also known as crystal polystyrene. It has good clarity and fairly good mechanical properties. Although GPPS can be processed into most types of objects, it is relatively brittle. It is used in applications where light transmission properties are important. On the other hand, HIPS contains an elastomer like polybutadiene rubber, which improves its mechanical properties. It has appreciable impact strength and hence, is milky or opaque in appearance. HIPS is tougher than GPPS, and also has good processing properties.

Q.126. List out the important products made from five important commodity plastics.

Ans. Important products made from five important commodity polymers/plastics namely LD/LLDPE, HDPE,PP, PVC and PS are tabulated below.

Table 34.2. Polymers/plastics and corresponding products

LD/LLDPE	HDPE	PP	PVC	PS
Cast and stretch cling films	Raffia/ Monofilaments	Raffia/FIBC	Irrigation pipes and fittings	Thermocol
Mono/ Multi layer films	Crates	Fish nets	Wire and cables	Foam
Wires and Cables	Multi layer films	Buckets, Tiffin boxes	Floorings/Tiles	Refrigerator trays
Water storage tanks	Water Conveyance pipes	Automobile bumpers	Shower curtains	CD/DVD cases
House wares	Pesticide bottles	BOPP films	Blister films	Structural insulation
Carrier bags	Woven sacks	Washing machine tubs	Profiles and Builders' ware	Bottles
Furniture	Caps and closures	Leno bags	Footwear	Disposable cups/ cutleries
Milk Packaging	Pallets	Fibres and Filaments	IV bottles and blood bags	Smoke detectors
Detergent Packaging	Thin wall moldings	Syringes	Bag liners	Films

Q.127. Describe polyethylene terephthalate (PET) production processes.

Ans. Polyethylene terephthalate (PET) is produced by esterifying terephthalic acid (TPA) and ethylene glycol or, more commonly, by the trans-esterification of dimethyl terephthalate(DMT) and ethylene glycol (EG). This route is favored because the free acid is not soluble in many organic solvents. The reaction occurs in two stages. Methanol is released in the first stage at approximately 200°C with the formation of bis(2-hydroxyethyl) terephthalate. In the second stage, polycondensation occurs, and excess ethylene glycol is driven away at approximately 280°C and at lower pressures ($\approx$ 0.01 atm):

Using excess ethylene glycol is the usual practice because it drives the equilibrium to near completion and terminates the acid end groups. This results in a polymer with terminal —OH. When the free acid is used (esterification), the reaction is self catalyzed. However, an acid catalyst is used to compensate for the decrease in terephthalic acid as the esterification nears completion. In addition to the catalyst and terminator, other additives are used such as color improvers and dulling agents. For example, PET is delustred by the addition of titanium dioxide.

The molecular weight of the polymer is a function of the extent of polymerization and could be monitored through the melt viscosity. The final polymer may be directly extruded or transformed to chips, which are stored.

Batch polymerization is still used. However, most new processes use continuous polymerization and direct spinning.

An alternative route to PET is by the direct reaction of terephthalic acid (TPA) and ethylene oxide. The product bis(2-hydroxyethyl)terephthalate reacts in a second step with TPA to form a dimer and ethylene glycol, which is released under reduced pressure at approximately 300°C.

This process differs from the direct esterification and the transesterification routes in that only ethylene glycol is released. In the former two routes, water or methanol are coproduced and the excess glycol released.

PET is a condensation polymer made by continuous melt-phase polymerization process followed by Solid State Polymerization (SSP). In the past, it was manufactured either by

DMT or PTA route:

- DMT Route (Di Methyl Terepthalate) DMT. + MEG = PET Polymer
- PTA Route (Purified Terephthalic acid) PTA + MEG = PET Polymer

DMT is seldom used for PET now a day. It is manufactured using PTA to help achieve better quality and consistency.

Q.128. Describe the properties and applications of polyethylene terephthalate (PET).

Ans. Polyethylene Terephthalate (PET) is a thermo plastic polyester. PET, the material used for the manufacturing of the plastic beverage bottles is of high strength and high barrier material. Pepsi-cola was the first company to commercially introduce its beverages products in PET.

PET containers when properly manufactured, offer excellent clarity and gloss, toughness and impact resistance, chemical resistance, low gas permeation, good process ability and good dimensional stability.

PET is used for the production of bottles, sheet, strapping and injection moulded products. PET bottles are used for packaging wide range of consumer products such as all types of Beverages including alcoholic, Edible Oil, Pharmaceutical formulations, Agro chemicals, Personal and Home Care etc. PET is also used for non-bottle applications such as thin/thick thermoformable sheet, dual-oven able, crystallised PET (CPET) containers, injection moulded components, fastening straps etc.

Injection Moulding of PET is becoming more prevalent in today's market place. PET can be best injection moulded into parts with wall thickness of up to 4.0 mm.

Q.129. Why there is a growing concern for the recycling of PET?

Ans. The concern for Recycled PET is continuously growing in recent times. PET bottles, which form the major market of PET packaging resin are the major concern from the point of recycling. More than 90% of PET is consumed in food packaging with drinks /beverages forming almost 80%.Since drinks and beverages are consumed mostly in residential houses, restaurants, stadia, railway stations, airports, entertainment venues and other public places, the organized collection and recycling of post-consumer PET bottles is important.

The PET bottle is 100% recyclable and most eco-friendly in nature. The need of the hour is proper handling system at all stages. The end consumers should take care to remove the cap and label from the used PET bottle before disposing it off and also should not throw the empty bottles in a public place. In order to ease the handling at different stages, the PET bottle should be 'flattened' and baled before sending it to the intended area of usage.

Recycled PET is becoming alternative for PET packaging options today and is getting serious consideration by the packaging world. There is a growing awareness among consumers towards their environment obligation in individual capacity and a marked preference thus for recycled content packaging.

PET consumption in India is growing rapidly. PET consumption in India is driven by Beverages, Health Care, Personal Care, Agro Chemicals, Cooking oil, Confectioneries and Non- bottle segments

In domestic market, beverages mostly covered Mineral water, Fortified water, flavored water, Sparkling water, Juices, Coke and Pepsi Cola and other alcoholic drinks. In non- bottle segment e.g., sheets, straps, lamipack, punnets and blisters presently PET is consumed with a steady growth.

Q.130. Describe the production processes for polycarbonates.

Ans. Polycarbonates (PC) are condensation thermoplastics used mainly for special engineering purposes. These polymers are considered polyesters of carbonic acid. They are produced by the condensation of the sodium salt of bisphenol A with phosgene in the presence of an organic solvent. Sodium chloride is precipitated, and the solvent is removed by distillation.

Another method for producing polycarbonates is by an exchange reaction between bisphenol A or a similar bisphenol with diphenyl carbonate.

Diphenol carbonate is produced by the reaction of phosgene and phenol. A new approach to diphenyl carbonate and non-phosgene route is by the reaction of CO and methyl nitrite using Pd/alumina. Dimethyl carbonate is formed which is further reacted with phenol in presence of tetraphenox titanium catalyst. Decarbonylation in the liquid phase yields diphenyl carbonate.

Q.131. Discuss the properties and uses of polycarbonates.

Ans. Polycarbonates, known for their toughness in molded parts, typify the class of polymers known as engineering thermoplastics. These materials, designed to replace metals and glass in applications demanding strength and temperature resistance, offer advantages of light weight, low cost, and ease of fabrication.

Materials made from polycarbonates are transparent, strong, and heat and break resistant. However, these materials are subject to stress cracking and can be attacked by weak alkalies and acids.

Polycarbonates are used in a variety of articles such as laboratory safety shields, street light globes, and safety helmets. The maximum usage temperature for polycarbonate objects is 125°C.

Polycarbonate is a high performance engineering resin which demonstrated phenomenal growth during the 1990s averaging close to 20 percent per year largely due to its use in optical media applications - primarily the production of CDs and DVDs.

Polycarbonate finds its application in Sheets/film, Automotive (window and non window), optical media, Appliances, Housewares, Electronics/Electrical, medical, ophthalmic, construction, sports, recreational, packaging etc. As blends, it finds usage as auto body.

Q.132. Write a short note on manufacture, properties and applications of acrylonitrile butadiene styrene (ABS).

Ans. Acrylonitrile butadiene styrene (ABS) is a copolymer of acrylonitrile, buta diene, and styrene (ABS). In a typical batch suspension process styrene is suspended in water by agitation and use of a stabilizer. The polymer forms beads. The bead/water slurry is separated by centrifugation, dried, and blended with additives. It is derived from Acrylonitrile, butadiene, and styrene either as such as graft polymer or SAN(Styrene Acrilo Nitrile) on Polybutadiene or as a blend of ABS plus SAN. Composition of the constituent monomers can be varied in order to alter the properties of the final resin.

ABS resins are well-established, high volume, amorphous engineering thermoplastics, which offer an excellent balance of heat, chemical and impact resistance with superior processing versatility.

ABS has a specific gravity of 1.03 to 1.06 and a tensile strength in the range of 6 to 7.5×10^3 psi. These polymers are tough plastics with outstanding mechanical properties. A wide variety of ABS modifications are available with heat resistance comparable to or better than polysulfones and polycarbonates. Another outstanding property of ABS is its ability to be alloyed with other thermoplastics for improved properties. For example, ABS is alloyed with rigid PVC for a product with better flame resistance.

The advantage of ABS is that this material combines the strength and rigidity of Acrylonitrile and styrene polymers with the toughness of the Poly-Butadiene rubber. Resistance and Toughness are the most important mechanical properties of this product.

Among the major applications of ABS are extruded pipes and pipe fittings, appliances such as refrigerator door liners, and in molded automobile bodies.

ABS also finds its application in Buildings/Construction,Transportation, Electronics/Electrical, Appliances,Alloys and other application like medical,toys and recreational.

Q.133. Write a short note on styrene acrylonitrile (SAN).

Ans. Styrene and acrylonitrile monomers can be copolymerized to form SAN which is a random, amorphous copolymer that has improved weatherability, stress crack resistance, and barrier properties. The SAN copolymer generally contains 70 to 80% styrene and 30 to 20% Acrylonitrile. This combination provides higher strength, rigidity and chemical resistance than polystyrene, but it is not quite as clear as crystal polystyrene and its appearance tends to yellow more quickly.

SAN ($T_g = 105°C$) is stiffer and has better chemical and heat resistance than the homopolymer. However, it is not as clear as polystyrene, and it is used in articles that do not require optical clarity, such as appliances and houseware materials. SAN has a high clarity with High Heat resistance and good chemical resistance besides good dimensional stability and food contact acceptability.

SAN finds its application in Electrical/Electronic, Automotive Applications, General Purpose Containers, Household Goods, Cosmetics and Compounding with ABS, Acrylics, etc.

Q.134. Write a short note on manufacture, properties and application of polyurathanes.

Ans. Urathanes are thermosetting resins obtained by condensation polymerisation of polyol (diols or triols) and a diisocyanate. Examples of monomers are toluene diisocyanate [$(CH_3)(C_6H_4)(NCO)_2$] and dihydric alcohol or glycol [$(CH_2OH)_2$]. The polymerisation is carried out in a stirred tank autoclave reactor at a temperature of around 200°C under nitrogen pressure.

Polyurethane polymers are either rigid or flexible, depending on the type of the polyol used. For example, triols derived from glycerol and propylene oxide are used for producing block slab foams.

Other polyhydric compounds with higher functionality than glycerol (three-OH) are commonly used. Examples are sorbitol (six-OH) and sucrose (eight-OH). Triethanolamine, with three OH groups, is also used. Diisocyanates generally employed with polyols to produce polyurethanes are 2,4-and 2,6-toluene diisocyanates prepared from dinitro-toluenes.

A different diisocyanate used in polyurethane synthesis is methylene diisocyanate (MDI), which is prepared from aniline and formaldehyde. The diamine product is reacted with phosgene to get MDI.

Improved polyurethane can be produced by copolymerization. Block copolymers of polyurethanes connected with segments of isobutylenes exhibit high temperature properties, hydrolytic stability, and barrier characteristics. The hard segments of polyurethane block polymers consist of $-(RNHCOO)_n-$, where R usually contains an aromatic moiety.

The physical properties of polyurethanes vary with the ratio of the polyol to the diisocyanate. For example, tensile strength can be adjusted within a range of 1,200-600 psi; elongation, between 150-800%.

The major use of polyurethanes is to produce foam. The density as well as the mechanical strength of the rigid and the flexible types varies widely with polyol type and reaction conditions. For example, polyurethanes could have densities ranging between 1-6 lb/ft^3 for the flexible types and 1–50 lb/ft^3 for the rigid types. Polyurethane foams have good abrasion resistance, low thermal conductance and good load bearing characteristics. However, they have moderate resistance to organic solvents and are attacked by strong acids. Flame retardency of polyurethanes could be improved by using special additives, spraying a coating material such as magnesium oxychloride, or by grafting a halogen phosphorous moiety to the polyol. Trichloro butylene oxide is sometimes copolymerized with ethylene and propylene oxides to produce the polyol.

Major markets for polyurethanes are furniture, transportation, and building and construction. Other uses include carpet underlay, textural laminates and coatings, footwear, packaging, toys, and fibers.

The largest use for rigid polyurethane is in construction and industrial insulation due to its high insulating property.

Molded urethanes are used in items such as bumpers, steering wheels, instrument panels, and body panels. Elastomers from polyurethanes are characterized by toughness and resistance to oils, oxidation, and abrasion. They are produced using short chain polyols such as polytetramethylene glycol from 1,4-butanediol. Polyurethanes are also used to produce fibers. Spandex (trade name) is a copolymer of polyurethane (85%) and polyesters.

Polyurethane networks based on triisocyante and diisocyanate connected by segments consisting of polyisobutylene are rubbery and exhibit high temperature properties, hydrolytic stability and barrier characteristics.

Q.135. What are epoxy resin? Describe their properties and uses.

Ans. Epoxy resins are produced by reacting epichlorohydrin and a diphenol. Bisphenol A is the diphenol generally used. The reaction, a ring opening polymerization of the epoxide ring, is catalyzed with strong bases such as sodium hydroxide.

The linear polymer formed is cured by cross-linking either with an acid anhydride, which reacts with the —OH groups, or by an amine, which opens the terminal epoxide rings. Cresols and other bisphenols are also used for producing epoxy resins.

Epoxy resins have a wide range of molecular weights (1,000-10,000). Those with higher molecular weights, termed phenoxy, are hydrolyzed to transparent resins that do not have the epoxide groups. These could be used in molding purposes, or crosslinked by diisocyanates or by cyclic anhydrides.

Important properties of epoxy resins include their ability to adhere strongly to metal surfaces, their resistance to chemicals, and their high dimensional stability. They can withstand temperatures up to 500°C. Epoxy resins with improved stress cracking properties can be made by using toughening agents, such as carboxyl-terminated butadiene-acrylonitrile liquid polymers. The carboxyl group reacts with the terminal epoxy ring to form an ester. The ester, with its pendant hydroxyl groups, reacts with the remaining epoxide rings, then more crosslinking occurs by forming ether linkages. This material is tougher than epoxy resins and suitable for encapsulating electrical units.

Major uses of epoxy resins are coatings for appliance finishes, auto primers, adhesive, and in coatings for cans and drums. Interior coatings of drums used for chemicals and solvents manifests its chemical resistance.

Q.136. Write a short note on thermoplastic polyester.

Ans. Thermoplastic polyesters are among the large volume engineering thermoplastics produced by condensation polymerization of terephthalic acid with ethylene glycol or 1,4-butanediol. These materials are used to produce film for magnetic tapes due to their abrasion and chemical resistance, low water absorption, and low gas permeability. Polyethylene terephthalate (PET) is also used to make plastic bottles (approximately 25% of plastic bottles are made from PET). Similar to nylons, the most important use of PET is for producing synthetic fibers. Polybutylene terephthalate (PBT) is another thermoplastic polyester produced by the condensation reaction of terephthalic acid and 1,4-butanediol.

The polymer is either produced in a bulk or a solution process. It is among the fastest growing engineering thermoplastics, and leads the market of reinforced plastics.

Q.137. What are unsaturated polyester?

Ans. Unsaturated polyesters are a group of polymers and resins used in coatings or for castings with styrene. These polymers normally have maleic anhydride moiety or an unsaturated fatty acid to impart the required unsaturation. A typical example is the reaction between maleic anhydride and ethylene glycol.

Phthalic anhydride, a polyol, and an unsaturated fatty acid are usually copolymerized to unsaturated polyesters for coating purposes. Many other combinations in variable ratios are possible for preparing these resins.

Q.138. Describe the manufacturing process of phenol-formaldehyde resins.

Ans. Phenol (C_6H_5OH) and formaldehyde (HCHO) in the presence of an aqueous ammoniacal medium condense to a phenol-formaldehyde resin. The reaction is carried out in a kettle-type reactor at a temperature of 160°C and at atmospheric pressure. The reaction time is less than 1 hour. The product from the reactor is usually blended with hexamethylene triamine and other fillers to make finished resins ready for moulding processes. Depending on the ratio of phenol to formaldehyde, a variety of grades of this resin may be made. This is a thermosetting type resin and is also termed as phenolic resin.

The reaction between phenol and formaldehyde is either base or acid catalyzed, and the polymers are termed resols (for the base catalyzed) and novalacs (for the acid catalyzed). The condensation reaction between the methylolphenols with the elimination of water forms the polymer. Crosslinking occurs by a reaction between the methylol groups and results in the formation of ether bridges. It occurs also by the reaction of the methylol groups and the aromatic ring, which forms methylene bridges. The formed polymer is a three-dimensional network thermoset.

Usually, the ratio of formaldehyde to phenol is kept less than unity to produce a linear fusible polymer in the first stage. Crosslinking of the formed polymer can occur by adding more formaldehyde and a small amount of hexamethylene tetramine [hexamine, $(CH_2)_6N_4$]. Hexamine decomposes in the presence of traces of moisture to formaldehyde and ammonia. This results in crosslinking and formation of a thermoset resin.

Q.139. Discuss the properties and uses of phenolic resins.

Ans. Important properties of phenolic resins are their hardness, corrosion resistance, rigidity, and resistance to water hydrolysis. They are also less expensive than many other polymers.

Many additives are used with phenolic resins such as wood flour, oils, asbestos, and fiberglass. Fiberglass piping made with phenolic resins can operate at 150°C and pressure up to 10 kg/cm^2.

Molding applications dominate the market of phenolic resins. Articles produced by injection molding have outstanding heat resistance and dimensional stability. Compression molded glass-filled phenolic resins disk brake pistons are replacing the steel ones in many automobiles because of their light weight and corrosion resistance. Phenolic resins are also used in a variety of other applications such as adhesives, paints, laminates for building, automobile parts, and ion exchange resins.

Q.140. Describe the manufacturing process of Amino resins (Aminoplasts) such as urea formaldehyde and urea melamine.

Ans. Amino resins are condensation thermosetting polymers of formaldehyde with either urea or melamine. Melamine is a condensation product of three urea molecules. It is also prepared from cyanamide at high pressures and temperatures.

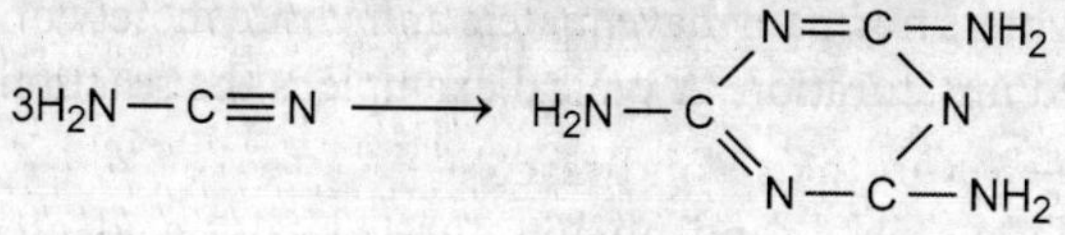

$$3H_2N{-}C{\equiv}N \longrightarrow \text{Melamine (ring: } H_2N{-}C,\ N{=}C{-}NH_2,\ N,\ N{-}C{-}NH_2)$$

Melamine

Urea-formaldehyde resin: The addition polymerization reaction of urea to formaldehyde produces mainly monomethylol urea and some dimethylol urea. When the mixture is heated in presence of an acid, condensation occurs, and water is released. This is accompanied by the formation of a crosslinked polymer:

Urea $(CONH_2)_2$ and formaldehyde (HCHO) readily react to polymerise into urea formaldehyde resin. Polymerisation is carried out in a glass-lined (or stainless steel) stirred tank reactor in an ammoniacal or weak alkaline medium. The polymerisation reaction takes place at room temperature and atmospheric pressure. This resin is a thermosetting resin.

Urea-Melamine Resins : A similar reaction occurs between melamine and formaldehyde and pro- duces methylolmelamines. A variety of methylols are possible due to the availability of six hydrogens in melamine. As with urea formaldehyde resins, polymerization occurs by a condensation reaction and the release of water.

Melamine formaldehyde: Melamine $[C_3N_3(NH_2)_3]$ and formaldehyde (HCHO) combine readily to form melamine-formaldehyde resin, which is also a thermosetting resin. The temperature of the reaction is 100°C in a neutral medium, i.e., at a pH of 7 in the aqueous medium at atmospheric pressure. Melamine formaldehyde resins are thermosets.

Q.141. Describe the properties and uses of amino resins (Aminoplasts).

Ans. Amino resins are characterized by being more clear and harder (tensile strength) than phenolic resins. However, their impact strength (breakability) and heat resistance are lower. Melamine resins have better heat and moisture resistance and better hardness than their urea analogs.

The most important use of amino resins is the production of adhesives for particleboard and hardwood plywood. Compression and injection molding are used with amino resins to produce articles such as radio cabinets, buttons, and cover plates. Because melamine resins have lower water absorption and better chemical and heat resistance than urea resins, they are used to produce dinnerware and laminates used to cover furniture. Almost all molded objects use fillers such as cellulose, asbestos, glass, wood flour, glass fiber and paper.

Q.142. What is poly methyl metha acrylate (PMMA) polymer?

Ans. This is a polymer of the methyl metha acrylate monomer, which is produced from acrylonitrile (CH_2CHCN) and sulfuric acid at a temperature below 90°C to yield methyl acrylate

(CH_2CONH_2) followed by reacting with methanol to convert methyl acrylate to methyl methacrylate $(CH_2CHCOOCH_3)$. This methyl metha acrylate monomer is then polymerised in a stirred tank autoclave under nitrogen pressure. Any of the solution, emulsion, mass or suspension polymerisation methods are used.

Q.143. What is polytetrafluoroethylene (PTEE) polymer? How it is prepared?

Ans. Polytetrafluoroethylene (PTFE) or *teflon* is a polymer of the tetrafluoroethylene (C_2F_4) monomer. This monomer is prepared from methanol chlorination followed by fluorination. The reactions are carried out in three reactors. Trichloromethane produced in the chlorinator is reacted with hydrofluoric acid in the first reactor at about 65°C in the presence of $SbCl_5$ as the catalyst and is converted to produce mono-chloro-difluoro- methane (CHF_2Cl) in the presence of $AlCl_3$ as the catalyst in the second reactor. In the third reactor, CHF_2Cl is then converted to C_2F_4 by catalytic pyrolysis at a temperature of 650 –800°C in the presence of platinum (Pt) as the catalyst. Polymerisation of C_2F_4 is then carried out in a batch reactor preferably by the suspension polymerisation method at a temperature of 200°C and at a pressure of 70 kg/cm^2. The reaction time is about 1 hour.

Q.144. What are synthetic rubber and natural rubber? Compare their properties.

Ans. Synthetic rubbers (elastomers) are long chain polymers with special chemical and physical as well as mechanical properties. These materials have chemical stability, high abrasion resistance, strength, and good dimensional stability. Many of these properties are imparted to the original polymer through crosslinking agents and additives. An important property of elastomeric materials is their ability to be stretched at least twice their original length and to return back to nearly their original length when released.

Synthetic rubbers include elastomers that could be crosslinked such as polybutadiene, polyisoprene, and ethylene-propylene-diene tere-polymer. It also includes thermoplastic elastomers that are not crosslinked and are adapted for special purposes such as automobile bumpers and wire and cable coatings. These materials could be scraped and reused. However, they cannot replace all traditional rubber since they do not have the wide temperature performance range of thermoset rubber.

Synthetic Rubber is broadly classified in two categories viz. General Purpose Rubber (GP) and Special Purpose Synthetic Rubber. GP includes Styrene Butadiene Rubber (SBR), Polybutadiene Rubber (PBR) and Polyisoprene Rubber (IR). Styrene Butadiene Rubber (SBR) is the largest tonnage synthetic rubber. Ethylene-Propylene Rubber (EPM and EDPM), Butyl Rubber (IIR) and Nitrile Rubber (NBR) are some of the other important types of synthetic rubbers.

Natural rubber is an elastomer constituted of isoprene units. These units are linked in a cis-1,4-configuration that gives natural rubber the outstanding properties of high resilience and strength. Natural rubber occurs as a latex (water emulsion) and is obtained from a tree. The latex can be vulcanized (crosslinked) by heating with sulfur or other agents. Vulcanization of rubber is a chemical reaction by which elastomer chains are linked together. The long chain molecules impart elasticity, and the crosslinks give load supporting strength.

The major use of rubber is for tire production. Non-tire consumption includes hoses, footwear, molded and extruded materials, and plasticizers.

Q.145. What are various synthetic rubbers produced from 1, 3-Butadiene (BD). List out properties and uses of different rubbers.

Ans. Butadiene is colourless easily liquefiable flammable gas used in the production of a wide variety of synthetic rubbers (SBR, BR, NBR etc.), Acrylonitrile-butadiene-styrene (ABS), polymer resins and few chemical intermediates.Chemical intermediates manufactured from butadiene are adiponitrile and chloroprene. Adiponitrile is used to make nylon fibres and polymers. Chloroprene is used to make polychloroprene (better known as Neoprene) which has uses such as wet suits, electrical insulation, car fan belts, gaskets, hoses, corrosion-resistant coatings and adhesives.

In India Butadiene (BD) is manufactured from C_4 hydrocarbons available from Naphtha and Gas Cracking units.

Table 34.3. Selected properties of some elastomers

Elastomers	Durometer hardness range	Tensile strength at room temp, psi	Elongation at room temp, %	Temp. range of service, °C	Weather resistance
Natural rubber	20-100	1,000-4,000	100-700	–55-80	Fair
Styrene-butadiene rubber (SBR)	40-100	1,000-3,500	100-700	–55-110	Fair
Polybutadiene	30-100	1,000-3,000	100-700	–60-100	Fair
Polyisoprene	20-100	1,000-4,000	100-750	–55-80	Fair
Polychloroprene	20-90	1,000-4,000	100-700	–55-100	Very good
Polyurethane	62-95 A 40-80 D	1,000-8,000	100-700	–70-120	Excellent
Polyisobutylene	30-100	1,000-3,000	100-700	–55-100	Very good

Uses of various rubbers:

Styrene butadiene rubber (**SBR**) is used mainly in automobile tyres and also in adhesives, sealants, coatings and rubber articles such as shoe soles.

Butadiene rubber (**BR**) is used mainly in tyres and also as an impact modifier of polymers such as high impact polystyrene (HIPS).

Nitrile rubbers are used mainly in the manufacture of hoses, gasket seals and fuel lines for the automobile industry as well as in gloves and footwear.

Acrylonitrile-butadiene-styrene (**ABS**) is used in business machines, computers, radios, televisions and telephone handsets and also in automobile industry.

Styrene-butadiene (**SB**) copolymer latex is used in paper coatings, carpet back coatings, foam mattresses and adhesives.

Q.146. Describe the manufacture, properties and applications of polybutadiene rubber (PBR).

Ans. Butadiene ($CH_2 = CH - CH = CH_2$) is obtained as the by-product from a Naphtha Cracker Unit (NCU). By catalytic (Ziegler–Nutta) polymerisation of butadiene, a product PBR that resembles natural rubber is obtained. A solvent is used to keep the catalyst and the monomer in the solution during the reaction.The product is separated from its unconverted monomer, which is recycled.The solvent is evaporated,stripped, and recycled. The polymer from the evaporator is then dried to get the final product(PBR).

Butadiene can be polymerized using free radical initiators or ionic or coordination catalysts. When butadiene is polymerized in emulsion using a free radical initiator such as cumene hydroperoxide, a random polymer is obtained with three isomeric configurations, the 1,4-addition configuration dominating.

Polymerization of butadiene using anionic initiators (alkyl lithium) in a nonpolar solvent produces a polymer with a high cis configuration. A high cis-polybutadiene is also obtained when coordination catalysts are used.

Manufacturing process of PBR from butadiene is shown below in figure.

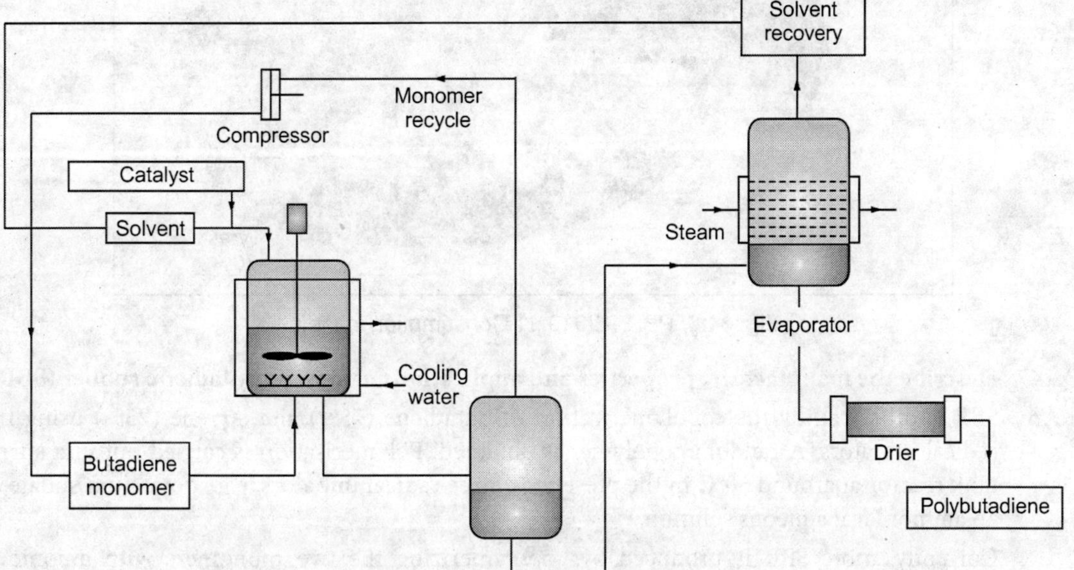

Fig. 34.4. FIGURE : Manufacturing process of polybutadiene rubber from butadiene.

cis-1,4-Polybutadiene rubber has high elasticity, low heat buildup, high abrasion resistance, and resistance to oxidation. However, it has a relatively low mechanical strength. This is improved by incorporating a cis- trans block copolymer or 1,2 (vinyl) block copolymer in the polybutadiene matrix. A small amount of natural rubber may be mixed with polybutadiene to improve its properties. trans 1,4-Polybuta- diene rubber has a higher glass transition temperature ($T_g = -14°C$) than the cis form ($T_g = -108°C$).The polymer has the toughness, resilience and abrasion resistance of natural rubber ($T_g = -14°C$).

PBR is used mainly in tyres and also as an impact modifier of polymers such as high impact polystyrene (HIPS).

Following figure shows the consumption pattern of PBR in India during 2010-2011.

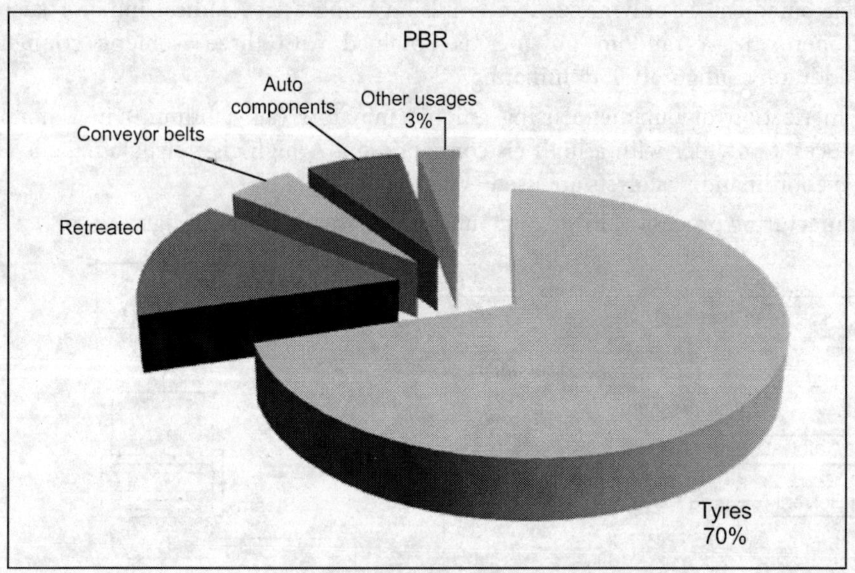

Fig. 34.5. PBR – 2010-11 Consumption in India

Q.147. Describe the manufacture, properties and applications of styrene butadiene rubber (SBR).

Ans. SBR is produced by the copolymerization of butadiene (75%) and styrene (25%) using free radical initiators. A random copolymer is obtained. Polymerisation is carried out in a stirred tank reactor at around 50°C in the presence of free radical initiators, e.g., potassium sulfate, in an ammoniacal aqueous solution.

Currently, more SBR is produced by copolymerizing the two monomers with anionic or coordination catalysts. The formed copolymer has better mechanical properties and a narrower molecular weight distribution. A random copolymer with ordered sequence can also be made in solution using butyl lithium, provided that the two monomers are charged slowly. Block copolymers of butadiene and styrene may be produced in solution using coordination or anionic catalysts. Butadiene polymerizes first until it is consumed, then styrene starts to polymerize. SBR produced by coordinaton catalysts has better tensile strength than that produced by free radical initiators.

Styrene butadiene rubber (SBR) is the most widely used synthetic rubber. The main use of SBR is for tire production. Other uses include footwear, coatings, carpet backing, and adhesives.

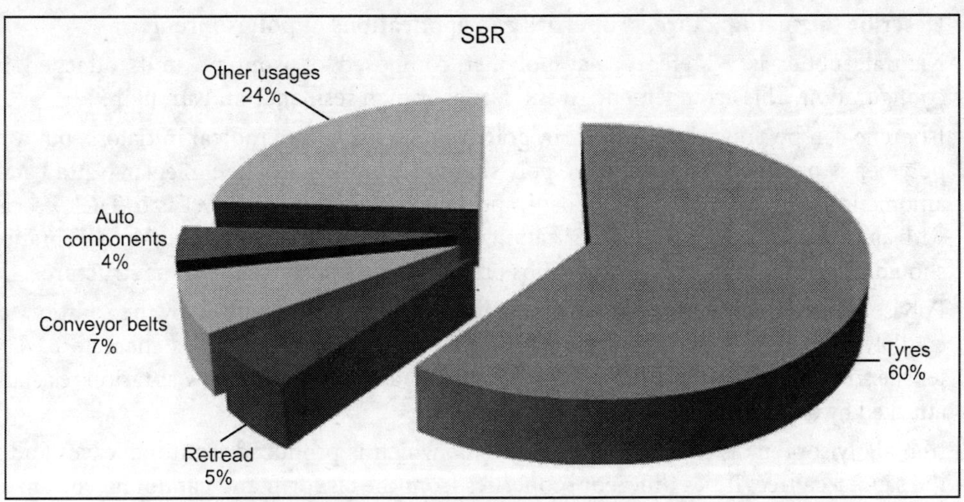

Fig. 34.6. Figure : SBR Consumption pattern in India during 2010-11.

Q.148. Describe the manufacture, properties and applications of nitrile rubber (NBR).

Ans. Nitrile rubber is a copolymer of butadiene and acrylonitrile. The copolymerization occurs in an aqueous emulsion. When free radicals are used, a random copolymer is obtained. Alternating copolymers are produced when a Zieglar-Natta catalyst is employed. Molecular weight can be controlled by adding modifiers and inhibitors. When the polymerization reaches approximately 65%, the reaction mixture is vacuum distilled in presence of steam to recover the monomer.

The ratio of acrylonitrile/butadiene can be adjusted to obtain a polymer with specific properties. Increasing the acrylonitrile ratio increases oil resistance of the rubber, but decreases its plasticizer compatibility. NBR has the special property of being resistant to hydrocarbon liquids.

NBR is produced in different grades depending on the end use of the polymer. Low acrylonitrile rubber is flexible at low temperatures and is generally used in gaskets, O-rings, and adhesives. The medium type is used in less flexible articles such as kitchen mats and shoe soles. High acrylonitrile polymers are more rigid and highly resistant to hydrocarbons and oils and are used in fuel tanks and hoses, hydraulic equipment, and gaskets.

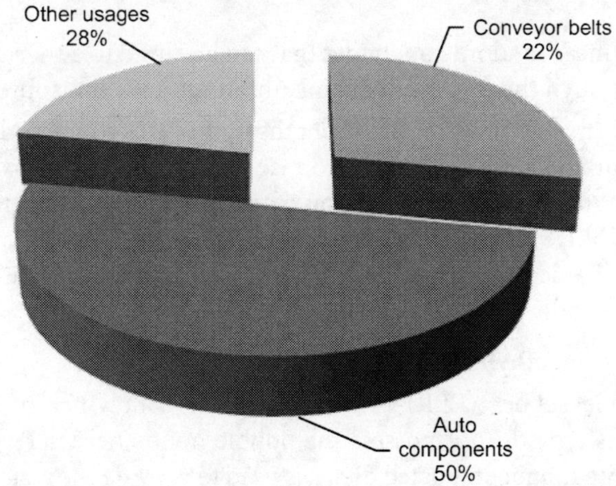

Fig. 34.7. NBR Consumption pattern in India during 2010-11.

Q.149. Describe the manufacture, properties and applications of polyisoprene.

Ans. Natural rubber is a stereoregular polymer composed of isoprene units attached in a cis configuration. This arrangement gives the rubber high resilience and strength.

Isoprene can be polymerized to form polyisoprene using free radical initiators, but a random polymer is obtained. Stereoregular polyisoprene is obtained when Zieglar-Natta catalysts or anionic initiators are used. The most important coordination catalyst is α-$TiCl_3$ co catalyzed with aluminum alkyls. The polymerization rate and cis content depends upon Al/Ti ratio, which should be greater than one. Lower ratios predominantly produce the trans structure.

Polyisoprene is a synthetic polymer (elastomer) that can be vulcanized by the addition of sulfur. cis-Polyisoprene has properties similar to that of natural rubber. It is characterized by high tensile strength and insensitivity to temperature changes, but it has low abrasion resistance. It is attacked by oxygen and hydrocarbons.

trans-Polyisoprene is similar to Gutta percha, which is produced from the leaves and bark of the sapotacea tree. It has different properties from the cis form and cannot be vulcanized. Few commercial uses are based on trans-polyisoprene.

Important uses of cis-polyisoprene include the production of tires, specialized mechanical products, conveyor belts, footwear, and insulation.

Q.150. Write a short note on polychloroprene (neoprene rubber).

Ans. Polychloroprene is the oldest synthetic rubber. It is produced by the polymerization of 2-chloro-1,3-butadiene in a water emulsion with potassium sulfate as a catalyst.

The product is a random polymer that is vulcanized with sulfur or with metal oxides (zinc oxide, magnesium oxide etc.). Vulcanization with sulfur is very slow, and an accelerator is usually required. Neoprene vulcanizates have a high tensile strength, excellent oil resistance (better than natural rubber), and heat resistance.

Neoprene rubber can be used for tire production, but it is expensive. Major uses include cable coatings, mechanical goods, gaskets, conveyor belts, and cables.

Q.151. Write a short note on butyl rubber.

Ans. Butyl rubber is a copolymer of isobutylene (97.5%) and isoprene (2.5%).The polymerization is carried out at low temperature (below –95°C) using $AlCl_3$ cocatalyzed with a small amount of water.

The product is a linear random copolymer that can be cured to a thermosetting polymer. This is made possible through the presence of some unsaturation from isoprene.

Butyl rubber vulcanizates have tensile strengths up to 140 kg/cm^2 and are characterized by low permeability to air and a high resistance to many chemicals and to oxidation. These properties make it a suitable rubber for the production of tire inner tubes and inner liners of tubeless tires. The major use of butyl rubber is for inner tubes. Other uses include wire and cable insulation, steam hoses, mechanical goods, and adhesives. Chlorinated butyl is a low molecular weight polymer used as an adhesive and a sealant.

Q.152. Write a short note on ethylene propylene rubber (EPR).

Ans. Ethylene-propylene rubber (EPR) is a stereoregular copolymer of ethylene and propylene. Elastomers of this type do not possess the double bonds necessary for crosslinking. A third monomer, usually a monoconjugated diene, is used to provide the residual double bonds needed for cross linking. 1,4-Hexadiene and ethylidene norbornene are examples of these dienes. The main polymer chain is completely saturated while the unsaturated part is pending from the main chain. The product elastomer, termed ethylene propylene terepolymer (EPT), can be crosslinked using sulfur. Crosslinking EPR is also possible without using a third component (a diene). This can be done with peroxides.

Important properties of vulcanized EPR and EPT include resistance to abrasion, oxidation, and heat and ozone; but they are susceptible to hydrocarbons attack.

The main use of ethylene propylene rubber is to produce automotive parts such as gaskets, mechanical goods, wire, and cable coating. It may also be used to produce tires.

Q.153. Write a short note on thermoplastic elastomers (TPE's).

Ans. Thermoplastic elastomers (TPE's), are plastic polymers with the physical properties of rubbers. They are soft, flexible, and possess the resilience needed of rubbers. However, they are processed like thermoplastics by extrusion and injection molding.

TPE's are more economical to produce than traditional thermoset materials because fewer steps are required to manufacture them than to manufacture and vulcanize thermoset rubber. An important property of these polymers is that they are recyclable.

Thermoplastic elastomers are multiphase composites, in which the phases are intimately depressed. In many cases, the phases are chemically bonded by block or graft copolymerization. At least one of the phases consists of a material that is hard at room temperature.

Important TPE's include blends of semicrystalline thermoplastic polyolefins such as propylene copolymers, with ethylene-propylene terepolymer elastomer. Block copolymers of styrene with other monomers such as butadiene, isoprene, and ethylene or ethylene/propylene are the most widely used TPE's. Styrene-butadiene-styrene (SBS) accounts for 70% of global styrene block copolymers (SBC). Polyurethane thermoplastic elastomers are relatively more expensive then other TPE's. However, they are noted for their flexibility, strength, toughness, and abrasion and chemical resistance. Blends of polyvinyl chloride with elastomers such as butyl are widely used in Japan.

Random block copolymers of polyesters (hard segments) and amorphous glycol soft segments, alloys of ethylene interpolymers, and chlorinated polyolefins are among the evolving thermoplastic elastomers.

Important properties of TPE's are their flexibility, softness, and resilience. However, compared to vulcanizable rubbers, they are inferior in resistance to deformation and solvents.

Important uses for TPE's include shoe soles, pressure sensitive adhesives, insulation, and recyclable bumpers.

Q.154. Write a short note on synthetic fibres.

Ans. Fibers are solid materials characterized by a high ratio of length to diameter. They can be manufactured from a natural origin such as silk, wool, and cotton, or derived from a natural fiber such as rayon. They may also be synthesized from certain monomers by polymerization (synthetic fibers). In general, polymers with high melting points, high crystallinity, and moderate thermal stability and tensile strengths are suitable for fiber production.

Man-made fibers include, in addition to synthetic fibers, those derived from cellulose (cotton, wood) but modified by chemical treatment such as rayon, cellophane, and cellulose acetate.

These are sometimes termed "regenerated cellulose fibers." Rayon and cellophane have shorter chains than the original cellulose due to degradation by alkaline treatment.

Major fiber-making polymers are those of polyesters, polyamides (nylons), polyacrylics, and polyolefins. Polyesters and polyamides are produced by step polymerization reactions, while polyacrylics and poly- olefins are synthesized by chain-addition polymerization.

Q.155. Describe the manufacture, properties and applications of polyester fibers.

Ans. Polyesters are the most important class of synthetic fibers. In general, polyesters are produced by an esterification reaction of a diol and a diacid. First polyester polymer was prepared by using terephthalic acid (an aromatic diacid) and ethylene glycol.

Polyesters can be produced by an esterification of a dicarboxylic acid and a diol, a transesterification of an ester of a dicarboxylic acid and a diol, or by the reaction between an acid dichloride and a diol.

Polyethylene terephthalate (PET) is produced by esterifying terephthalic acid (TPA) and ethylene glycol or, more commonly, by the trans- esterification of dimethyl terephthalate and ethylene glycol. This route is favored because the free acid is not soluble in many organic solvents. The reaction occurs in two stages. Methanol is released in the first stage at approximately 200°C with the formation of bis(2-hydroxyethyl) terephthalate. In the second stage, polycondensation occurs, and excess ethylene glycol is driven away at approximately 280°C and at lower pressures ($\approx$ 0.01 atm).

Batch polymerization is still used. However, most new processes use continuous polymerization and direct spinning.

An alternative route to PET is by the direct reaction of terephthalic acid and ethylene oxide. The product bis(2-hydroxyethyl) terephthalate reacts in a second step with TPA to form a dimer and ethylene glycol, which is released under reduced pressure at approximately 300°C.

This process differs from the direct esterification and the transesterification routes in that only ethylene glycol is released. In the former two routes, water or methanol are coproduced and the excess glycol released.

Polyethylene terephthalate is an important thermoplastic. However, most PET is consumed in the production of fibers. Polyester fibers contain crystalline as well as noncrystalline regions. The degree of crystallinity and molecular orientation are important in determining the tensile strength of the fiber (between 18–22 denier) and its shrinkage. The degree of crystallinity and molecular orientation can be determined by X-ray diffraction techniques.

Important properties of polyesters are the relatively high melting temperatures ($\approx$ 265°C), high resistance to weather conditions and sunlight, and moderate tensile strength.

Melt spinning polyesters is preferred to solution spinning because of its lower cost. Due to the hydrophobic nature of the fiber, sulfonated terephthalic acid may be used as a co monomer to provide anionic sites for cationic dyes. Small amounts of aliphatic diacids such as adipic acid may also be used to increase the dyeability of the fibers by disturbing the fiber's crystallinity.

Polyester fibers can be blended with natural fibers such as cotton and wool. The products have better qualities and are used for men's and women's wear, pillow cases, and bedspreads. Fiberfill, made from polyesters, is used in mattresses, pillows, and sleeping bags. High tenacity polymers for tire cord reinforcement are equivalent in strength to nylon tire cords and are superior because they do not "flat spot." V-belts and fire hoses made from industrial filaments are another market for polyesters.

Q.156. Write a short note on nylon resin.

Ans. Nylon resins are important engineering thermoplastics. Nylons are produced by a condensation reaction of amino acids, a diacid and a diammine, or by ring opening lactams such as caprolactam. The polymers, however, are more important for producing synthetic fibers.

Important properties of nylons are toughness, abrasion and wear resistance, chemical resistance, and ease of processing. Reinforced nylons have higher tensile and impact strengths and lower expansion coefficients than metals. They are replacing metals in many of their applications. Objects made from nylons vary from extruded films used for pharmaceutical packaging to bearings and bushings, to cable and wire insulation.

Q.157. Write a short note on polyamides (nylon fibers).

Ans. Polyamides/Nylons fibres are the second largest group of synthetic fibers after polyesters. Polymers of amine or polyamides are excellent for making fibres. These are well known as nylons.

Polyamides are produced by the reaction between a dicarboxylic acid and a diamine (e.g., nylon 66), ring openings of a lactam, (e.g., nylon 6) or by the polymerization of *w*-amino acids (e.g., nylon 11). Condensation polymerisation is carried out with the monomer/monomers in autoclaves or stirred tank batch reactors.

Numbers that follow the word "nylon" denote the number of carbons present within a repeating unit and whether one or two monomers are being used in polymer formation. For nylons using a single monomer such as nylon 6 or nylon 12, the numbers 6 and 12 denote the number of carbons in caprolactam and laurolactam, respectively. For nylons using two monomers such as nylon 610, the first number, 6, indicates the number of carbons in the hexamethylene diamine and the other number,10, is for the second monomer sebacic acid.

Q.158. Describe the properties and uses of nylons.

Ans. Nylons have relatively high melting points due to the presence of the amide linkage. They are also highly crystalline, and the degree of crystallinity depends upon factors such as the polymer structure and the distance between the amide linkages. An increase in polymer crystallinity increases its tensile strength, abrasion resistance, and modulus of elasticity.

Hydrogen bonding in polyamides is fairly strong and has a pronounced effect on the physical properties of the polymer such as the crystallinity, melting point, and water absorption. For example, nylon 6, with six carbons, has a melting point of 223°C, while it is only 190°C for nylon 11. This reflects the higher hydrogen bonding in nylon 6 than in nylon 11.

Moisture absorption of nylons differs according to the distance between the amide groups. For example, nylon 4 has a higher moisture absorption than most other nylons, and it is approximately similar to that of cotton. This is a result of the higher hydrophilic (water loving) character of nylon 4. Nylons, however, are to some extent subject to deterioration by light.

This has been explained on the basis of chain breaking and crosslinking. Nylons are liable to attack by mineral acids but are resistant to alkalies. They are difficult to ignite and are self-extinguishing.

In general, most nylons have remarkably similar properties, and the preference of using one nylon over the other is usually dictated by economic considerations except for specialized uses.

Melting points of various nylons and the monomer formula are tabulated below.

Monomer	Nylon Type	Monomer Formula	Approximate M.P. °C
Caprolactam	6	$CH_2-(CH_2)_4-C=O$ (ring closed through NH)	223
Adipic acid	66	$HOOC-(CH_2)_4COOH$	265
Hexamethylenediamine		$H_2N-(CH_2)_6-NH_2$	
Sebacic acid	610	$HOOC-(CH_2)_8COOH$	215
Hexamethylenediamine		$H_2N-(CH_2)_6-NH_2$	
11-Aminoundecanoic acid	11	$H_2N-(CH_2)_{10}-COOH$	190
Laurolactam	12	$CH_2-(CH_2)_{10}-C=O$ (ring closed through NH)	119
Acrylamide	3	$CH_2=CHCONH_2$	320
2-pyrrolidone	4	$CH_2-(CH_2)_2-C=O$ (ring closed through NH)	265
Valerolactam	5	$CH_2-(CH_2)_3-C=O$ (ring closed through NH)	260
7-aminoheptanoic acid or Ethyl 7-aminoheptanoate	7	$N_2H-(CH_2)_6-COOH$ $H_2H-(CH_2)_6-COOC_2H_5$	233
Capryllactam	8	$CH_2-(CH_2)_6-C=O$ (ring closed through NH)	209
9-aminopelargonic acid	9	$H_2N-(CH_2)_8-COOH$	209
10-aminodecanoic acid	10	$H_2N-(CH_2)_9-COOH$	188

Nylons have a variety of uses ranging from tire cord to carpet to hosiery. The most important application is cord followed by apparel. Nylon staple and filaments are extensively used in the carpet industry. Nylon fiber is also used for a variety of other articles such as seat belts, monofilament finishes, and knitwear. Because of its high tenacity and elasticity, it is a valuable fiber for ropes, parachutes, and underwear.

Q.159. Describe the manufacturing process of Nylon 66 (Poly hexamethylene adipate).

Ans. *Nylon 6,6* is produced by polymerising adipic acid and hexamethylene diamine in an autoclave in an aqueous solution at about 200°C and a pressure of 16 kg/cm^2 for a reaction time of 1-2 hour. The reaction of hexamethylene diamine and adipic acid produces hexamethylene diammonium adipate salt. The product is a dilute salt solution which is concentrated to approximately 60% and charged with acetic acid to a reactor where water is continuously removed. The presence of a small amount of acetic acid limits the degree of polymerization to the desired level.

The temperature is then increased to 270-300°C and the pressure to 17 atmospheres, which favors the formation of the polymer. The pressure is finally reduced to atmospheric to permit further water removal. After a total of three hours, nylon 66 is extruded under nitrogen pressure.

Q.160. Describe the manufacture of Nylon 6 (Polycaproamide).

Ans. Nylon 6 is produced by the polymerization of caprolactam. The monomer is first mixed with water, which opens the lactam ring and gives *w*-amino acid. The formed amino acid reacts with itself or with at approx- imately 250-280°C to form the polymer. The reaction time is about 8-12 hours.Temperature control is important, especially for depolymerization, which is directly proportional to reaction temperature and water content.

Q.161. Describe the production process of Nylon 12 (Poly laurylamide).

Ans. *Nylon 12* is a product of the polymerisation of laurolactam at a temperature of about 260°C in an autoclave. Nylon 12 is produced in a similar way to nylon 6 by the ring opening polymerization of laurolactam. The polymer has a lower water capacity than nylon 6 due to its higher hydrophobic (water hating) properties. The polymerization reaction is slower than for caprolactam. Higher temperatures are used to increase the rate of the reaction.

The monomer (laurolactam) can be produced from 1,5,9-cyclododecatriene, a trimer of butadiene. The trimer is epoxidized with peracetic acid or acetaldehyde peracetate and then hydrogenated. The saturated epoxide is rearranged to the ketone with MgI2 at 100°C. This is then changed to the oxime and rearranged to laurolactam.

Q.162. Describe the production process of Nylon 4 (Polybutyramide).

Ans. Nylon 4 is produced by ring opening 2-pyrrolidone. Anionic polymerization is used to polymerize the lactam. Co catalysts are used to increase the yield of the polymer. Carbon dioxide is an excellent polymerization activator.

Nylon 4 has a higher water absorption capacity than other nylons due to its lower hydrophobic property.

Q.163. Describe the production process of Nylon 11 (Poly undecanylamide).

Ans. Nylon 11 is produced by the condensation reaction of 11- amino undecanoic acid, which is obtained from castor oil, by polymerising in an aqueous medium at a temperature of 220°C in an autoclave. This is an example of the self condensation of an amino acid where only one monomer is used. The monomer is first suspended in water, then heated to melt the monomer and to start the reaction. Water is continuously removed. The polymer is finally withdrawn for storage.

Many other nylons could be produced such as nylon nylon 5, nylon 7, nylon 610, and nylon 612. These have properties generally similar to those nylons described earlier.

Q.164. Describe the manufacturing processes for the various monomers used for producing different types of of nylons.

Ans. Manufacturing processes for the various monomers used for producing different types of of nylons are described below.

Adipic acid: Commercial adipic acid is made from benzene. Benzene is catalytically hydrogenated to cyclohexane (C_6H_{12}) by Raney nickel as the catalyst at a temperature of 200-300°C under a hydrogen pressure of 3-4 MPa. The reaction takes place in a stirred tank reactor. Cyclohexane is then oxidised to a mixture of cyclohexanol ($C_6H_{11}OH$) and cyclohexanone ($C_6H_{10}O$) by cobalt naphthenate as the catalyst at a temperature of 145-175°C and at a pressure of 0.8-1 MPa in an oxygen atmosphere. Both the cyclohexanol and cyclohexanone are then further oxidised to adipic acid in the presence of copper and ammonium metavandate as the catalyst in the presence of nitric acid. The oxidation is carried out in a stirred tank reactor at a temperature of 60-80°C and atmospheric pressure. The reaction time is about 5-7 minutes. The product adipic acid [$(CH_2)_4(COOH)_2$] is then further purified by steam distillation.

***Hexamethylene diamine*:** This is obtained from adipic acid in two stages. In the first stage, adipic acid is converted to adiponitrile [$(CH_2)_4(CN)_2$] by reacting with gaseous ammonia in the presence of the catalyst, boron phosphate, and at a temperature of about 340°C in a fixed bed reactor. The product adiponitrile is then catalytically hydrogenated to hexamethylene diamine [$(CH_2)_6(NH_2)_2$] by a cobalt catalyst at a temperature of 100-135°C and at a hydrogen pressure of 60-65 MPa in a fixed bed reactor.

***Sebacic acid*:** This acid is commercially produced from castor oil saponification with a caustic solution at about 250°C. In the reactor, the castor oil (glyceride) is split into ricinoleic acid [$CH_3(CH_2)_{13}CHOH(CH)_2COOH$] and glycerine [$(CH_2OH)_2CHOH$]. Ricinoleic acid is separated from glycerine and further heated in a separate reactor to yield sebacic acid [$(CH_2)_8(COOH)_2$].

***11-Amino-undecanoic acid*:** This is also produced from castor oil. The oil is first esterified by methanol to yield methyl ester of recinolic acid, followed by pyrolysis at about 500°C. A mixture of products of heptyl alcohol, ester of undecylenic acid, heptaldehyde, etc., is obtained. Hydrolysis yields the undecylenic acid [$CH(CH_2)_9COOH$], which is then brominated by hydrobromic acid to bromo-undecanoic acid [$CHBr(CH_2)_{10}COOH$]. Treating bromo-undecanoic acid with ammonia yields 11-amino undecanoic acid [$NH_2(CH_2)_{10}COOH$].

***Caprolactam*:** Benzene is converted to cyclohexane followed by oxidation to cyclohexanol and cyclohexanone, as described in the production of adipic acid. Cyclohexanol from the mixture is further catalytically dehydrogenated to cyclo hexanone in the presence of zinc at a temperature of 400°C at 1 atm. The cyclo hexanone is then converted to cyclohexanone oxime [$C_6H_{10}NOH$] by reacting with hydroxylamine sulfate [$(NH_2OH)_2H_2SO_4$] at about 18-25°C. Cyclohexanone oxime is then further treated with sulfur trioxide in the presence of sulfuric acid to yield caprolactum [$(CH_2)_5(CO)(NH)$] by a reaction mechanism known as the Beckmann rearrangement. The temperature of this reaction is maintained at 140-150°C.

***Laurolactam or dodecyl lactam*:** Butadiene is first catalytically trimerised in a liquid phase reaction by the Ziegler catalyst (TiCl4) in a stirred tank reactor. The product stream containing the desired trimer, 1,5,9-cyclo-dodecatriene, is separated from the by-products polybutadiene, cyclo-octadiene, and vinyl cyclohexene by distillation. This trimer is then catalytically hydrogenated using a nickel catalyst at 200°C and a hydrogen pressure of 1.4 MPa. The reaction is carried out in the liquid phase in a stirred tank autoclave. The hydrogenated product is cyclododecane [$(CH_2)_{12}$]. Cyclododecane is then oxidised in the liquid phase at about 150°C in jacketed stirred tank reactors in the presence of boric acid as the catalyst. The oxidised product stream, containing cyclododecanol [$(CH_2)_{11}OH$] and cyclododecanone [$(CH_2)_{10}O$], is separated by distillation. Cyclododecanol is further dehydrogenated to yield cyclodo decanone in a separate reactor at 200°C in the liquid phase and in the presence of a copper catalyst supported on alumina. Finally, cyclododecanone is oximated by hydroxylamine hydrogen sulfate [$(NH_2OH)_2H_2SO_4$] at a temperature of 100°C and converted to cyclododecanone oxime [$(CH_2)_{10}NOH$]. Finally, cyclododecanone oxime is converted to laurolactam [$(CH_2)_{11}(CO)(NH)$] ammonium sulfate and concentrated sulfuric acid. The Beckmann rearrangement reaction takes place during the formation of laurolactam.

Q.165. What are acrylic and modacrylic fibers? Describe the properties and uses of polyacrylic fibres.

Ans. Modacrylic fibers are copolymers containing between 35–85% acrylonitrile. Acrylic fibers contain at least 85% acrylonitrile. Orlon is an acrylic fiber developed by DuPont in 1949; Dynel is a modacrylic fiber developed by Union Carbide in 1951.

Polyacrylic fibres (acrylic and modacrylic fibers) are produced by copolymerizing acrylonitrile with other monomers such as vinyl acetate, vinyl chloride, and acrylamide. Solution polymerization may be used where water is the solvent in the presence of a redox catalyst. Free radical or anionic initiators may also be used. The produced polymer is insoluble in water and precipitates.

Copolymers of acrylonitrile are sensitive to heat, and melt spinning is not used. Solution spinning (wet or dry) is the preferred process where a polar solvent such as dimethyl formamide is used. In dry spinning the solvent is evaporated and recovered.

Dynel, a modacrylic fiber, is produced by copolymerizing vinyl chloride with acrylonitrile. Solution spinning is also used where the polymer is dissolved in a solvent such as acetone. After the solvent is evaporated, the fibers are washed and subjected to stretching, which extends the fiber 4-10 times of the original length.

Important properties of acrylics are resistance to solvents and sunlight, resistance to creasing, and quick drying. Acrylic fiber breaking strength ranges between 1500 and 2650 kg/cm^2 and they have a water absorption of approximately 5%. Dynel, due to the presence of chlorine, is less flammable than many other synthetic fibers.

Major uses of acrylic fibers are woven and knitted clothing fabrics (for apparel), carpets, and upholstery. Acrylic fibers are characterized by having properties similar to wool and have replaced wool in many uses such as blankets, carpets, and sweaters.

Q.166. Describe the important characteristics and uses of polypropylene fibers.

Ans. Polypropylene fibers represent a small percent of the total polypropylene production. (Most polypropylene is used as a thermoplastic.) The fibers are usually manufactured from isotactic polypropylene.

Important characteristics of polypropylene are high abrasion resistance, strength, low static buildup, and resistance to chemicals. Crystallinity of fiber grade polypropylene is moderate and when heated, it starts to soften at 145°C and then melts at 170°C. Melt spinning is normally used to produce the fibers.

Polypropylene fibers are used for face pile of needle felt, tufted carpets, upholstery fabrics, etc.

Q.167. Describe the merits and demerits of emulsion polymerization w.r.t. other polymerization processes.

Ans. One of the major advantages of emulsion polymerization is that it has good heat control over the entire polymerization process. The heat control occurs by transfer to the aqueous phase, and this enables little change in the viscosity of the emulsion medium. Secondly, in this process, the reaction mass is emulsified with soap in a continuous aqueous medium. Soap has a stabilizing action, and hence, the emulsion obtained is stable and no agitation is necessary. Due to this, extra cost incurred in employing agitators is avoided. Moreover, Emulsion Polymerization overcomes the difficulty of low degree of polymerization, as encountered in Solution polymerization and Bulk Polymerization. Since the degree of polymerization directly depends upon the droplet sizes, and in this process, spherical monomer particles (1 – 10 μ) are yielded ; hence, the degree of polymerization is sufficiently higher than in the other two processes. However, Emulsion Polymerization has certain disadvantages. It blocks the emulsifiers used in the process. In addition, the process has poor capacity and low electrical resistance, which sometimes create major problems.

Q.168. What are the additives used in polymer products?

Ans. Polymer additives include materials which make and modify polymers. These include surfactants, chain transfer agents, modifiers, plasticizers, stabilizers, initiators and cross linking agents.

Q.169. What is glass transition temperature?

Ans. Glass Transition Temperature of a non crystalline material is the temperature at which the material changes its behaviour from its glassy state to a rubbery state. Glassy state refers to the state at which a material is hard and brittle; whereas, rubbery state refers to the state at which the material is flexible and elastic.

Q.170. Mention the advantages of blow molding as compared to injection molding.

Ans. Some of the advantages of blow molding as compared to injection molding are-

- Possibility of variable wall thickness.
- Low stresses and favorable cost factor.

Q.171. In molding process, how can we prevent the polymer material sticking to the wall of the mold (container)?

Ans. The polymeric material can be prevented from sticking to the walls of the mold by changing the viscosity of the polymer. This can be achieved by selecting the MOC of the container (mold) in such a way that it will create substantial viscosity gradient between the walls and the polymeric material. It can also be done by heating the walls of the mold externally.

Q.172. What are the various commercial uses of formaldehyde?

Ans. Formaldehyde is commercially available as aqueous solution with concentration ranging from 30-56 wt.% HCHO. It is also sold in solid form as paraformaldehyde or trioxane. Some major intermediates derived from formaldehyde are chelating agents, acetal resins, 1,4-butanediol, polyols, methylene diisocynate. It is also used for the manufacture of wide variety of chemicals, including sealant, herbicides, fertilisers, coating, and pharmaceutical. Product profile of formaldehyde is tabulated below.

Product	Uses
Formaldehyde	Thermosetting resin: Phenol, Urea Melamine, Formaldehyde resins Hexamethylenetetramine, Plastic and pharmaceuticals 1,4-Butadiol Methylene diisocyanate Fertiliser, Disinfectant, Biocide Preservative, Reducing agent, Corrosion inhibitor Polyaceta resin p-formaldehyde Pentaerythritol (Explosive-PETN), Alkyl resins

Q.172. Describe the industrial manufacturing process of formaldehyde.

Ans. The main industrial route for producing formaldehyde (HCHO) is the catalyzed air oxidation of methanol (CH_3OH).

$$CH_3OH\,(g) + 1/2O_2\,(g) \longrightarrow H-\overset{\overset{\displaystyle O}{\|}}{C}-H\,(g) + H_2O\,(g) \qquad \Delta H^\circ = -148.0 \text{ KJ/mol}$$

A silver-gauze catalyst is still used in some older processes that operate at a relatively higher temperature (about 500°C). Newtion is exothermic and occurs at 400-425°C and atmospheric pressure. Excess air is used to keep the methanol air ratio below the explosion limits. Various industrial processes for manufacture of formaldehyde using processes use an iron-molybdenum oxide catalyst. Chromium or cobalt oxides are sometimes used to dope the catalyst.

Process diagram for manufacture of formaldehyde using silver and iron- molybdenum catalyst is shown below in figure.

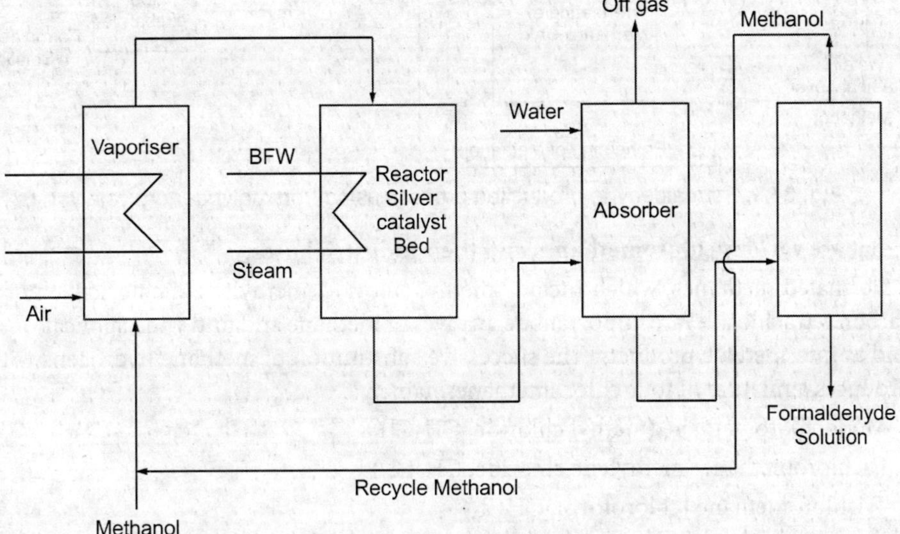

Fig. 34.8. Formaldehyde production process using silver catalyst

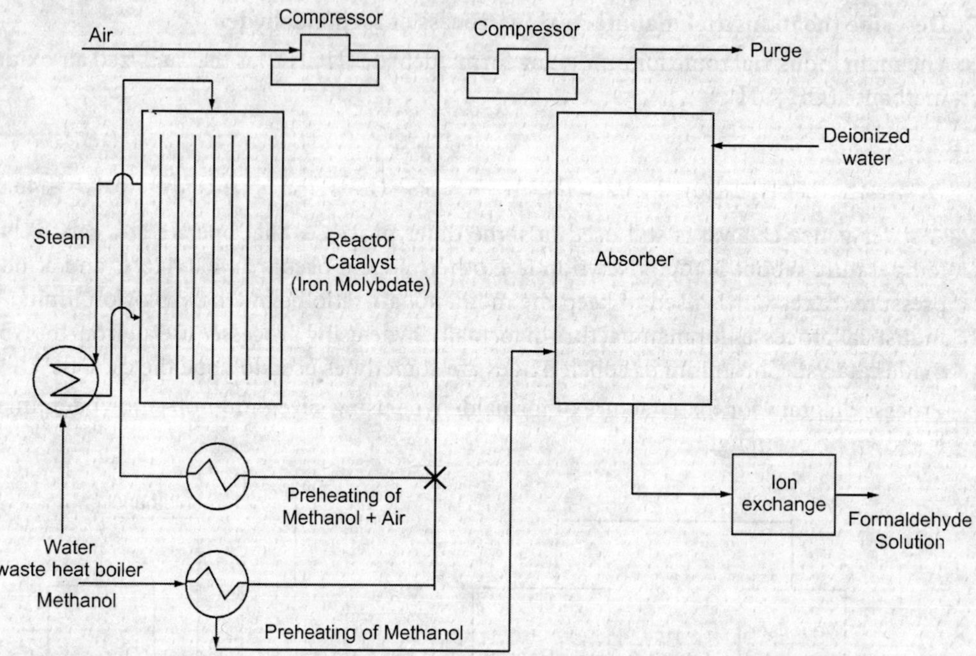

Fig. 34.9. Formaldehyde production process using Iron molybdenum catalyst

Q.173. What are various chloromethanes and their industrial uses.

Ans. Chlorinated methanes, which include methyl chloride, methylene dichloride, chloroform and carbon tetrachloride, are important derivatives of methane and find wide application as solvents and as intermediate products. The successive substitution of methane hydrogens with chlorine produces a mixture of four chloromethanes namely:

- Monochloromethane (methyl chloride, CH_3Cl)
- Dichloromethane (methylene chloride, CH_2Cl_2)
- Trichloromethane (chloroform, $CHCl_3$)
- Tetrachloromethane (carbon tetrachloride, CCl_4)

Each of these four compounds has many industrial applications as given below :

The major use of methyl chloride is to produce silicon polymers. Other uses include the synthesis of tetramethyl lead as a gasoline octane booster, a methylating agent in methyl cellulose production, a solvent, and a refrigerant.

Methylene chloride's major use is as a paint remover. It is also used as a degreasing solvent, a blowing agent for polyurethane foams, and a solvent for cellulose acetate.

Chloroform is mainly used to produce chloro difluoro methane (Fluoro-carbon 22) by the reaction with hydrogen fluoride. This compound is used as a refrigerant and as an aerosol propellent. It is also used to synthesize tetrafluoroethylene, which is polymerized to a heat resistant polymer Polytetrafluoroethylene(PTFE) which is commercially known as Teflon).

Carbon tetrachloride is used to produce chlorofluorocarbons by the reaction with hydrogen fluoride using an antimony pentachloride ($SbCl_5$) catalyst. The formed mixture is composed of trichlorofluoromethane (Freon-11) and dichlorodifluoromethane (Freon-12). These compounds are used as aerosols and as refrigerants. Due to the depleting effect of chlorofluorocarbons (CFCs) on the ozone layer, the production of these compounds has been reduced.

Q.174. What are the various feedstocks and routes of ethylene production. What are various products made from ethylene?

Ans. Ethylene, the key petrochemical building block is produced through multiple routes using different feedstocks like Naphtha, Natural Gas (Ethane), Propane, Butane, etc. However, Naphtha and Ethane are the most commonly used feedstocks, accounting for 82% of global ethylene production. Today, 50% of the ethylene produced in the world is through Naphtha cracking while one-third is produced through the Ethane route.

Ethylene is one of the most versatile petrochemicals and its production has steadily increased over the years. Ethylene is called as king of chemicals and surpasses all organic chemicals in production and in amount sold. Ethylene is the basic building block for petrochemicals.

Because of its ready availability at low cost and high purity and reactivity, ethylene has become one of the important raw materials for large number of petrochemicals and products. Ethylene has replaced the earliest route of production of vinyl chloride, acetaldehyde, vinyl acetate and other chemicals through acetylene route. Large tonnage of ethylene is being used for the manufacture of polyethylene, ethylene oxide, ethylene glycol and styrene.

Product profile of ethylene is given below in Table.

Table 34.4. Product profile of ethylene

Product	Uses
Polyethylene LDPE, LLDPE, HDPE	Films, moldings, pipes, cable covering, netting
Ethylene oxide and Ethylene glycol	Antifreeze, polyester, solvents, detergent, textile, ethanol amine
Styrene	Synthetic rubber and polystyrene
Ethyl alcohol	Industrial solvent and Chemical intermediates
Acetaldehyde (from ethyl alcohol)	Acetic acid (Peracetic acid), Acetic anhydride, Cellulose acetate, Vinyl acetate, Pyridine, Butyraldehyde (ethyl hexanol)
Olefin	n-Butenes, Synthetic detergent, Oxo alcohols, Synthetic lubricants
Chlorinated Solvents	Trichloroethylene, Perchloroethylene
Ethyl Chloride	Tetraethyl lead, Chemical intermediates
Vinyl Acetate	Polyvinyl acetate, Polyvinyl alcohol
Vinyl Chloride	Polyvinyl chloride(PVC)

Over 60% of global Ethylene production is used for manufacturing Polyethylenes (PE), 13% for Ethylene oxide (EO), 11% for production of ethylene dichloride (EDC), and the residual Ethylene goes in producing a wide array of petrochemicals.

Q.175. What are the major chemicals produced from ethylene?

Ans. Major chemicals produced from ethylene is shown below in figure.

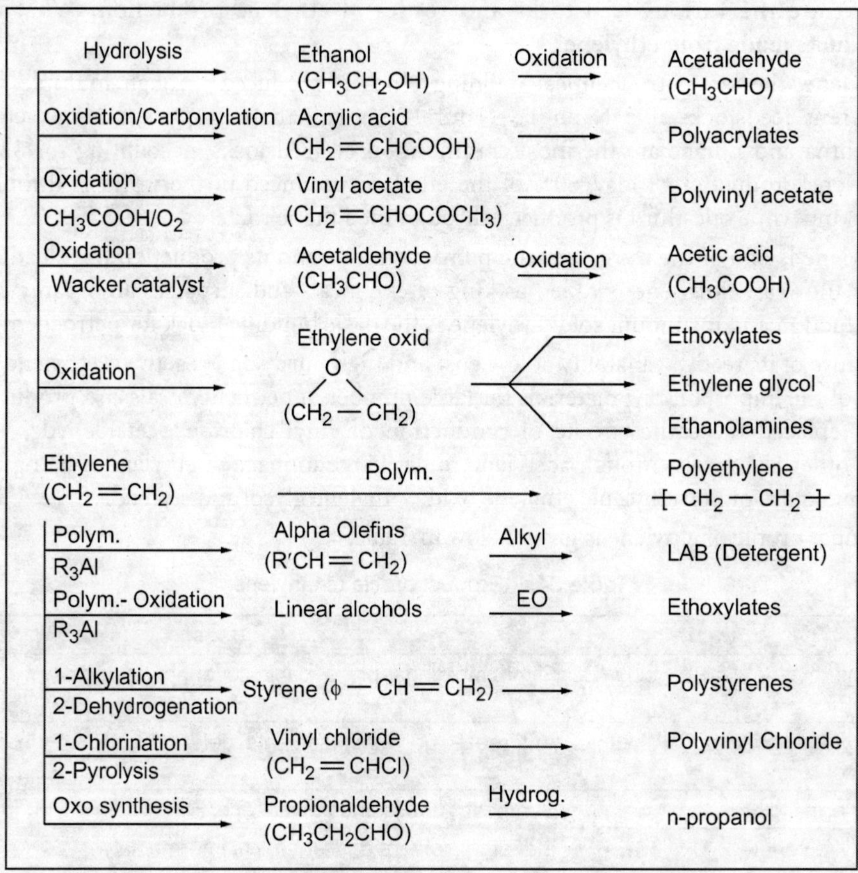

Fig. 34.10. Major chemicals produced from ethylene

Q.176. How is (CH_2-CH_2) produced? (O bridging the two CH_2 groups)

Ans. Ethylene oxide (EO) is a colorless gas that liquefies when cooled below 12°C. It is highly soluble in water and in organic solvents.The main route to ethylene oxide is oxygen or air oxidation of ethylene over a silver catalyst. The reactionis exothermic, so heat control is important.

A concomitant reaction is the complete oxidation of ethylene to carbon dioxide and water:

$$CH_2 == CH_2 + 3\,O_2 \longrightarrow 2\,CO_2 + 2\,H_2O \qquad (\Delta H = -1{,}421\ KJ/mol))$$

This reaction is highly exothermic; the excessive temperature increase reduces ethylene oxide yield and causes catalyst deterioration. Over oxidation can be minimized by using modifiers such as organic chlorides. It seems that silver is a unique epoxidation catalyst for ethylene.

In the process, compressed oxygen, ethylene, and recycled gas are fed to a multitubular reactor. The temperature of oxidation is controlled by boiling water in the shell side of the reactor. Effluent gases are cooled and passed to the scrubber where ethylene oxide is absorbed as a dilute aqueous solution. Unreacted gases are recycled. Epoxidation reaction occurs at 200–300°C with a short residence time of one second. A selectivity of 70–75% can be reached for the oxygen based process. Selectivity is the ratio of moles of ethylene oxide produced per mole of ethylene reacted. Ethylene oxide selectivity can be improved when the reaction temperature is lowered and the conversion of ethylene is decreased (higher recycle of unreacted gases).

Q.177. What are the properties and uses of ethylene oxide?

Ans. Ethylene oxide is a highly active intermediate. It reacts with all compounds that have a labile hydrogen such as water, alcohols, organic acids, and amines. The epoxide ring opens, and a new compound with a hydroxyethyl group is produced. The addition of a hydroxyethyl group increases the water solubility of the resulting compound. Further reaction of ethylene oxide produces polyethylene oxide derivatives with increased water solubility.

Ethylene oxide is a precursor for many chemicals of great commercial importance, including ethylene glycols, ethanol amines, and alcohol ethoxylates. Ethylene glycol is one of the monomers for polyesters, the most widely used synthetic fiber polymers.

Ethylene oxide is one of the important petrochemical intermediate used for the manufacture of large number of products; some of the major uses are in the manufacture monoethylene glycol, glycol ethers which is made by reaction of ethylne oxide and alcohols, ethanol amines. Surfactant industry is one of largest user of EO, both for industrial and house hold applications. Product profile of ethylene oxide is tabulated below.

Table 34.5. Product Profile of Ethylene Oxide(EO)

Product	Uses
Ethanol Amines	Detergent, Soap solvent, Cosmetics, Morpholine
Mono Ethylene Glycol	Polyester, Staple fiber yarn, PET Bottles, Film
Diethylene Glycol	Coolants, Pesticides, Rubber Compounding, Plasticizer, Polyurethane, Alkyl Resin
Tri ethylene Glycol	Natural gas conditioning agents plasticizer
Polyethylene Glycol	Pharmaceuticals; Brake Fluid, cosmetics
Non-ionic surfactants, Ethoxylates	Textile auxiliaries, Binders, Dyes, Pesticides, Pharmaceuticals, Cosmetics
Glycol Ethers	Brake Fluid and Protective Coating

Q.178. Describe the manufacturing process of Mono-, Di- and Tri-Ethylene Glycols (MEG, DEG, TEG).

Ans. The main route for producing ethylene glycol is the hydration of ethylene oxide in presence of dilute sulfuric acid.

$$\overset{O}{CH_2-CH_2} + H_2O \xrightarrow{H^+} HO-CH_2-CH_2OH$$

The hydrolysis reaction occurs at about 200 °C.Contact time is about 45 minutes. Di- and triethylene glycols are coproducts with the monoglycol. Increasing the water/ethylene oxide ratio and decreasing the contact time decreases the formation of higher glycols. A water/

ethylene oxide ratio of 10 is normally used to get approximately 90% yield of the monoglycol. However, the di- and the triglycols are not an economic burden, because of their commercial uses. Ethylene oxide is preheated to 195°C. EO:H_2O ratio is maintained at 1:10 to maximize MEG production. By Products are DEG and TEG.

Chemical reactions involved are as follows :

$$C_2H_4O + H_2O \longrightarrow HO{-}CH_2{-}CH_2{-}OH \quad \text{(MEG)}$$

$$\underset{\text{MEG}}{HO{-}CH_2{-}CH_2{-}OH} + C_2H_4O \longrightarrow \underset{\text{DEG}}{HO(C_2H_4O)_2H}$$

$$\underset{\text{DEG}}{HO(C_2H_4O)_2H} + C_2H_4O \longrightarrow \underset{\text{TEG}}{HO(C_2H_4O)_3H}$$

Figure below gives detail manufacturing of MEG, DEG and TEG from ethylene oxide.

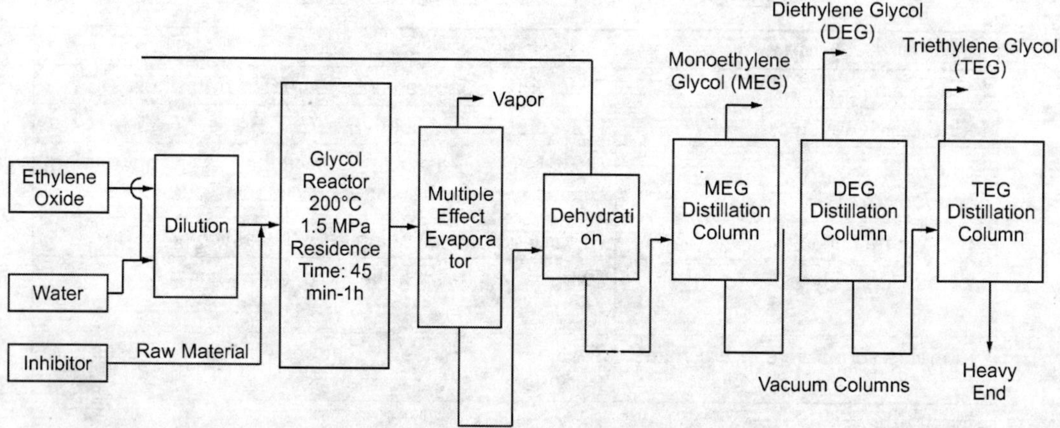

Fig. 34.11. MEG, DEG and TEG from ethylene oxide

A new route to ethylene glycol from ethylene oxide via the intermediate formation of ethylene carbonate has recently been developed by Texaco. Ethylene carbonate may be formed by the reaction of carbon monoxide, ethylene oxide, and oxygen. Alternatively, it could be obtained by the reaction of phosgene and methanol.

Ethylene carbonate is a reactive chemical. It reacts smoothly with methanol and produces ethylene glycol in addition to dimethyl carbonate. The reaction occurs at approximately 80-130°C using the proper catalyst. Many catalysts have been tried for this reaction, and the best catalyst types are those of the tertiary amine and quaternary ammonium functionalized resins. This route produces ethylene glycol of a high purity and avoids selectivity problems associated with the hydrolysis of ethylene oxide.

The coproduct dimethyl carbonate is a liquid soluble in organic solvents. It is used as a specialty solvent, a methylating agent in organic synthesis, and a monomer for polycarbonate resins. It may also be used as a gasoline additive due to its high oxygen content and its high octane rating.

Q.179. Write a short note on properties and uses of ethylene glycols.

Ans. Ethylene glycol (EG) is colorless syrupy liquid, and is very soluble in water. The boiling and the freezing points of ethylene glycol are 197.2° and –13.2°C, respectively.

Ethylene Glycol (CH_2OHCH_2OH) which finds major application in manufacture of polyester and as antifreeze compound account for about 70% of Ethylene oxide production.

Most of EG is used for producing polyethylene terephthalate (PET) resins (for fiber, film, bottles), antifreeze, and other products. Approximately 50% of the world EG is consumed in the manufacture of polyester fibers and another 25% for the antifreeze.

Q.180. Describe the manufacturing process and uses of ethylene dichloride (EDC).

Ans. The direct addition of chlorine to ethylene produces ethylene dichloride (1,2-dichloroethane). Ethylene dichloride is the main precursor for vinyl chloride, which is an important monomer for polyvinyl chloride plastics and resins.

Other uses of ethylene dichloride include its formulation with tetraethyl and tetramethyl lead solutions as a lead scavenger, as a degreasing agent, and as an intermediate in the synthesis of many ethylene derivatives.

The reaction of ethylene with hydrogen chloride, on the other hand, produces ethyl chloride. This compound is a small volume chemical with diversified uses (alkylating agent, refrigerant, solvent etc).

Ethylene reacts also with hypochlorous acid, yielding ethylene chlorohydrin:

$$CH_2 == CH_2 + HOCl \longrightarrow ClCH_2CH_2OH$$

Ethylene chlorohydrin via this route was previously used for producing ethylene oxide through an epoxidation step. Currently, the catalytic oxy- chlorination route is an alternative for producing ethylene glycol where ethylene chlorohydrin is an intermediate. In organic synthesis, ethylene chlorohydrin is a useful agent for introducing the ethyl hydroxy group. It is also used as a solvent for cellulose acetate.

Q.181. Describe the manufacturing process of vinyl chloride ($CH_2 = CHCl$).

Ans. Vinyl chloride monomer (VCM) was originally produced by the reaction of hydrochloric acid and acetylene in the presence of $HgCl_2$ catalyst at 100-180°C. The reaction is straight forward and proceeds with high conversion (96% on acetylene).

However, ethylene as a cheap raw material from cracker plant has replaced acetylene for obtaining vinyl chloride. The production of vinyl chloride via ethylene is a three-step process. The first step is the direct chlorination of ethylene to produce ethylene dichloride. The exothermic reaction occurs at approximately 4 atmospheres and 50°C in the presence of $FeCl_3$, $CuCl_2$ or $SbCl_3$ catalysts. Ethylene bromide may also be used as a catalyst.

The second step is the dehydrochlorination of ethylene dichloride (EDC) to vinyl chloride and HCl. The pyrolysis reaction occurs at approximately 500°C and 25 atmospheres in the presence of pumice on charcoal.

The third step, the oxychlorination of ethylene, uses by-product HCl from the previous step to produce more ethylene dichloride.

$$CH_2{=}CH_2 + 2HCl + 1/2O_2 \longrightarrow ClCH_2{-}CH_2Cl + H_2O$$

Ethylene dichloride from this step is combined with that produced from the chlorination of ethylene and introduced to the pyrolysis furnace.

The reaction conditions are approximately 225°C and 2-4 atmospheres. In practice the three steps, direct chlorination, oxychlorination, and dehydrochlorination, are integrated in one process so that no chlorine is lost.

Direct chlorination of ethylene: The process of vinyl chloride manufacture takes place in two stages.

- **First stage:** Ethylene is reacted with chlorine in either liquid or vapor phase in presence of ferric chloride. However, the liquid phase process is more common and the reaction takes place at around 50°C and 4 atm pressure.
- **Second stage:** Vinyl chloride is produced by pyrolysis of vaporised ethylene dichloride in a set of tubular furnaces at temperature of about 500 ^{0}C.

$$CH_2{=}CH_2 + Cl_2 = ClCH_2{-}CH_2Cl$$

Oxychlorination: The original process of manufacture of vinyl chloride by ethylene chlorination and cracking of ethylene dichloride had been replaced by oxychlorination process in which no hydrochloric acid is formed as byproduct. The process involves production of ethylene dichloride by exothermic reaction of ethylene, hydrochloric acid and oxygen.

Liquid phase: at about 170-180°C in at 15-20 atm pressure in presence of copper chloride in either fixed or fluidised bed reactor.

Vapor phase reaction: The temperature and pressure are 200-220°C and 20-50 atm pressure.

- Direct chlorination:

$$CH_2{=}CH_2 + Cl_2 \longrightarrow ClCH_2{-}CH_2Cl$$

- Oxychlorination:

$$CH_2{=}CH_2 + 2\,HCl + \tfrac{1}{2}\,O_2 \longrightarrow ClCH_2{-}CH_2Cl + H_2O$$

- Ethylene dichloride pyrolysis:

$$ClCH_2{-}CH_2Cl \longrightarrow CH_2{=}CHCl + HCl$$

- Overall reaction:

$$2\,CH_2{=}CH_2 + Cl_2 + \tfrac{1}{2}\,O_2 \longrightarrow 2\,CH_2{=}CHCl + H_2O$$

Process diagram of vinyl chloride production from oxychlorination process is shown below in Figure.

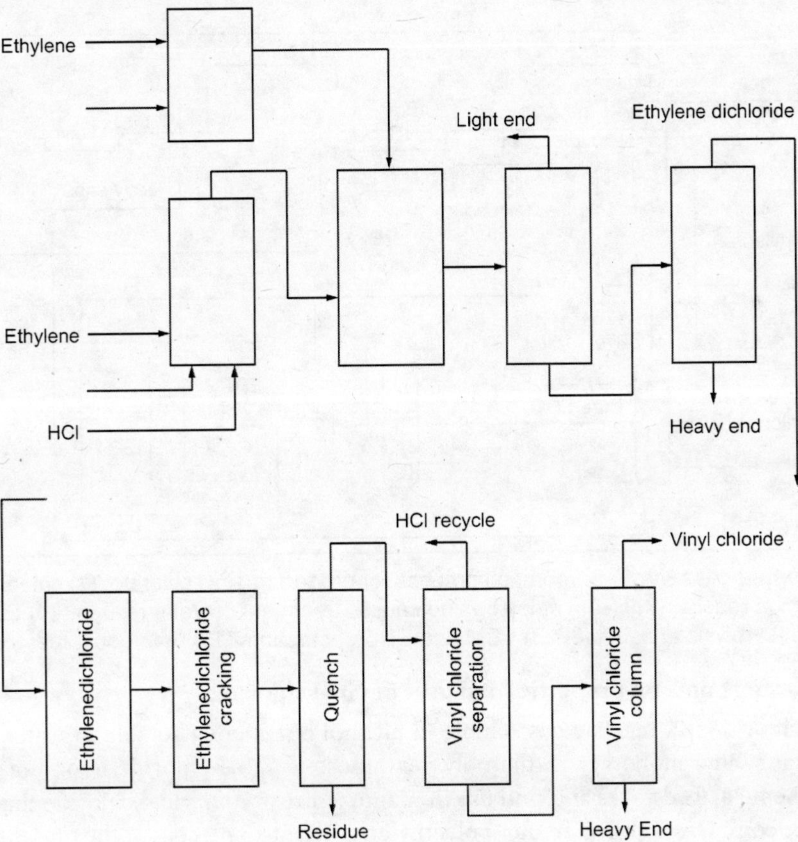

Fig. 34.12. Vinyl chloride production from oxychlorination process

Figure below shows the European Vinyls Corporation process for producing vinyl chloride.

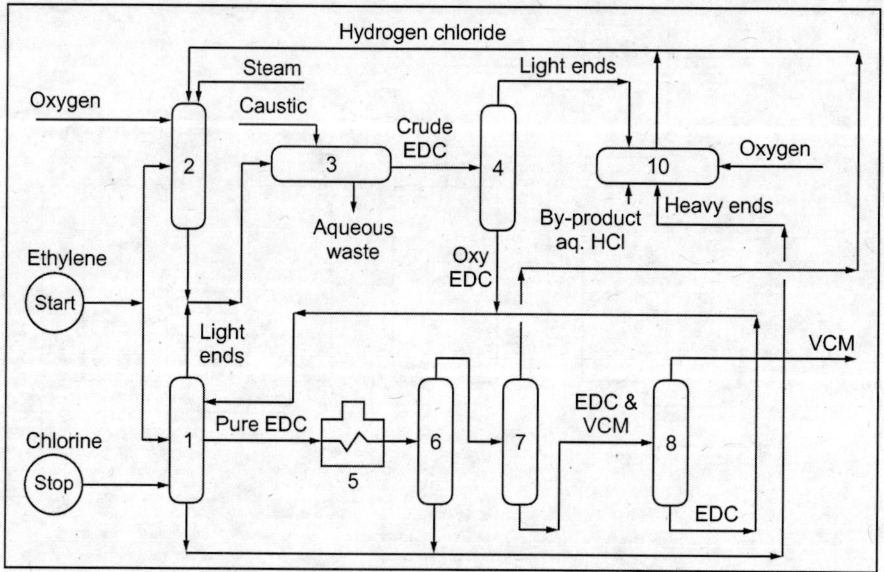

Fig. 34.13. The European Vinyls Corporation process for producing vinyl chloride: (1) chlorination section, (2) oxychlorination reactor, (3) steam stripping and caustic treatment of water effluent, (4) EDC distillation, (5) pyrolysis furnace, (6,7,8) VCM and EDC separation, (10) by-product reactor.

Q182. Write a short note on properties and uses of vinyl chloride.

Ans. Vinyl chloride is a reactive gas soluble in alcohol but slightly soluble in water. It is the most important vinyl monomer in the polymer industry. Vinyl chloride is one of the important petrochemical feedstocks and find use in manufacture of poly vinyl chloride the second largest tonnage commercial polymer after polyethylene. About 95 percent of the present vinyl chloride production worldwide is used in polymer production or copolymer application. Another important use of vinyl chloride is in the production of vinylidiene chloride.

Q.183. Write a short note on manufacture and uses of vinyl acetate.

Ans. The production of vinyl acetate through traditional acetylene route, which was developed by Wacker in 1930, involves reaction of acetylene and acetic acid in liquid phase at 60-80°C and 1-2 atm pressure in presence of mercury salt catalyst. The ethylene route has now replaced the traditional process of manufacture of vinyl acetate.

Vinyl acetate from Ethylene route: Vinyl acetate is made by reaction of ethylene with acetic acid by liquid phase process or by vapor phase process in presence of palladium and cupric chloride catalyst.

Vinyl acetate is one of the important derivatives of ethylene which is used as intermediate for manufacture of polyvinyl alcohol, polyvinyl acetate, polyvinyl butyral, etc.

Use of vinyl acetate based products has been tabulated below.

Vinyl Acetate	Polyvinyl Acetate	Surface coating adhesives, Textile resins
	Polyvinyl Alcohol	Textile size, Grease proofing paper, Vinyl emulsifier, Thickener, Viscosity regulators, Adhesives
	Acrylo- nitrile Copolymer	Acrylic Fibers
	Polyvinyl Formate	Water resistant insulation enamel
	Ethylene Vinyl Acetate Copolymers	Textile and Paper Coating
	Vinyl chloride comonomers	VC-VAC, LP Records, VC-VAC Coating
	Polyvinyl Butryaldehyde	Safety Glass

Global end use pattern of vinyl acetate is:- Adhesives (23%), paints and coating (29%), textiles (21%), plastics (17%), paper and board (10%). End use pattern of vinyl acetate in India is :- Adhesives (35-40%), textiles (30-35%), paints and coating (15-20%), others (10-15%).

Consumption pattern of vinyl acetate in India is polyvinyl acetate emulsions and resins (50%), polyvinyl alcohol (25%), ethylene vinyl acetate (10%), others (15%).

Q.184. Describe the manufacturing processes, properties and uses of acetaldehyde (CH_3CHO).

Ans. There are many ways to produce acetaldehyde. Historically, it was produced either by the silver-catalyzed oxidation or by the chromium activated copper-catalyzed dehydrogenation of ethanol. Currently, acetaldehyde is obtained from ethylene by using a homogeneous catalyst (Wacker catalyst). The catalyst allows the reaction to occur at much lower temperatures (typically 130°C) than those used for the oxidation or the dehydrogenation of ethanol (approximately 500°C for the oxidation and 250°C for the dehydrogenation).

Ethylene oxidation is carried out through oxidation-reduction (redox). The overall reaction is the oxidation of ethylene by oxygen as represented by:

$$CH_2{=}CH_2 + \frac{1}{2}O_2 \longrightarrow CH_3-\overset{\overset{\displaystyle O}{\|}}{C}-H \qquad \Delta H = -218.6 \text{ KJ/mol}$$

The Wacker process uses an aqueous solution of palladium(II) chloride, copper(II) chloride catalyst system.

In the course of the reaction, the Pd^{2+} ions are reduced to Pd metal, and ethylene is oxidized to acetaldehyde:

$$CH_2{=}CH_2 + PdCl_2 + H_2O \longrightarrow CH_3CHO + 2HCl + Pd^{\circ}$$

The oxidation reaction may be carried out in a single-stage or a two- stage process. In the single-stage, ethylene, oxygen, and recycled gas are fed into a vertical reactor containing the catalyst solution. Heat is controlled by boiling off some of the water. The reaction conditions are approximately 130°C and 3 atmospheres. In the two-stage process, the reaction occurs under relatively higher pressure (approximately 8 atmospheres) to ensure higher ethylene conversion. The reaction temperature is approximately 130°C. The catalyst solution is then withdrawn from the reactor to a tube-oxidizer to effect the oxidation of the catalyst at approximately 10 atmospheres. The yield of acetaldehyde from either process is about 95%. By-products from this reaction include acetic acid, ethyl chloride, chloro acetaldehyde, and carbon dioxide.

Acetaldehyde is a colorless liquid with a pungent odor. It is a reactive compound with no direct use except for the synthesis of other compounds. For example, it is oxidized to acetic acid and acetic anhydride. It is a reactant in the production of 2-ethyl hexanol for the synthesis of plasticizers and also in the production of pentaerithritol, a polyhydric compound used in alkyd resins. Acetaldehyde is an intermediate for many chemicals such as acetic acid, n-butanol, pentaerithritol, and polyacetaldehyde.

Products made from acetaldehyde and their uses are tabulated below.

Table 34.6. Product profile of acetaldehyde

Product	Uses
Pyridine, Picoline	Solvent, Drugs, Dyes, Agricultural chemicals
Chlor-aldehydes	Insecticides, Fungicides, Disinfectants
Acetaldol	1,3-Butylene Glycol (Polyesters),Urethane coating, Humcetant, Printing ink, Crotonaldehyde, n-butyl alcohol, n-butyric acid anhydride, 2-ethyl hexanol,Rubber accelerator, Sorbic acid
Paraldehyde	Rubber accelerator, Antioxidant dyestuff
Per-acetic acid	Epoxidation reaction, Reagent in caprolactam, Synthetic glycerols
Penta-erythritol	Alkyl resin, Stabilizer, Plasticizers, Chlorinated polyether resin, Intumescents
Acetic, anhydride	Acetyl salicylic acid, Cellulose acetate, Esters
Acetic acid	Cellulose acetate, Vinyl acetate Chloro-acetic acid Ammonium acetate
Lactic acid	Food and, beverages, lactates, adhesives, Leather processing

Q.185. Describe the various manufacturing processes of propylene.

Ans. Propylene is second to ethylene as the largest volume hydrocarbon intermediate for the production of chemicals. Propylene is a byproduct of steam crackers and varying amount of olefins is produced from steam crackers depending on the type of feedstock. Other sources of propylene may be recovery of propylene from Fluid Catalytic Cracker (FCC) light ends, Propane dehydrogenation, Metathesis.

Some of the major processes for production of propylene is given below in Table.

Table 34.7. Propylene production technologies

Technology	Process description
Olefin conversion technology	This process involves production of propylene from ethylene and 2-butenes in a fixed bed metathesis reactor containing proprietary catalyst, which promotes reaction of ethylene and 2-butene to form propylene and simultaneously isomerises 1-butene to 2-butene.
Superflex process	The process uses a fluidised bed catalytic reactor system using proprietary catalyst which converts low value feedstock to predominantly propylene and ethylene products. Low value light hydrocarbon streams from ethylene plant and refineries can be used, e.g., C_4 and C_5 olefin rich stream from ethylene plants, FCC naphtha, C_4 stream, thermally cracked naphtha from visbreakers or cokers.

Propylur process	This process produces propylene beside ethylene from low value rich feeds ranging from C_4-C_8 from ethylene plant and refineries in a fixed bed reactor using proprietary catalyst. The process offers high selectivity towards propylene.
UOP Oleflex process	This process produces polymer grade propylene from propane and the process consist of a reactor, catalyst regeneration section and product separation and fractionation section. The process uses platinum catalyst.
UOP/Hydro MTO process	This process converts crude methanol (produced from synthesis gas using natural gas) to ethylene and propylene and can be operated either a maximum ethylene or a maximum propylene production mode using MTO-100 silicoaluminophosphate synthetic molecular sieve based catalyst. The process utilizes fluidised bed reactor and regenerator.
Methanol to propylene (MTP) technology	This process produces propylene through methanol route using natural gas. In this process propylene is produced in two steps. First methanol is converted to dimethyl ether in reactor followed by reaction of methanol/ DME in second reactor. Methanol can be produced from methane from conventional method.
C_4 hydrogenation and Meta-4 process	This process involves production of polymer grade propylene plus an isobutylene rich stream or MTBE by upgrading low value C_4 stream pyrolysis C_4 cuts or butene rich cut. The process steps involve – butadiene and C_4 acetylenes selective hydrogenation and butadiene hydroisomerisation, isobutylene removal or MTBE production and metathesis step for conversion of butene and ethylene to propylene. The two main equilibrium reactions taking place are metathesis and isomerisation.
Olefin Ultra TM	A new ultra high activity ZSM-5 additive that provides the highest activity has been developed by Davision catalysts.
KBR's MAXOFIN-3 technology	KBR's MAXOFIN process is based on fluidised bed cracking of gas oils and residue feeds using ZSM catalyst and proprietary MAXOFIN-3 catalyst additive. The process gives 15% or higher propylene yield from gas oil.

A Typical Composition of FCC Gas Stream is tabulated below.

Table 34.8. Typical composition of FCC gas stream

Products	Yield, weight percent
Dry gas (including ethylene)	12.7
Propane	6.5
Propylene	21.0
Butene	35.8

Propylene production by catalytic dehydrogenation of propane: Catalytic dehydrogenation of light paraffins is of increasing importance because of the growing demand of olefins such as propylene and isobutene and n- butenes. Propane dehydrogenation accounts for 2 % of the total world propylene production.

Catalytic dehydrogenation takes place at high temperature (650°C) using platinum based or chromium-alumina or Fe, Cr/Al_2O_3 as catalyst. Reactor effluent treatment for the separation of hydrogen, propylene, and propane is not simple and total investment is high. These production units can be installed only in areas where field propane is available at low costs.

Methanol to propylene: This process produces propylene from natural gas via methanol by converting methanol to dimethyl ether in adiabatic reactor using high activity, high selectivity catalyst. The methanol, water, DME stream is then feed to series of MTP reactor where steam is added. The product stream is first processed for removal of traces of water, CO2 and DME, followed by further processing for yielding polymer grade propylene.

Q.186. Describe the properties and uses of propylene.

Ans. Propylene is a reactive compound that can react with many common reagents used with ethylene such as water, chlorine, and oxygen. However, structural differences between these two olefins result in different reactivities toward these reagents. For example, direct oxidation of propylene using oxygen does not produce propylene oxide as in the case of ethylene. Instead, an unsaturated aldehyde, acrolein, is obtained. This could be attributed to the ease of oxidation of allylic hydrogens in propylene. Similar to the oxidation reaction, the direct cat- alyzed chlorination of propylene produces allyl chloride through substitution of allylic hydrogens by chlorine. Substitution of vinyl hydrogens in ethylene by chlorine, however, does not occur under normal conditions. Many important chemicals are produced from propylene.

Propylene is used in many of the world's largest and fastest growing synthetic materials and thermoplastics. The demand of propylene has increased rapidly during the last twenty years and primarily driven by polypropylene demand. Product profile of propylene is given below in table.

Table 34.9. Product profile of propylene

Product	Uses
Miscellaneous chemicals	1 butanol, 2-ethyl hexanol, Allyl chloride, Epichlorohydrin
Polymer	Polypropylene,, Polyacrylamide, nylon 66, acrylic sheets
Propylene oxide	Polyether-polyols, glycol ethers, isopropyl amines, propylene carbamate, surfactants
Propylene glycol	Unsaturated polyester resins, food additives, cellophane, paints and coating, plasticisers, functional fluids, antifreeze, tobacco treatment
Acrylonitrile	Acrylic fiber, acrylic acid, acrylates, methyl methacrylates, adiponitrile
Isopropanol	Acetone, cosmetics, solvents, pharmaceuticals, isopropyl acetate
Polyols	Polyurethane and Polyester

Consumption of refinery grade propylene made up 9% of total consumption in 2009, chemical grade 23% and polymer grade 68%. Refinery grade propylene is consumed mainly for production of cumene and isopropyl alcohol. Chemical grade propylene mostly goes into oxo alcohol, propylene oxide and acrylonitrile production.

Figure below shows the important chemicals produced from propylene.

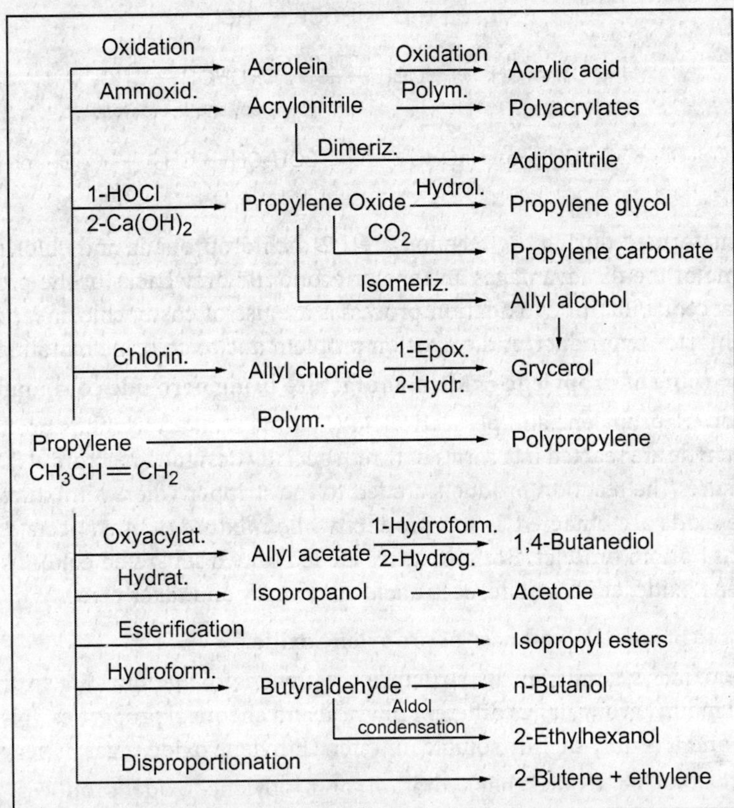

Fig. 34.14. Important chemicals produced from propylene

Q.187. Describe the manufacturing process of propylene oxide.

Ans. There are two major processes for the manufacture of propylene oxide: Propylene chlorohydrin process and propylene oxidation process using peroxides.

Propylene chlorohydrin process route: The chlorohydrination process consists of formation of propylene chlorohydrin by the reaction between hypochlorous acid and propylene. The propylene chlorohydrin is epoxidised to propylene oxide by a 10% solution of milk of lime or NaOH. Various steps involved are

- ***Propylene hypochlorination:*** Propylene is reacted with aqueous chlorine resulting in the formation of propylene chlorohydrins. Unreacted propylene is recycled.
- ***Neutralisation:*** Neutrialisation of propylene chlorohydrins containing hydrochloric acid which is formed during the process.
- ***Dehydrochlorination:*** Reaction of propylene chlorohydrin with milk of lime or caustic soda to produce propylene oxide.
- ***Purification:*** Distillation of crude propylene oxide for separation heavy ends.

Reactions involved are:

$$Cl_2 + H_2O == HOCl + HCl$$

$$H—CH=CH + HOCl \xrightarrow{350°C} CHCHOH—CHCl$$

Propylene chlorohydrin

$$2CH_3CHOH—CH_2Cl + Ca(OH)_2 \text{ or } NaOH \longrightarrow 2CH_3CH—CH_2 (\text{epoxide, } -O-) + CaCl_2 \text{ or } NaCl + 2H_2O$$

Byproducts formed during the reaction are 1,2-dichloropropane and chlorinated di-isopropyl ether. Some of the disadvantages and major economic drawbacks of the process which led to the wide acceptability of epoxidation processes are use of costly chlorine, production of weak calcium chloride byproduct, and corrosion problem due to chlorine handling.

Oxidation route of propylene oxide manufacture using peroxide compounds:

In this process, propylene and peracetic acid (in ethyl acetate) which is produced by oxidation of acetaldehyde are reacted in a series of three specially designed reactors at 50-80°C and 90-120 MPa pressure. The reaction products are fed to the stripper where a mixture of propylene and propylene oxide are obtained as top product while mixture of ethyl acetate and acetic acid is obtained as bottom product. Both mixtures are fed to two separated columns where separation of propylene oxide, ethyl acetate, acetic acid, and heavy end takes place.

Q.188. Describe the properties and uses of propylene oxide.

Ans. Propylene oxide is similar in its structure to ethylene oxide, but due to the presence of an additional methyl group, it has different physical and chemical properties. It is a liquid that boils at 33.9°C, and it is only slightly soluble in water. (Ethylene oxide, a gas, is very soluble in water).

Similar to ethylene oxide, the hydration of propylene oxide produces propylene glycol. Propylene oxide also reacts with alcohols, producing polypropylene glycol ethers, which are used to produce polyurethane foams and detergents. Isomerization of propylene oxide produces allyl alcohol, a precursor for glycerol.

Propylene oxide, propylene glycols and polyols are important derivatives of propylene. Propylene oxide is used for the manufacture of propylene glycol and polyols. Major consumption of propylene oxide is manufacture of polyurethane and polyester resins. Propylene glycol find major application in the manufacture of unsaturated polyester resins, food additives, pharmaceuticals and personal care, tobacco humectants, cellophane, paints and coatings. Polyols major use is in the manufacture of polyurethane.

Multiple Choice Questions and Answers

1. Molecular weight of polymers are in the range of
   (a) $10 - 10^3$ (b) $10^2 - 10^7$
   (c) $10^7 – 10^9$ (d) $10^9 - 10^{11}$
2. Poly Vinyl chloride (PVC) is a............material.
   (a) thermoplastic (b) thermosetting
   (c) fibrous (d) chemically active
3. Dacron is a/an
   (a) polyester
   (b) unsaturated polyester
   (c) polyamide
   (d) inorganic polymer
4. Phenol formaldehyde
   (a) employs addition polymerisation.
   (b) employs condensation polymerisation.
   (c) is a monomer.
   (d) is an abrasive material.
5. Nylon-6 is manufactured from
   (a) caprolactum.
   (b) adipic acid and hexamethylene diamine.
   (c) maleic anhydride and hexamethylene diamine.
   (d) sebasic acid and hexamethylene diamine.
6. Poly tetra flouro ethylene (PTFE) is known as
   (a) dacron (b) teflon
   (c) bakelite (d) celluloid
7. Styrene butadiene rubber (SBR) is
   (a) a natural rubber.
   (b) another name of silicone rubber.
   (c) a synthetic polymer.
   (d) a synthetic monomer.
8. Nylon-66 is manufactured from
   (a) adipic acid and hexamenthylene diamine.
   (b) caprolactum.
   (c) maleic anhydride and hexamethylene diamine.
   (d) dimethyl terephthalate (DMT) and ethylene glycol.
9. Epoxy resin is
   (a) not used for surface coating.
   (b) a good abrasive.
   (c) an elastomer.
   (d) a polyester.
10. Thermosplastic materials
    (a) do not soften on application of heat.
    (b) are heavily branched molecules.
    (c) are solvent insoluble.
    (d) none of these.
11. Vulcanisation of rubber
    (a) decreases its tensile strength.
    (b) increases its ozone & oxygen reactivity,
    (c) increases its oil & solvent resistance.
    (d) converts its plasticity into elasticity.
12. Thermosetting materials
    (a) are cross-linked molecules.
    (b) soften on application of heat.
    (c) are solvent soluble.
    (d) none of these.

Answers

1. (b)	2. (a)	3. (a)	4. (b)	5. (a)	6. (b)	7. (c)	8. (a)
9. (b)	10. (d)	11. (d)	12. (a)				

13. Most commonly used rubber vulcanisation agent is
(*a*) sulphur (*b*) bromine
(*c*) platinum (*d*) alumina

14. Zeigler process
(*a*) produces high density polythene.
(*b*) uses no catalyst.
(*c*) produces low density polythene.
(*d*) employs very high pressure.

15. Celluloid is
(*a*) cellulose acetate
(*b*) regenerated cellulose
(*c*) cellulose nitrate
(*d*) cellulose acetate butytrate

16. Viscose rayon is
(*a*) cellulose nitrate
(*b*) regenerated cellulose nitrate
(*c*) regenerated cellulose acetate
(*d*) none of these

17. Thermosetting plastic materials
(*a*) can be repeatedly melted.
(*b*) is useful for melt casting.
(*c*) can not be melted after forming.
(*d*) is useful for spinning.

18. Phthalic anhydride is used
(*a*) in making PVC (*b*) as plasticisers
(*c*) in insecticides manufacture
(*d*) for making nylon-6

19. Bakelite is a/an
(*a*) addition polymer (*b*) elastomer
(*c*) thermoplastic (*d*) none of these

20. The repeating units of PTFE are
(*a*) $C1_2CH - CH_3$ (*b*) $F_2C = CF_2$
(*c*) $F_3C - CF_3$ (*d*) $FClC = CF_2$

21. ________ is a homopolymer.
(*a*) Neoprene (*b*) Bakelite
(*c*) Nylon-66 (*d*) Terylene

22. Synthetic polymer produced by using terephthalic acid and ethylene glycol is
(*a*) terylene (*b*) nylon-66
(*c*) PVC (*d*) polyesterene

23. Nylon-66 is a polyamide of
(*a*) hexamethylene diamine and adipic acid.
(*b*) adipic acid and methyl amine.
(*c*) vinyl chloride and formaldeyde.
(*d*) none of these.

24. ________ is a natural fibre.
(*a*) Cellulose (*b*) Dacron
(*c*) Nylon-6 (*d*) none of these

25. ________ is a copolymer.
(*a*) styrene butadiene rubber (SBR)
(*b*) Neoprene (*c*) PVC
(*d*) None of these

26. Which of the following types of polymers has the strongest inter particle forces?
(*a*) Elastomers (*b*) Fibres
(*c*) Thermoplastics
(*d*) Thermosetting polymers

27. ________ is not a condensation polymer.
(*a*) Teflon (*b*) Nylon-66
(*c*) Dacron (*d*) Polystyrene

28. ________ is a copolymer.
(*a*) Nylon-66 (*b*) Polyrophylene
(*c*) PVC
(*d*) Poly tetra flouro ethylene

29. The monomer of poly vinyl chloride (PVC) is
(*a*) chloroethane (*b*) ethylene dichloride
(*c*) ethyl chloride (*d*) chloroform

30. Which of the following is not a natural fibre?
(*a*) Silk (*b*) Viscose rayon
(*c*) Wool (*d*) Cotton

31. Pick out the wrong statement.
(*a*) Polystyrene is a thermoplastic polymer.
(*b*) Protein is a natural polymer.
(*c*) Neoprene is a natural rubber.
(*d*) Polythene is a copolymer, while SBR is a homopolymer.

32. Polycaprolactum is
(*a*) nylon-6 (*b*) nylon-66
(*c*) dacron (*d*) rayon

Answers

13. (*a*)	14. (*a*)	15. (*c*)	16. (*d*)	17. (*c*)	18. (*b*)	19. (*d*)	20. (*b*)
21. (*a*)	22. (*a*)	23. (*a*)	24. (*a*)	25. (*a*)	26. (*d*)	27. (*a*)	28. (*a*)
29.(*a*)	30. (*b*)	31. (*c*)	32. (*a*)				

33. Acrylonitrile is mainly used in the ________ industry.

(*a*) polymer (*b*) printing
(*c*) dyeing (*d*) photographic

34. The main use of butadiene is

(*a*) as a plasticiser for unsaturated polyester.
(*b*) in the manufacture of synthetic rubber.
(*c*) as an anti-skimming agent in paint.
(*d*) none of these.

35. Silicone is a/an

(*a*) monomer (*b*) inorganic polymer
(*c*) thermoplastic material
(*d*) a natural polymer

36. Polyvinyl chloride (PVC) is

(*a*) a thermosetting material.
(*b*) a condensation polymerisation product.
(*c*) made by employing emulsion polymerisation.
(*d*) none of these.

37. Flexible foam (for mattresses) are usually made of

(*a*) PVC (*b*) silicone rubber
(*c*) polyurethanes (*d*) polyamides

38. Which of the following is not an elastomer?

(*a*) Polyisoprene (*b*) Neoprene
(*c*) Nitrile-butadiene (*d*) None of these

39. Nylon-6 is a

(*a*) polyamide (*b*) thermosetting resin
(*c*) polyester (*d*) none of these

40. Styrene butadiene rubber (SBR) as compared to natural rubber has

(*a*) poor tensile strength.
(*b*) poorer resistance.
(*c*) greater amount of heat build up on heavv loading,
(*d*) all (*a*), (*b*) and (*c*).

41. Nylon-6 as compared to nylon-66 is

(*a*) harder.
(*b*) more abrasion resistant.
(*c*) having higher melting point.
(*d*) none of these.

42. Plasticisers are added to synthetic plastics to

(*a*) impart flexibility.
(*b*) improve workability during fabrication.
(*c*) develop new unproved properties not present in the original resin.
(*d*) all (*a*), (*b*) and (*c*),

43. Polythene a/an

(*a*) addition polymerisation product.
(*b*) condensation polymerisation product.
(*c*) thermosetting material
(*d*) none of these.

44. Dacron is a

(*a*) condensation polymerisation product of hexamethylene diamine and adipic acid.
(*b*) condensation polymerisation product of dimethyl terephthalate (DMT) and ethylene glycol.
(*c*) thermosetting material.
(*d*) none of these.

45. Bakelite is

(*a*) same as polytetra fiouro ethylene (PTFE).
(*b*) an inorganic polymer.
(*c*) same as thermoset phenol formaldehyde.
(*d*) not a polymer.

46. The process involved in converting rubber into a thin sheet or coating it on fabric is called

(*a*) extrusion (*b*) mastication
(*c*) calendering (*d*) vulcanisation

47. Terylene is

(*a*) same as dacron (*b*) a polyester
(*c*) both (*a*) & (*b*) (*d*) neither (*a*) nor (*b*)

48. Nitrile rubber is produced by the polymerisation of

(*a*) acrylonitrile & butadiene.
(*b*) acrylonitrile & styrene.
(*c*) isobutylene & isoprene.
(*d*) none of these.

49. Main constituent of natural rubber is

(*a*) polystyrene (*b*) polyisoprene
(*c*) polybutadiene (*d*) poly chloroprene

Answers

33. (*a*)	34. (*b*)	35. (*b*)	36. (*c*)	37. (*c*)	38. (*d*)	39. (*a*)	40. (*d*)
41. (*d*)	42. (*d*)	43. (*a*)	44. (*d*)	45. (*c*)	46. (*c*)	47. (*c*)	48. (*a*)
49. (*b*)							

50. Caprolactum, a raw material for the manufacture of nylon-6, is produced from

(*a*) phenol (*b*) naphthalene
(*c*) benzene (*d*) pyridine

51. Polymethyl methacrylate (PMMA) is known as

(*a*) bakelite (*b*) teflon
(*c*) perspex (*d*) nylon–6

52. ________ is a thermosetting plastic.

(*a*) PVC (*b*) Polythene
(*c*) Bakelite (*d*) Polystyrene

53. Phenol formaldehyde is produced by condensation polymerisation. It is also known as

(*a*) teflon (*b*) bakelite
(*c*) polyester (*d*) nylon-66

54. Which of the following is not a thermoplastic material?

(*a*) Epoxy polymer (*b*) PVC
(*c*) Polystyrene (*d*) Polythene

55. Polymerisation product of C_2F_4 (carbon tetraflouride) is called P.T.F.E. (poly chloro tetra fiouro ethylene). It is also called

(*a*) polyurethane (*b*) silicone rubber
(*c*) teflon (*d*) epoxy resin

56. Starting material for the production of butadiene in India is

(*a*) naphthalene (*b*) benzol
(*c*) ethyl alcohol (*d*) phthalic anhydride

57. Neoprene is chemically known as

(*a*) polybutandiene
(*b*) styrene butadiene rubber (SBR)
(*c*) polyurethane
(*d*) polychloroprene

58. Dacron (or terylene) fibres as compared to nylon fibres have

(*a*) better heat & acid resistant properties.
(*b*) poorer resistance to alkalis.
(*c*) poorer dyeability.
(*d*) all (*a*), (*b*) and (*c*).

59. Teflon is

(*a*) phenol formaldehyde
(*b*) an inorganic polymer
(*c*) polytetra floroueth ylene (PTFE)
(*d*) a monomer

60. Styrene which is a monomer for the production of polystyrene, is commercially produced by the

(*a*) catalytic dehydrogenation of ethyl benzene.
(*b*) dehydration of ethyl alcohol followed by hydrogenation.
(*c*) reacting ethylene oxide with acetaldehyde.
(*d*) fermentation of starch.

61. Adipic acid is an intermediate in the manufacture of

(*a*) perspex (*b*) nylon-66
(*c*) polystyrene (*d*) bakelite

62. Reaction of dimethyl terephthalate (DMT) and ethylene glycol produces

(*a*) nylon-6 (*b*) dacron
(*c*) polyester (*d*) PVC

63. Thermal pyrolysis of ethylene dichloride produces

(*a*) trichlorethylene (*b*) vinyl chloride
(*c*) ethanol amine (*d*) ethylene oxide

64. Neoprene is a

(*a*) monomer (*b*) synthetic rubber
(*c*) polyester (*d*) none of these

65. In a cross linked polymer, the monomeric units are linked together to constitute a three dimensional network. Which of the following is a cross-linked polymer?

(*a*) Bakelite (phenol formaldehyde)
(*b*) Polyester (*c*) Polythene
(*d*) Nylon-6

66. In a linear polymer, the monomeric units are linked together to form long straight chains The cross linked or branched chain polymers compared to linear polymers have higher

(*a*) densities (*b*) melting point
(*c*) tensile strength
(*d*) hardness, rigidity & brittleness

67. Dacron is a/an

(*a*) addition polymer
(*b*) condensation polymer
(*c*) polyester (*d*) both (*b*) and (*c*)

Answers

50. (*c*)	51. (*c*)	52. (*c*)	53. (*b*)	54. (*a*)	55. (*c*)	56. (*c*)	57. (*d*)
58. (*d*)	59. (*c*)	60. (*a*)	61. (*b*)	62. (*b*)	63. (*b*)	64. (*b*)	65. (*a*)
66. (*d*)	67. (*d*)						

68. Pick out the wrong statement.
(*a*) Addition polymers are generally formed by chain growth polymerisation.
(*b*) Condensation polymers are generally formed by step growth polymerisation.
(*c*) Teflon is formed by step growth polymerisation.
(*d*) Bakelite is formed by step growth polymerisation.

69. Poymers are classified into four categories namely thermosetting, thermoplastic, elastomer and fibre depending upon their
(*a*) molecular sizes.
(*b*) magnitude of intermolecular forces
(*c*) resistance to heat.
(*d*) polymerisation mechanism.

70. Which of the following has the weakest intermolecular forces?
(*a*) Polyisoprene (*b*) Nylon–66
(*c*) Polystyrene (*d*) Bakelite

71. Low density polythene as compared to high density polythene is
(*a*) harder (*b*) tougher
(*c*) chemically inert (*d*) more flexible

72. Pick out the wrong statement.
(*a*) Polymers made of only one monomer is called homopolymer.
(*b*) Polymers made of more than one monomer is called copolymer or mixed polymer.
(*c*) Vulcanised rubber is more elastic than natural rubber.
(*d*) The intermoleculer forces in thermosplastic polymers are intermediate to that of elastomers & fibres.

73. Condensation polymerisation of caprolactum is carried out in the reactor maintained at a temperature of ________ °C for producing nylon-6.
(*a*) –20 to 25 (*b*) 50 to 75
(*c*) 100 to 150 (*d*) 250 to 280

74. The conversion of caprolactum in the above case is about ________ percent.
(*a*) 25 (*b*) 50
(*c*) 70 (*d*) 90

75. Low pressure Zeigler process for the manufacture of polythene uses a catalyst which is
(*a*) Ni (*b*) V_2O_5
(*c*) Fe
(*d*) aluminium triethyl combined with titanium tetrachloride

76. Temperature and gage pressure maintained during the manufacture of hot SBR (styrene butediene rubber) are
(*a*) 50°C and 3 – 4 kg/cm^2
(*b*) 50°C and 1 kgf/cm^2
(*c*) 250°C and 10 kgf/cm^2
(*d*) 250 °C and 1 kgf/cm^2

77. Temperature and gage pressure maintained during the manufacture of cold SBR (styrene butadiene rubber) are
(*a*) 50°C and 1 kgf/cm^2.
(*b*) – 20°C and 1 kgf/cm^2.
(*c*) 0°C and 1 kgf/cm^2.
(*d*) 0°C and 3 kgf/cm^2.

78. Which of the following is a copolymer?
(*a*) Polystyrene (*b*) SBR
(*c*) P.T.F.E. (*d*) Polypropylene

79. Polypropylene compared to polythene is
(*a*) harder (*b*) stronger
(*c*) lighter (*d*) all (*a*), (*b*) and (*c*)

80. Polystyrene is a light, transparent, thermoplastic material used for making
(*a*) toys and combs (*b*) packaging bags
(*c*) non-sticking utensils
(*d*) electrical insulation

81. Rain coats are made of
(*a*) neoprene (*b*) PVC
(*c*) polyurathane (*d*) SBR

82. Neoprene which is used for making shoe heels & belts is superior to natural rubber in its stability to aerial oxidation and resistance to oils & other solvents. The monomer used for making neoprene is
(*a*) chloroethane (*b*) chlorophrene
(*c*) isoprene (*d*) none of these

Answers

68. (*c*)	69. (*b*)	70. (*a*)	71. (*d*)	72. (*a*)	73. (*d*)	74. (*d*)	75. (*d*)
76. (*a*)	77. (*a*)	78. (*b*)	79. (*d*)	80. (*a*)	81. (*b*)	82. (*b*)	

83. Buna-S is also known as

(*a*) teflon (*b*) PTFE
(*c*) SBR (*d*) polycrylates

84. _______ tubes are good substitude for human blood vessels in heart by-pass operation.

(*a*) PVC (*b*) Polythene
(*c*) Teflon/dacron (*d*) Polystyrene

85. Which of the following is a polymer of hexamethylene diamine and adipic acid?

(*a*) Nylon-6 (*b*) Nylon-66
(*c*) Nylon-6,10 (*d*) Epoxy resin

86. The starting material used for the manufacture of caprolactum is,

(*a*) ethyl benzene (*b*) cyclohexane
(*c*) ethylene glycol (*d*) DMT

87. The synthetic fibres produced from............are known as rayon.

(*a*) lignin (*b*) cellulose
(*c*) polyamides (*d*) ethylene glycol

88. Which of the following is a natural polyamide fibre?

(*a*) Wool (*b*) Silk
(*c*) Cotton (*d*) None of these

89. Main constituent of cotton fibre is

(*a*) lignin (*b*) cellulose
(*c*) starch (*d*) gelatine

90. Cellulose content in bomboo fibre is about _______ percent.

(*a*) 10 (*b*) 20
(*c*) 50 (*d*) 85

91. Which of the following is not present in bagasse fibre?

(*a*) Cellulose (*b*) Lignin
(*c*) Pentogens (*d*) None of these

92. 90% of the caprolactum is converted to nylon-6 on its condensation polymerisation in the reactor maintained at a temperature of _______ °C.

(*a*) –5 (*b*) 10-30
(*c*) 250-280 (*d*) 500-600

93. In nylon-66, the first and second numbers (*i.e.*, 6) respectively designate the number of carbon atoms present in the

(*a*) hexamethylene diamine and the ring.
(*b*) hexamethylene diamine and the adipic acid.
(*c*) adipic acid and the ring.
(*d*) none of these.

94. Linear polymers are normally

(*a*) thermosetting. (*b*) thermosplastic.
(*c*) elastometric.
(*d*) having extremely high softening point.

95. Polyesters are manufactured by the polycondensation reaction of

(*a*) dibasic acid with dihydric alcohol.
(*b*) a single monomer.
(*c*) carboxylic acid with ethylene.
(*d*) inorganic acid with ethylene glycol.

96. Nylon-66 compared to nylon-6 has

(*a*) lower melting point.
(*b*) more abrasion resistant properties.
(*c*) higher hardness.
(*d*) all (*a*), (*b*) and (*c*).

97. In nylon-6, the number 6 represents the total number of

(*a*) carbon atoms in the ring.
(*b*) carbon atoms in the linear polymer chain.
(*c*) nitrogen atoms in the ring.
(*d*) hydrogen atoms in the ring.

98. Polycaprolactum is also known as

(*a*) nylon-66 (*b*) nylon-6
(*c*) teflon (*d*) SBR

99. Polyhexamethylene adipimide is also known as

(*a*) bakelite (*b*) nylon-66
(*c*) epoxy resin (*d*) silicone rubber

100. The major component of acrylic fibres is

(*a*) polyamides (*b*) poly olefins
(*c*) polyacrylonitrile (*d*) polyesters

101. Elastomers are

(*a*) thermosetting material.
(*b*) exemplified by protein derivatives.

Answers

83. (*c*)	84. (*c*)	85. (*b*)	86. (*b*)	87. (*b*)	88. (*a*)	89. (*b*)	90. (*c*)
91. (*d*)	92. (*b*)	93. (*b*)	94. (*b*)	95. (*a*)	96. (*c*)	97. (*a*)	98. (*b*)
99. (*b*)	100. (*c*)	101. (*c*)					

(*c*) having high flexural strength.
(*d*) having very high tensile strength and heat resistance.

102. Maximum consumption of polymers is in
(*a*) electrical insulation (*b*) toys making
(*c*) coating and films (*d*) packaging

103. Low pressure Zeigler process of polythene manufacture
(*a*) employs a pressure of 30 kgf/crn .
(*b*) achieves an yield of 95-98% based on ethylene.
(*c*) produces very low density polythene.
(*d*) does not use any catalyst for polymerisation.

104. Polyvinyl chloride (PVC)
(*a*) is produced by polyconaensation reaction.
(*b*) uses either emulsion or suspension polymerisation methods.
(*c*) can be made thermosetting by adding a plasticiser.
(*d*) softening temperature is 200°C.

105. Cross linked Polymers formed from bi and trifuctional groups in which cross-linkage in three dimensions *via* few chemical bonding across linear chains occur imparts to the polymer ________ properties.
(*a*) thermoplastic (*b*) thermosetting
(*c*) elastometric (*d*) brittleness

106. Pick out the wrong statement.
(*a*) Linear polymers are formed from bifunctional groups only and are normally thermoplastic.
(*b*) Cross-linked branched chain polymers are either elastometric or thermosetting.
(*c*) Branching in case of cross-linked polymers caused by small amount of impurities in bifunctional monomer formulation reduces its solubility and increases the softening point.
(*d*) Dibasic acids reacts with dihydric alcohols to give polyesters using addition polymerisation reaction.

107. Typical solvent polymerisation reaction conditions for the production of high density polythene by Zeigler process is
(*a*) 7 kgf/cm^2 and 70°C.
(*b*) 1000 kgf/cm^2 and 100°C.
(*c*) 7 kgf/cm^2 and 700°C.
(*d*) 1 kgf/cm^2 (gage) and 70°C.

108. Temperature maintained in the emulsion polymerisation reactor for PVC manufacture is about ________ °C.
(*a*) –20 (*b*) 50
(*c*) 250 (*d*) 500

109. Polycondensation reaction of polymerisation
(*a*) does not produce linear polymers.
(*b*) produces only thermoplastic material.
(*c*) produces epoxy polymers.
(*d*) does not need any catalyst.

110. Epoxy resins (i.e., epoxide polymers)
(*a*) are made by addition polymerisation reaction only.
(*b*) contain an epoxy group at the ends of the polymer.
(*c*) are cross-linked polymers only.
(*d*) use emulsion polymerisation methods.

111. Which of the following is generally not used as cord for synthetic rubber tyre casing?
(*a*) Dacron (*b*) Nylon
(*c*) Cellulose (*d*) None of these

112. Which of the following is an inorganic polymer?
(*a*) Teflon (*b*) perspex
(*c*) Silicones (*d*) Bakelite

113. The monomer of natural rubber is
(*a*) DMT (*b*) isoprene
(*c*) 2 methyl-l-propene (*d*) both (*b*) and (*c*)

114. ________ is an addition polymer.
(*a*) Nylon (*b*) Bakelite
(*c*) Polythene (*d*) None of these

115. Which of the following polymers is used for making a non stick coating on frying pans?
(*a*) Bakelite (*b*) Teflon
(*c*) Perspex (*d*) PVC

116. Cross linked polymers are
(*a*) thermoplastic (*b*) thermosetting
(*c*) either (*a*) or (*b*) (*d*) fibres only

Answers

102. (*c*)	103. (*b*)	104. (*b*)	105. (*c*)	106. (*d*)	107. (*a*)	108. (*b*)	109. (*c*)
110. (*b*)	111. (*c*)	112. (*c*)	113. (*d*)	114. (*c*)	115. (*b*)	116. (*b*)	

117. Due to its excellent permeability to air/gas and oxidation resistance, the tubes of automobile tyres is made of

(*a*) cold SBR (*b*) butyl rubber
(*c*) Buna *N* (*d*) Buna S

118. ________ is not a polyester fibre.

(*a*) Terylene (*b*) Dacron
(*c*) Nylon (*d*) Polyacrylonitrile

119. Plasticisers are high boiling liquids added to plastic polymers to impart toughness and flexibility at ordinary temperature. Which of the following is not a plasticiser?

(*a*) Ethylene glycol
(*b*) Stearic acid estors
(*c*) Tricresyl phosphate
(*d*) Esters of phthalic acid

120. Orlan fibre which is used as a wool substitute is

(*a*) an amorphous polymer.
(*b*) a natural polymeric fibre.
(*c*) polyacrylonitrile.
(*d*) polymethylmethacrylate (PMMA).

121. The major constituent of laminate of safety glass, which holds the broken glass, pieces in their places during accident (and thus minimises the danger from flying glass fragments) is

(*a*) polyvinyl alcohol (*b*) polyvinyl acetate
(*c*) polyvinyl butyral (*d*) PVC

122. Polyvinyl alcohol is used as a

(*a*) cation/anion exchanger.
(*b*) water soluble adhesive.
(*c*) textile fibre.
(*d*) non-sticky coating on frying pans.

123. Perspex can be used as a substitute of glass. Its monomer is

(*a*) methyl methacrylate
(*b*) DMT (*c*) butadiene
(*d*) tetrafluroethylene

124. Which of the following is a copolymer?

(*a*) PVC (*b*) Bakelite
(*c*) Polythene (*d*) Teflon

125. Which of the following is not a condensation polymer?

(*a*) Bakelite (*b*) Melamine polymer
(*c*) Polymethylmethacrylate (PMMA)
(*d*) None of these

126. ________ polymer is used for making unbreakable crockery.

(*a*) Thermoplastic (*b*) Melamine
(*c*) Addition (*d*) none of these

127. Which of the following polymers belong to the class of formaldehyde resin?

(*a*) Melamine resins (*b*) Teflon
(*c*) Dacron (*d*) None of these

128. Ion exchange resins are made of

(*a*) lucite (*b*) sulphonated bakelite
(*c*) polystyrene (*d*) teflon

129. Diphenylamine is added to rubber to

(*a*) valcanise it.
(*b*) protect it from deterioration on exposure to air.
(*c*) make it non-inflammable.
(*d*) make it thermosetting.

130. Neoprene is rendered non-inflammable, because of

(*a*) its cross-linked structure.
(*b*) its linear chain structure.
(*c*) the presence of chlorine atoms in its monomer.
(*d*) the absence of chlorine atoms in its monomer.

131. Branched chain polymers as compared to linear polymers have

(*a*) higher melting point.
(*b*) higher tensile strength.
(*c*) lower density. (*d*) none of these.

132. In condensation polymerisation as compared to addition polymerisation

(*a*) the monomers are unsaturated compounds.
(*b*) no co-product is lost.
(*c*) the monomers contain two functional groups.
(*d*) generally only one monomer is involved

133. Chain growth polymerisation is a process, in which the monomers are added in a chain fashion, and it requires an initiator to produce the free radical. An

Answers

117. (*b*) 118. (*c*) 119. (*a*) 120. (*c*) 121. (*c*) 122. (*b*) 123. (*a*) 124. (*b*)
125. (*c*) 126. (*b*) 127. (*a*) 128. (*b*) 129. (*b*) 130. (*c*) 131. (*c*) 132. (*c*)
133. (*b*)

example of chain growth polymerisation products is

(*a*) nylon-66 (*b*) teflon
(*c*) polyester (*d*) bakelite

134. In step growth polymerisation, condensation occurs in a stepwise manner with or witnout the elimination of smaller molecules. An example of step growth polymerisation product is

(*a*) terylene (*b*) Polybutadiene
(*c*) PVC (*d*) polypropylene

135. Properties of a polymer is affected by the

(*a*) chain length.
(*b*) intermolecular forces.
(*c*) branching & cross-linking
(*d*) all (*a*), (*b*) & (*c*)

136. Rayon is superior to cotton in making gauge for wound treatment, because rayon

(*a*) is a synthetic polymer.
(*b*) does not stick to the wound unlike cotton.
(*c*) can absorb over *90%* of its own weight of water.
(*d*) both (*b*) and (*c*).

137. Vulcanisation of rubber does not increase its

(*a*) elasticity (*b*) plasticity
(*c*) ductility (*d*) none of these

138. Thermosetting polymers as compared to thermoplastic polymers

(*a*) are formed by addition polymerisation.
(*b*) have three dimensional structure.
(*c*) have linear structure.
(*d*) none of these.

139. Polypropylene is preferred to polythene because the former is

(*a*) non-inflammable (*b*) harder
(*c*) stronger (*d*) both (*b*) & (*c*)

140. Branched chain polymers compared to linear polymers have higher

(*a*) density.
(*b*) tensile strength.
(*c*) melting point.
(*d*) degree of irregularity in atomic packing.

141. Hot drink (*e.g.*, tea) cups are usually made of

(*a*) polystyrene (*b*) polythene
(*c*) polypropylene (*d*) PVC

142. Liners of bags are usually made of

(*a*) polythene (*b*) PVC
(*c*) polypropylene (*d*) polyesters

143. Glyptal used in the manufacture of paints & lacquers is a ________ polymer.

(*a*) polyamide (*b*) polystyrene
(*c*) polyester (*d*) polyacrylonitrile

144. Which of the following is not a polyolefin?

(*a*) Polystyrene (*b*) Polypropylene
(*c*) Neoprene (*d*) None of these

145. Which of the following polymers does not belong to the class of polyacrylates polymer?

(*a*) PMMA (*b*) Polyacrylonitrile
(*c*) Polyethyl acrylate (*d*) None of these

146. Which of the following is not the commercial name of polymethylmethacrylate (PMMA)?

(*a*) Perspex (*b*) Lucite
(*c*) Plexiglass (*d*) Teflon

147. Acrilan fibres used for making cloth, carpet & blankets, which is a hard, horny & high melting polymeric material is nothing but

(*a*) polyacrylonitrile (*b*) polyamide
(*c*) saturated polyester (*d*) alkyd resin

148. Vinyl flooring is done using ________ sheets.

(*a*) polypropylene (*b*) PVC
(*c*) polythene (*d*) polyvinyl acetate

149. Bristles of tooth brushes are made of

(*a*) nylon-6 (*b*) nylon-66
(*c*) polystyrene (*d*) PVC

150. Non sulphonated hard bakelites are not used for making

(*a*) ion-exchange resins (*b*) fountain pen barrels
(*c*) formica table tops (*d*) combs

151. Contact lenses for eyes are made of perspex, which is nothing but

(*a*) polymethylmethacrylate
(*b*) polystyrene
(*c*) unsaturated polyester
(*d*) polypropylene

Answers

134. (*a*)	135. (*d*)	136. (*b*)	137. (*b*)	138. (*b*)	139. (*d*)	140. (*d*)	141. (*a*)
142. (*c*)	143. (*c*)	144. (*c*)	145. (*d*)	146. (*d*)	147. (*a*)	148. (*b*)	149. (*b*)
150. (*a*)	151. (*a*)						

152. Addition polymerisation takes place either by a free radical mechanism or ionic maehanism depending on the reagents used. Free radical polymerisation is catelyzed by ________, which decompose to give free radicals.

(*a*) organic peroxides (*b*) sulphuric acid
(*c*) hydroflouric acid (*d*) none of these

153. Fillers such as zinc oxide and carbon black are added to the crude natural rubber before vulcanisation in order to improve its

(*a*) elasticity (*b*) strength
(*c*) plasticity
(*d*) weathering characteristics

154. The monomer for the production of neoprene rubber is

(*a*) acetylene (*b*) chloroprene
(*c*) isoprene (*d*) none of these

155. Orion is

(*a*) a copolymer.
(*b*) a condensation polymer.
(*c*) obtained by polymerising vinyl cyanide,
(*d*) all (*a*), (*b*) and (*c*).

156. Size range of polymer molecules varies from ________ metre.

(*a*) 10^{-2} to 10^{-5} (*b*) 10^{-4} to 10^{-7}
(*c*) 10^{-1} to 10^{-2} (*d*) 10^{-8} to 10^{-10}

157. Polymerisation process in which two or more monomers of chemically different nature take part is called

(*a*) copolymerisation
(*b*) addition polymerisation
(*c*) chain polymerisation
(*d*) none of these

158. Condensation polymerisation of ________ produces bakelite.

(*a*) propylene
(*b*) phenol & formaldehyde
(*c*) phenol & acetaldehyde
(*d*) urea & formaldehyde

159. Vulcanisation of raw rubber makes it

(*a*) soft (*b*) less elastic
(*c*) plastic (*d*) tacky

160. Ebonite is a/an

(*a*) highly vulcanised rubber.
(*b*) natural rubber.
(*c*) unvulcanised raw rubber.
(*d*) adhesive.

161. Buna-S is also called

(*a*) polyurathane (*b*) SBR
(*c*) teflon (*d*) bakelite

162. Thermocol is a spongy, porous, rigid or flexible foamed plastic, obtained by blowing gas/air through molten

(*a*) saturated polyester (*b*) polyurathane
(*c*) polystyrene (*d*) either (*b*) or (*c*)

163. A copolymer is formed by the combination of two or more monomer molecules

(*a*) in a chain without the elimination of water.
(*b*) with the elimination of small amount of water.
(*c*) of the same monomer by elimination of small molecules of water.
(*d*) none of these.

164. Thermocol (expanded polystyrene) is not used for

(*a*) low temperature thermal insulation as in refrigerator and air conditioners.
(*b*) accoustic control and ceiling for building.
(*c*) high temperature thermal insulation in furnaces.
(*d*) packing of delicate electronic gadgets.

165. Rexin (also called artificial leather), which is used for making table cover, automobile seat cover, shoes etc. is made by coating thick cloth with molten

(*a*) teflon (*b*) bakelite
(*c*) SBR (*d*) PVC

166. Buna-N is also called

(*a*) butyl rubber (*b*) nitrile rubber
(*c*) neoprene (*d*) thiokol

167. ________ is produced by polymerisation of chloroprene.

(*a*) Thiokol (a polysulphide rubber)
(*b*) Butyl rubber
(*c*) Neoprene
(*d*) Polyurathane rubber

Answers

152. (*a*)	153. (*d*)	154. (*b*)	155. (*c*)	156. (*a*)	157. (*a*)	158. (*b*)	159. (*b*)
160. (*a*)	161. (*b*)	162. (*d*)	163. (*a*)	164. (*c*)	165. (*d*)	166. (*b*)	167. (*c*)

168. Addition of plasticisers to polymers results in partial neutralisation of intermolecular forces of attraction between the macromolecules thereby increasing its

(*a*) tensile strength (*b*) chemical resistance
(*c*) flexibility (*d*) all (*a*), (*b*) & (*c*)

169. Pick out the wrong statement regarding the solubility characteristics of high polymers.

(*a*) Greater the degree of cross-linking in the polymer, lesser is its solubility.
(*b*) Polymers having more aliphatic character are more soluble in aliphatic solvents, while those polymers having more aromatic character are more soluble in aromatic solvents.
(*c*) Swelling tendency or solubility of polymers in a particular solvent decreases with increase in molecular weight of the solvent.
(*d*) High molecular weight polymers on dissolving gives solution of very low viscosity.

170. Which of the following low molecular weight (<10^4), soft & waxy polymer is used in *chewing gum*?

(*a*) Cellulose acetate (*b*) Polyvinyl acetate
(*c*) Thiokol (*d*) PVC

171. Thermosetting resins/polymers as compared to thermoplastic ones are

(*a*) soluble in all organic solvents.
(*b*) more brittle.
(*c*) formed by addition polymerisation only.
(*d*) easily reshaped & reused.

172. Phosphates (*e.g.,* triceresyl, tributyl, tetrabutyl, tripheyl etc.) are added to polymers to act as

(*a*) hardeners
(*b*) anti-shrinkage agents
(*c*) plasticisers
(*d*) transparency improver

173. ________ moulding is used for shaping of thermosetting plastics exclusively.

(*a*) Compression (*b*) Injection
(*c*) Transfer (*d*) Extrusion

174. The only natural thermoplastic resin, which is a product of animal life is

(*a*) rosin (*b*) shellac
(*c*) amber (*d*) copal

175. Automobile steering wheels are normally made of

(*a*) cellulose acetate (*b*) cellulose nitrate
(*c*) PVC
(*d*) high density polythene

176. Density of low density polythene is about ________ gm/c.c.

(*a*) 0.38 (*b*) 0.56
(*c*) 0.81 (*d*) 0.91

177. Melt spinning of polymers involves the forcing of polymer melt through spinnerettes (fine holes) into an atmosphere kept at a temperature lower than the melting point of the polymer, which causes the fine diameter polymer melt to harden into filaments. Melt spinning is not used in case of the.........fibres.

(*a*) acrylic (*b*) polyester
(*c*) nylon 6:6 (*d*) polypropylene

178. Density of high density polythene is about ________ gm/c.c.

(*a*) 1.18 (*b*) 1.05
(*c*) 0.95 (*d*) 0.99

179. In case of dry spinning of polymers, the polymer solution in a volatile solvent is forced through the spinnerates into a warm air chamber, where the solvent evaporates leaving behind the polymer in the filament form. Dry spinning is used for ________ fibres.

(*a*) polythene (*b*) PVC
(*c*) rayon (*d*) polyvinyl acetate

180. Which of the following is generally not drawn into fibre?

(*a*) Polyamide
(*b*) Unsaturated polyesters
(*c*) Saturated polyesters
(*d*) Polyacrylonitrile

181. In case of wet spinning of polymers, the polymer solution is forced through spinnerates into a coagulating bath to give a filament form. Wet spinning is not used in case of the....................fibres.

(*a*) viscose rayon (*b*) acrylic
(*c*) polyvinyl acetate (*d*) saturated polyester

Answers

168. (*c*)	169. (*d*)	170. (*b*)	171. (*b*)	172. (*c*)	173. (*c*)	174. (*b*)	175. (*a*)
176. (*d*)	177. (*a*)	178. (*c*)	179. (*b*)	180. (*b*)	181. (*d*)		

182. Zeigler-Natta catalyst (AIR_3 – $AlCl_3$) is used in the polymerisation of

(*a*) vinyl acetate (*b*) vinyl chloride
(*c*) propylene (*d*) styrene

183. Which of the following polymers are produced by employing all bulk polymerisation, solution polymerisation & suspension polymerisation technique of addition (chain) polymerisation?

(*a*) PVC (*b*) Bakelite
(*c*) PTFE (*d*) Epoxy resin

184. Which of the following is an elastomer?

(*a*) Thiokol
(*b*) Phenol formaldehyde
(*c*) Urea formaldehyde
(*d*) Polystyrene

185. Which of the following additives are added to plastics to make it impervious to X-rays?

(*a*) Asbestos (*b*) Barium salt
(*c*) Carborundum (*d*) Phthalic acid

186. Softening point of high density polythene is about ________ °C.

(*a*) 85 (*b*) 135
(*c*) 165 (*d*) 205

187. Plexiglass (also called lucite) because of its high optical transparency is used for making lenses. It is chemically

(*a*) polymethyl methacrylate (PMMA).
(*b*) Polytetraflouroethylene (PTFE).
(*c*) polycarbonates.
(*d*) phenolic resins.

188. ________ are produced by reacting polybasic acid (*e.g.*, phthalic anhydride) with polyhydric alcohol (*e.g.*, glycerol).

(*a*) Unsaturated polyester
(*b*) Alkyd resins (*c*) Saturated polyester
(*d*) Amino resins

189. Viscosity of a polymer solution or melt

(*a*) decreases with increase in molecular weight.
(*b*) decreases with increase in temperature.
(*c*) increases with increase in temperature.
(*d*) does not vary with temperature rise.

190. Flexible plastic pipes are made of

(*a*) high density polyethylene (HDPE).
(*b*) low density polyethylene (LDPE).
(*c*) polypropylene.
(*d*) unsaturated polyester.

191. Valcunisation of rubber decreases its

(*a*) tensile strength.
(*b*) resistance to organic solvents.
(*c*) tackiness.
(*d*) working temperature range.

192. Tubeless tyres are made of rubber, which is a co-polymer of isoprene & isobutylene.

(*a*) nitrile (*b*) silicone
(*c*) neoprene (*d*) butyl

193. Which of the following rubbers has the widest service temperature range (–75 to 275°C)?

(*a*) Butyl rubber (*b*) Silicon rubber
(*c*) Nitrile rubber (*d*) Silicone rubber

194. Cellulose is the main constituent of most ________ fibres.

(*a*) acrylic (*b*) spandex
(*c*) synthetic (*d*) natural

195. Pick out the wrong statement.

(*a*) Denier is defined as weight in gm of 9000 metres length of a fibre, while tex is defined as weight in gm of 1000 metres length of fibre.
(*b*) Crimp is a measure of the difference between the length of the straightened and unstraightend fibre.
(*c*) Dacron fibres, acrylic fibres, viscose rayon fibres & cellulose acetate fibres are all prepared by wet spinning.
(*d*) Nylon fibres have greater water absorption capacity than polyester fibres.

196. Tyres are made by

(*a*) injection moulding (*b*) extrusion
(*c*) rotational moulding
(*d*) compression moulding

197. Antioxidants are added to rubber to protect it from the attack of light, heat & atmospheric ozone. Which of the following is an antioxident used in rubber?

Answers

182. (*c*)	183. (*a*)	184. (*a*)	185. (*b*)	186. (*b*)	187. (*a*)	188. (*b*)	189. (*c*)
190. (*b*)	191. (*c*)	192. (*d*)	193. (*d*)	194. (*d*)	195. (*c*)	196. (*c*)	197. (*b*)

(*a*) Carbon
(*b*) Alkylated diphenyl amine
(*c*) Thiokol (*d*) Magnesium

198. Cellulose acetate has very high film permeability among all the polymers of the order of 5000 gm/100 m^2. Which of the following polymers has the maximum film elongation (of the order of 100%)?
(*a*) Water impermeable cellophane
(*b*) Polythene (*c*) Cellulose acetate
(*d*) Teflon

199. Plastic tubes & pipes are generally made by ________ moulding.
(*a*) injection (*b*) transfer
(*c*) extrusion (*d*) compression

200. Which of the following is stretched into fibres?
(*a*) Saturated polyester
(*b*) Unsaturated polyester
(*c*) Isoprene (*d*) Bakelite

201. Mastication of rubber means
(*a*) its softening.
(*b*) a treatment to retard its deterioration due to oxidation.
(*c*) improving its curing rate.
(*d*) depression of its freezing point.

202. Polyvinyl acetate is never used for making
(*a*) moulded articles (*b*) fibres
(*c*) adhesives (*d*) all (*a*), (*b*) & (*c*)

203. Synthetic rubber
(*a*) deforms, if stretched to double of its original dimension.
(*b*) is brittle at low temperature.
(*c*) is softer at higher temperature.
(*d*) is highly permeable to air & water and is readily attacked by chemicals & atmospheric gases.

204. Styrene butadiene rubber (SBR) is never used for making
(*a*) coated fabrics
(*b*) front wheel tyres of aeroplanes (*i.e.*,heavy duty tyres)
(*c*) gaskets
(*d*) soles of shoes

205. Peptizers like aromatic mercaptans (*e.g.* thiophenes) in rubber are added to
(*a*) protect rubber goods from attack by oxygen & ozone present in the atmosphere.
(*b*) reduce its viscosity to permit easier processing.
(*c*) reduce the time of vulcanisation and quantity of vulcanising agent.
(*d*) increase its viscosity.

206. Which of the following is the lowest cost plastic commercially available?
(*a*) Polythene (*b*) Teflon
(*c*) Bakelite (*d*) PVC

207. Paper like thin plastic articles can be produced by
(*a*) blow moulding
(*b*) vacuum thermo forming
(*c*) injection moulding
(*d*) none of these

208. Polyurathane can not be used to make
(*a*) automobile cushion
(*b*) thermal insulation in refrigerator
(*c*) coating and adhesive
(*d*) fibre reinforced plastic (FRP)

209. Most of the plastics are safe to be used upto a maximum temperature of ________ °C.
(*a*) 100 (*b*) 150
(*c*) 350 (*d*) 450

210. Which of the following polymers has the tendency of decomposing before melting?
(*a*) Polystyrene (*b*) Nylon
(*c*) PVC (*d*) None of these

211. Polycondensation of saturated dicarboxylic acid with polyhydric alcohol produces
(*a*) epoxy resin (*b*) polyamide
(*c*) alkyd resin (*d*) phenolic resin

212. Silicone resins, which are highly water repellant and has good heat resistance can not be used
(*a*) as room temperature adhesive.
(*b*) as grease & lubricant.
(*c*) hydraulic fluid for heat transfer.
(*d*) resin for lamination.

Answers

198. (*b*)	199. (*c*)	200. (*a*)	201. (*a*)	202. (*a*)	203. (*d*)	204. (*b*)	205. (*b*)
206. (*a*)	207. (*b*)	208. (*d*)	209. (*b*)	210. (*c*)	211. (*c*)	212. (*a*)	

213. Epoxy resin is a polymer containing two or more —C—C— (with O bridging, epoxide ring) groups called epoxide group or ethoxyline group. It is a

(*a*) polyamide & an elastomer.
(*b*) good adhesive.
(*c*) surface coating agent.
(*d*) both (*b*) and (*c*).

214. Polycaprolactum is nothing but

(*a*) orlon (*b*) nylon 6,6
(*c*) Nylon 6 (*d*) saran

215. Out of all the elastomers, natural rubber has the longest elongation range & flexibility of the order of ________ percent.

(*a*) 1–1000 (*b*) 1000–1500
(*c*) 1500–2000 (*d*) 2000–2500

216. A copolymer of vinyl and vinylidine chloride is called

(*a*) treylene (*b*) orlon
(*c*) saran (*d*) dacron

217. Addition polymerisation is not involved in the manufacture of

(*a*) low density polythene.
(*b*) poly vinyl chloride.
(*c*) polystyrene.
(*d*) polyhexamethylene adipamide.

218. ________ polythene is most prone to stress cracking.

(*a*) High density (*b*) Low density
(*c*) Cross–linked
(*d*) Linear low density

219. Commercial production of polypropylene employs ________ polymerisation.

(*a*) emulsion (*b*) suspension
(*c*) solution (*d*) bulk

220. Addition of stabiliser during PVC manufacture is done to

(*a*) improve its impact strength.
(*b*) improve its elasticity.
(*c*) reduce the melt viscosity & glass transition temperature.
(*d*) prevent its thermal degradation.

221. ________ polymer is produced by the copolymerisation of vinyl chloride-vinyl acetate.

(*a*) Fibrous (*b*) Leathery
(*c*) Rubbery (*d*) Hard

222. ________ is normally used for the manufacture of refrigerator components and transistor parts.

(*a*) Polyproylene (*b*) Polystyrene
(*c*) Polyester (*d*) Polyurathane

223. Nylon 6 as compared to nylon 6,6 has lower

(*a*) abrasion resistance. (*b*) thermal stability.
(*c*) adhesion to rubber. (*d*) hardness.

224. Nylon 6, 10 which is used for bristles making is superior to nylon 6, 6 due to its lower-water absorption capacity, is a/an

(*a*) polyester (*b*) polyamide
(*c*) polyisoprene (*d*) polystyrene

225. Melamine formaldehyde resin which has a very high anti tacking properties, is not used for the

(*a*) electrical insulation purpose.
(*b*) tanning of leather.
(*c*) strengthening of plaster of paris.
(*d*) decorative laminates.

226. Amino resins are used in paper treatment to improve its

(*a*) wet tear and bursting strength.
(*b*) folding endurance
(*c*) wet rub resistance
(*d*) all (*a*), (*b*) & (*c*)

227. Cation exchange resins (0.3 to 1 mm size) used in water treatment is prepared from......resins.

(*a*) epoxy
(*b*) phenol formaldehyde
(*c*) urea formaldehyde
(*d*) melamine formaldehyde

228. Alkyd resin *e.g.*, glyptal resin formed by phthalic anhydride and glycerene is not used

(*a*) for surface coating of automobiles & aircrafts.
(*b*) for fibre making.
(*c*) as plasticiser for PVC & nitrocellulose.
(*d*) for film forming materials.

Answers

213. (*d*)	214. (*c*)	215. (*a*)	216. (*c*)	217. (*d*)	218. (*a*)	219. (*c*)	220. (*d*)
221. (*b*)	222. (*b*)	223. (*a*)	224. (*b*)	225. (*c*)	226. (*d*)	227. (*c*)	228. (*b*)

229. Molecular weight of a polymer is equal to the molecular weight of the repeat unit multiplied by the degree of polymerisation. What is the molecular weight of poly vinyl chloride (PVC), if its degree of polymerisation is 800?

(*a*) 50000 (*b*) 51600
(*c*) 49000 (*d*) 50800

230. Scouring is a finishing operation during the manufacture of fibre, which aims at ________ of fibres.

(*a*) improving the stretchability
(*b*) dyeing/colouring
(*c*) detergent solution washing
(*d*) glycol soaking

231. Most of the fibre forming polymers are crystalline in nature, one of the exceptions being which is amorphous in nature.

(*a*) nylon (*b*) polyacrylonitrile
(*b*) polypropylene (*d*) polyester

232. All thermoplastic, thermosetting & elastic materials can be processed in a extrusion machine, however it can not be used for the production of plastic

(*a*) filaments (*b*) pipes
(*c*) buckets (*d*) tubings

233. ________ scrap can be recycled & reutilised.

(*a*) Bakelite (*b*) Epoxy resin
(*c*) Polythene (*d*) None of these

234. The physical state in which polymers exist is

(*a*) melts & rubber like state.
(*b*) amorphous glassy state.
(*c*) partially crystalline state.
(*d*) all (*a*), (*b*) and (*c*).

235. Lavatory cisterns are normally made of

(*a*) expanded polystyrene
(*b*) saturated polyester
(*c*) perspex
(*d*) PVC

236. ________ resins are produced by the condensation polymerisation of formaldehyde with urea or melamine.

(*a*) Epoxy (*b*) Amino
(*c*) Alkyd (*d*) Phenolic

237. Condensation of bisphenol A with phosgene produces ________ which possess very good heat resistance.

(*a*) polyurathane (*b*) polysulphone
(*c*) polycarbonate (*d*) polyester

238. The visco-elastic behaviour exhibited by plastics is a ________ like behaviour.

(*a*) solid (*b*) liquid
(*c*) combination of solid & liquid
(*d*) neither solid nor liquid

239. Pick out the wrong statement.

(*a*) Polymeric fibres are never produced by addition polymerisation.
(*b*) Property of tackiness is exhibited by uncured rubber.
(*c*) Sharp melting point is not observed in thermoplastic polymers.
(*d*) Polythene generally has an excellent resistance to ultra violet rays.

240. Thermoplastic resins usually

(*a*) remain hard as long as they are hot.
(*b*) can not be reclaimed from waste.
(*c*) permanent setting resins.
(*d*) less brittle than thermosetting resins.

241. Gutta parcha rubber is

(*a*) soft & tacky at room temperature.
(*b*) an isomer of natural rubber.
(*c*) a thermosetting resin.
(*d*) recovered by coagulation of rubber latex.

242. Vulcanisation of rubber does not increase its

(*a*) softness
(*b*) oxidation resistance
(*c*) weight & strength
(*d*) elasticity & water solubility

243. Condensation polymerisation of formaldehyde with ________ does not produce phenolic resin.

(*a*) resorcinol (*b*) phenol
(*c*) para-cresol (*d*) melamine

244. ________ of SBR is adversely affected, if more quantity of styrene is added to butadiene during its co-polymerisation to produce SBR.

Answers

229. (*a*)	230. (*c*)	231. (*b*)	232. (*c*)	233. (*c*)	234. (*d*)	235. (*a*)	236. (*b*)
237. (*c*)	238. (*c*)	239. (*a*)	240. (*d*)	241. (*b*)	242. (*a*)	243. (*d*)	244. (*b*)

(*a*) Percent elongation (*b*) Resilience
(*c*) Freezing point (*d*) Strength

245. ________ practically possess no elasticity.
(*a*) Vulcanite or ebonite
(*b*) Spandex fibre
(*c*) Polysulphide rubber
(*d*) Epoxy resin

246. Polycaprolactum (Nylon-6) is produced by the condensation polymerisation of caprolactum at 240–280° C in which the conversion of caprolactum is about ________ percent.
(*a*) 50 (*b*) 75
(*c*) 90 (*d*) 99

247. In a co-polymer, the repeating units contain two difierent monomers. Which of the following is a copolymer?
(*a*) PTFE (*b*) Buna-S
(*c*) PMMA (*d*) Polycaprolactum

248. Thiokol is nothing but
(*a*) polysulphide rubber
(*b*) polyamide fibre
(*c*) engineering plastic
(*d*) expanded polystyrene

249. Polymethyl methacrylate (PMMA) which is also known as perspex and is produced by bulk polymerisation of methyl methacrylate is not a ________ polymer.
(*a*) thermoplastic (*b*) thermosetting
(*c*) linear
(*d*) glass like transparent

250. Acrylonitrile-butadiene-styrene (ABS) copolymer, which is produced by blending styrene-acrylonitrile copolymer with butadiene based elastomer, is a/an
(*a*) rigid foam
(*b*) engineering plastic
(*c*) thermosetting polymer
(*d*) spongy rubber

251. Natural rubber is obtained from latex which is a colloidal dispersion of rubber in water. Which of the following is used as a coagulant in latex?
(*a*) Ammonium alum (*b*) Potassium alum
(*c*) both *a* & *b* (*d*) neither *a* nor *b*

252. Bakelite (phenol-formaldehyde) resin can not be used as
(*a*) decorative paint.
(*b*) decorative laminates.
(*c*) electrical insulation.
(*d*) glass reinforced plastics.

253. Catalyst used in the production of high density polythene by low pressure Ziegler process is
(*a*) aluminium triethyl activated with $TiCl_4$
(*b*) platinum (*c*) molybdenum
(*d*) Nickel

254. SBR is produced by the copolymerisation of butadiene & styrene by employing emulsion polymerisation. The weight ratio of styrene and butadiene is maintained at
(*a*) 1:3 (*b*) 3:1
(*c*) 1:2 (*d*) 2:1

255. Alkyd resin is a/an
(*a*) polyamide (*b*) polyester
(*c*) polyolefin (*d*) addition polymer

256. Which of the following is the most important rubber compounding ingradient which is used to improve wearing qualities of both natural rubber & SBR by imparting toughness?
(*a*) Phosphorous (*b*) Carbon black
(*c*) Pine oil (*d*) Rosin

257. Rayon is a ________ fibre.
(*a*) cellulosic (*b*) polyamide
(*c*) polyester (*d*) natural

258. In step growth polymerisation, generally only one type of reaction & some basic mechanism is involved. Step growth polymerisation reaction is not involved in the manufacture of
(*a*) cross linked polystyrene.
(*b*) phenol formaldehyde resins.
(*c*) polyesters.
(*d*) polyamides.

259. A chain growth polymerisation reaction consists of three different types of reaction namely initiation reaction, propagation reaction & termination reaction. Chain growth polymerisation reaction is not involved in the manufacture of

Answers

245. (*a*)	246. (*c*)	247. (*b*)	248. (*a*)	249. (*b*)	250. (*b*)	251. (*c*)	252. (*a*)
253. (*a*)	254. (*a*)	255. (*b*)	256. (*b*)	257. (*a*)	258. (*a*)	259. (*d*)	

(*a*) siloxane elastomers.
(*b*) polyamides.
(*c*) vinyl polymers.
(*d*) urea-formaldehyde resins.

260. Which of the following is not a polymer of two monomes?
(*a*) Teflon (*b*) Bakelite
(*c*) SBR (*d*) None of these

261. Transistor Parts and refrigerator components are normally made of
(*a*) polystyrene
(*b*) polyester
(*c*) high density polythene
(*d*) polyurathane

262. Polyurathane can not be used for making
(*a*) mattresses & foam (*b*) coating material
(*c*) adhesives (*d*) bottles

263. Alkyd resin can not be used for making
(*a*) plasticiser (*b*) paint & varnish
(*c*) fibres
(*d*) film forming materials

264. Which of the following polymers shows the highest anti-tacking properties?
(*a*) Melamine formaldehyde resin
(*b*) Phenolic resin
(*c*) Epoxy resin
(*d*) Alkyd resin

265. Trade name of ________ is neoprene.
(*a*) polychloroprene
(*b*) polyisoprene
(*c*) polytetraflouroethylene
(*d*) poly vinyl acetate

266. ________ fibres are made of polyamides.
(*a*) Dacron (*b*) Nylon
(*c*) Rayon (*d*) Orlon

267. The generic chemical name for the class of polymers which are commerically known as nylons is
(*a*) polyolefins (*b*) polyamide
(*c*) polyacrylate (*d*) polyurathane

268. Plastic articles are normally produced by ________ moulding.
(*a*) green sand (*b*) injection
(*c*) shell (*d*) dry sand

269. Collapsible tubes for tooth paste are produced by ________ extrusion.
(*a*) direct (*b*) indirect
(*c*) impact (*d*) none of these

270. Pick out the correct statement.
(*a*) Plastics are good conductors of heat and electricity.
(*b*) All the polymers are highly crystalline in nature.
(*c*) Polymers can be vaporised by heating to a very high temperature.
(*d*) The liquid polymer becomes greasy, then waxy and finally solid on increasing the degree of polymerisation.

271. Polystyrene is a ________ plastic at room temperature.
(*a*) ductile (*b*) brittle
(*c*) malleable (*d*) none of these

272. Buna-S is a ________ material.
(*a*) fibrous (*b*) plastic
(*c*) resinous (*d*) rubbery

273. Polymerisation of poly functional monomers produces polymers having
(*a*) good machanical strength
(*b*) low viscosity
(*c*) low melting point
(*d*) none of these

274. Unbreakable crockeries are made from ________ polymers.
(*a*) polystyrene (*b*) melamine
(*c*) polyester (*d*) polyurathane

275. Crystallisation of polymers is an undesirable property. Crystallisation of celluloid is prevented by adding
(*a*) glycerol (*b*) nitrocellulose
(*c*) camphor (*d*) none of these

Answers

260. (*a*)	261. (*a*)	262. (*d*)	263. (*c*)	264. (*a*)	265. (*a*)	266. (*b*)	267. (*b*)
268. (*b*)	269. (*c*)	270. (*d*)	271. (*b*)	272. (*d*)	273. (*a*)	274. (*b*)	275. (*c*)

276. Pick out the wrong statement.

(*a*) For the manufacture of styrene, the major raw materials are benzene and ethylene.

(*b*) One inportant copolymer of styrene is SBR, which is widely used in the manufacture of automobile tyres.

(*c*) Manufacture of phenol by chlorobenzene-caustic process involves the chlorination of benzene, causticisation and hydrolysis.

(*d*) Phenol manufacture by chlorobenzene-caustic process is competitive even when low cost chlorine is not available.

277. Butyl rubber is a copolymer of

(*a*) 1-butene with a small amount of isobutene.

(*b*) isobutene with a small amount of 2-methylbutadiene (isoprene).

(*c*) butadiene with a small amount of propylene.

(*d*) 1-butene with a small amount of butadiene.

278. Which of the following variety of rubber has a very low permeability for gases and finds application in tyre inner tubes?

(*a*) Neoprene.

(*b*) Styrene butadiene rubber.

(*c*) Nitrile rubber (copolymer of butadiene and acrylonitrile).

(*d*) Butyl rubber.

279. Methyl teritiary butyl ether (MTBE) produced by an acid catalyzed reaction of methanol with isobutene is a/an

(*a*) octane improver of gasoline

(*b*) plasticizer

(*c*) defoaming reagent

(*d*) liquid detergent

280. Nylon 6 is produced by polymerizing caprolactam in presence of catalytic amount of

(*a*) ammonia (*b*) hydrocyanic acid

(*c*) water (*d*) hydrofluoric acid

281. The properties of Nylon 6 are ________ nylon 66.

(*a*) very similar to

(*b*) somewhat similar to those of

(*c*) very different from

(*d*) unpredictable

282. Bakelite, introduced by Bakeland in 1910, is basically

(*a*) polycarbonates

(*b*) phenol-formaldehyde resins

(*c*) polyethylene

(*d*) poly vinyl chloride

283. Which of the following is a thermosetting polymer?

(*a*) Polyethylene (*b*) Polystyrene

(*c*) Polyvinyl chloride (*d*) Phenolic resin

284. PVC is manufactured by employing emulsion polymerisation. Solution polymerisation is carried out for ________ polymerisation reactions.

(*a*) slow rate (*b*) highly exothermic

(*c*) highly endothermic (*d*) high pressure

285. During manufacture of polythene by low pressure Zeigler process, vacuum is maintained in flash drum

(*a*) to facilitate production of polythene in pellet form.

(*b*) to facilitate production of polythene in sheet form.

(*c*) because the catalyst is highly flammable at high temperatures.

(*d*) to evaporate the solvent in solution polymerisation.

286. Chemical name of nylon 6, 6 is

(*a*) polyhexamethylene diamine,

(*b*) polyhexadiamine.

(*c*) polyhexamethylene adipimide.

(*d*) none of these

287. Match the examples of various types of polymers

List I

A. Thermosetting

B. Thermoplastic

C. Elastomer

D. Fibres.

List II

I. Polystyrene II. Rayon

III. Epoxy resin IV. Neoprene

(*a*) A-III, B-I, C-IV, D-II

(*b*) A-III, B-IV, C-I, D-II

(*c*) A-IV, B-III, C-II, D-I

(*d*) A-I, B-II, C-III, D-IV

Answers

276. (*d*)	277. (*b*)	278. (*d*)	279. (*a*)	280. (*d*)	281. (*a*)	282. (*b*)	283. (*d*)
284. (*b*)	285. (*c*)	286. (*c*)	287. (*a*)				

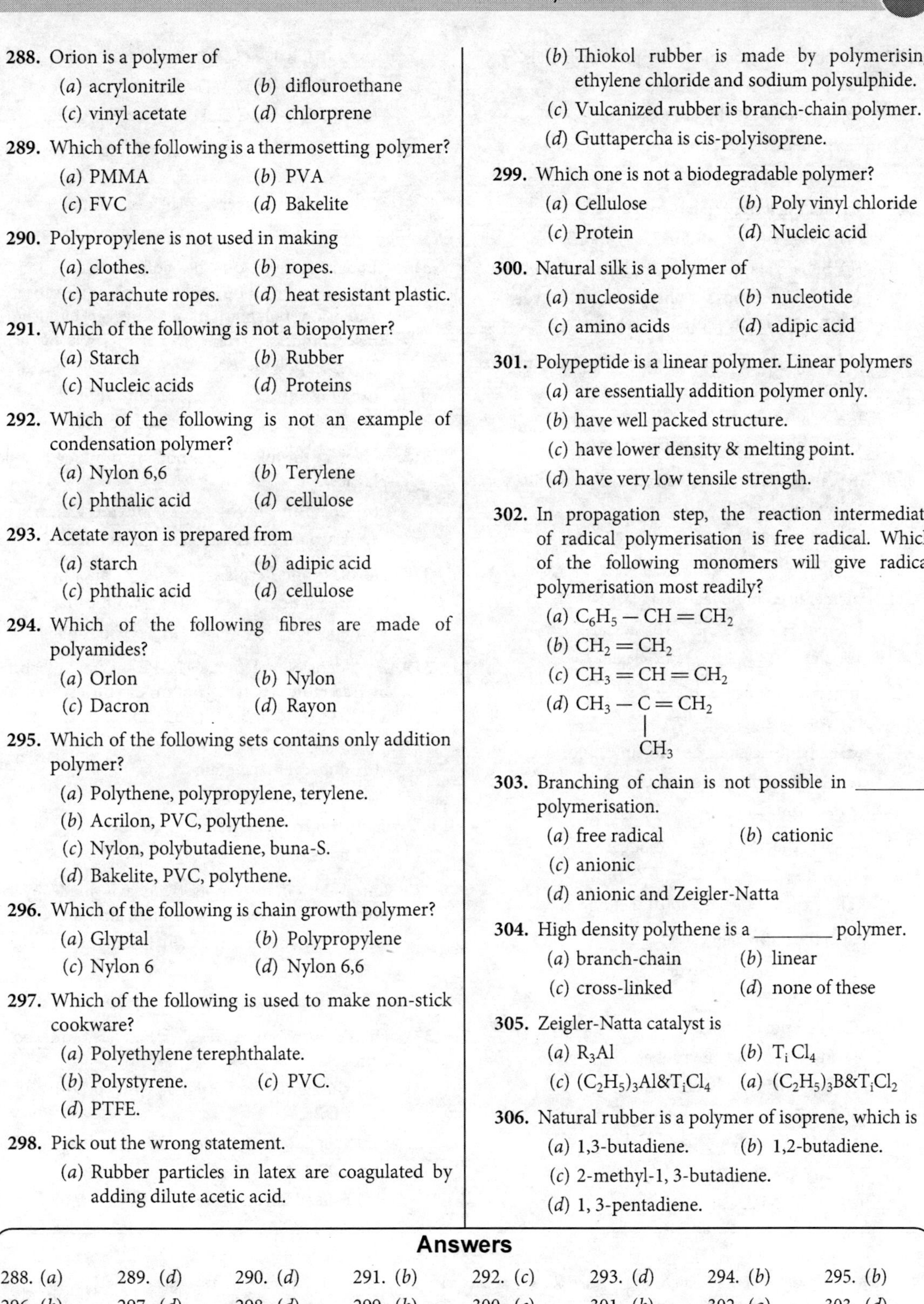

288. Orion is a polymer of

(*a*) acrylonitrile (*b*) diflouroethane
(*c*) vinyl acetate (*d*) chlorprene

289. Which of the following is a thermosetting polymer?

(*a*) PMMA (*b*) PVA
(*c*) FVC (*d*) Bakelite

290. Polypropylene is not used in making

(*a*) clothes. (*b*) ropes.
(*c*) parachute ropes. (*d*) heat resistant plastic.

291. Which of the following is not a biopolymer?

(*a*) Starch (*b*) Rubber
(*c*) Nucleic acids (*d*) Proteins

292. Which of the following is not an example of condensation polymer?

(*a*) Nylon 6,6 (*b*) Terylene
(*c*) phthalic acid (*d*) cellulose

293. Acetate rayon is prepared from

(*a*) starch (*b*) adipic acid
(*c*) phthalic acid (*d*) cellulose

294. Which of the following fibres are made of polyamides?

(*a*) Orlon (*b*) Nylon
(*c*) Dacron (*d*) Rayon

295. Which of the following sets contains only addition polymer?

(*a*) Polythene, polypropylene, terylene.
(*b*) Acrilon, PVC, polythene.
(*c*) Nylon, polybutadiene, buna-S.
(*d*) Bakelite, PVC, polythene.

296. Which of the following is chain growth polymer?

(*a*) Glyptal (*b*) Polypropylene
(*c*) Nylon 6 (*d*) Nylon 6,6

297. Which of the following is used to make non-stick cookware?

(*a*) Polyethylene terephthalate.
(*b*) Polystyrene. (*c*) PVC.
(*d*) PTFE.

298. Pick out the wrong statement.

(*a*) Rubber particles in latex are coagulated by adding dilute acetic acid.
(*b*) Thiokol rubber is made by polymerising ethylene chloride and sodium polysulphide.
(*c*) Vulcanized rubber is branch-chain polymer.
(*d*) Guttapercha is cis-polyisoprene.

299. Which one is not a biodegradable polymer?

(*a*) Cellulose (*b*) Poly vinyl chloride
(*c*) Protein (*d*) Nucleic acid

300. Natural silk is a polymer of

(*a*) nucleoside (*b*) nucleotide
(*c*) amino acids (*d*) adipic acid

301. Polypeptide is a linear polymer. Linear polymers

(*a*) are essentially addition polymer only.
(*b*) have well packed structure.
(*c*) have lower density & melting point.
(*d*) have very low tensile strength.

302. In propagation step, the reaction intermediate of radical polymerisation is free radical. Which of the following monomers will give radical polymerisation most readily?

(*a*) $C_6H_5-CH=CH_2$
(*b*) $CH_2=CH_2$
(*c*) $CH_3=CH=CH_2$
(*d*) $CH_3-\underset{\underset{CH_3}{|}}{C}=CH_2$

303. Branching of chain is not possible in ________ polymerisation.

(*a*) free radical (*b*) cationic
(*c*) anionic
(*d*) anionic and Zeigler-Natta

304. High density polythene is a ________ polymer.

(*a*) branch-chain (*b*) linear
(*c*) cross-linked (*d*) none of these

305. Zeigler-Natta catalyst is

(*a*) R_3Al (*b*) T_iCl_4
(*c*) $(C_2H_5)_3Al\&T_iCl_4$ (*a*) $(C_2H_5)_3B\&T_iCl_2$

306. Natural rubber is a polymer of isoprene, which is

(*a*) 1,3-butadiene. (*b*) 1,2-butadiene.
(*c*) 2-methyl-1, 3-butadiene.
(*d*) 1, 3-pentadiene.

Answers

288. (*a*)	289. (*d*)	290. (*d*)	291. (*b*)	292. (*c*)	293. (*d*)	294. (*b*)	295. (*b*)
296. (*b*)	297. (*d*)	298. (*d*)	299. (*b*)	300. (*c*)	301. (*b*)	302. (*a*)	303. (*d*)
304. (*b*)	305. (*c*)	306. (*c*)					

307. Monomer of neoprene chloroprene, which is
(*a*) 1-chloro-1, 3-butadiene.
(*b*) 2-chloro-1, 3-butadiene.
(*c*) 2-methyl-1, 3-butadiene.
(*d*) none of these.

308. Pick out the wrong statement.
(*a*) Plexiglases stands for PMMA.
(*b*) Buna-S stands for styrene butadiene rubber.
(*c*) Bakelite stands for phenol-formaldehyde.
(*d*) Teflon stands for poly vinyl acetate.

309. Which of the following polymers has an ester linkage?
(*a*) Nylon 6, 6 (*b*) Nylon 6, 10
(*c*) Bakelite (*d*) Dacron

310. Glyptal polymer is an alkyd resin obtained by reaction of glycerol with
(*a*) maleic acid (*b*) malonicacid
(*c*) phthalic acid (*d*) acetic acid

311. Polymer used in bullet proof glass is
(*a*) nomex (*b*) kevlar
(*c*) plexiglass (*d*) lexan

312. Saran is a co-polymer of
(*a*) vinyl acetate and styrene.
(*b*) vinylidene chloride and vinyl chloride.
(*c*) acrylonitrile and styrene.
(*d*) acrylonitrile.

313. Which of the following is a biodegradable polymer?
(*a*) Nylon 2,6 (*b*) Neoprene
(*c*) Perspex (*d*) Teflon

314. Match the following:
A. Phenol formaldehyde
B. Nylon
C. High purity PVC
D. Styrene butadiene rubber
I. Suspension polymerisation
II. Emulsion polymerisation
III. Solution polymerisation
IV. Bulk polymerisation
(*a*) A-III, B-IV, C-1, D-II
(*b*) A-III, B-IV, C-II, D-1
(*c*) A-IV, B-III, C-1, D-II
(*d*) A-IV, B-III, C-II, D-1

315. Carbon black used in rubber manufacture acts as a/an
(*a*) vulcanising agent. (*b*) accelerator.
(*c*) extender. (*d*) reinforcing agent.

316. Though styrene can be polymerised using all bulk polymerisation, solution polymerisation & suspension polymerisation; however in German Tower Process, styrene is polymerised by using ________ polymerisation.
(*a*) solution (*b*) emulsion
(*c*) bulk (*d*) suspension

317. Which of the following is not a reinforcing fibre for plastic?
(*a*) Nichrome fibres (*b*) Polymeric fibres
(*c*) Carbons fibres (*d*) Glass fibres

318. Novolaks and Resols are ________ plastics.
(*a*) phenolic (*b*) amino
(*c*) epoxy (*d*) epoxide

319. Low density polythene (LDPE) is manufactured by free radical polymerisation of ethylene (at 800-3000 °C & 1000-3000 atm). Initiator used in the polymerisation reaction is
(*a*) dodecyl mercaptans.
(*b*) ethylene oxide or fluorine.
(*c*) oxygen or an organic peroxide.
(*d*) none of these.

320. Linear low density polyethylene is made by co-polymerizing ethylene with small amount of
(*a*) propylene. (*b*) isobutene.
(*c*) isoamylene.
(*d*) 1-alkenes such as 1-butene or 1-hexene.

321. High density polyethylene (HDPE) is produced by a process-
(*a*) similar to that used for the production of LDPE.
(*b*) called co-ordination or Ziegler polymerization.
(*c*) called Clans process.
(*d*) called Halcon process.

Answers

307. (*b*)	308. (*d*)	309. (*d*)	310. (*c*)	311. (*d*)	312. (*b*)	313. (*a*)	314. (*a*)
315. (*d*)	316. (*c*)	317. (*a*)	318. (*a*)	319. (*c*)	320. (*d*)	321. (*b*)	

322. Which of the following ethylene polymers will you recommend for making buckets and washing-up bowls?

(*a*) LDPE. (*b*) HDPE.
(*c*) Linear LDPE.
(*d*) None of the foregoing; I will recommend polystyrene for the said purpose.

323. The major uses of ethylene glycol are as

(*a*) antifreeze agents in automobiles and as a raw material in the manufacture of poly ethyleneterephthalate.
(*b*) a plasticizer
(*c*) an antioxidant
(*d*) all *a*,*b* & *c*

324. Raw materials for the production of styrene are

(*a*) benzene and ethylene
(*b*) toluene and propylene
(*c*) toluene and ethanol
(*d*) m-xylene and ethylene

325. Gas phase dehydrogenation of ethyl-benzene to styrene occurs over catalyst based on

(*a*) iron oxide. (*b*) silica-alumina.
(*c*) titanium dioxide. (*d*) sodium silicate.

326. Acrylonitrile is produced by "oxidation" of

(*a*) ethylene (*b*) propylene
(*c*) isobutylene (*d*) styrene

327. Propylene, on hydration in presence of acidic ion exchange resin catalyst, produces isopropyl alcohol. This alcohol on dehydrogeneration gives

(*a*) propylene oxide. (*b*) ethylene oxide.
(*c*) maleic anhydride (*d*) methanol.

328. The major use of propylene tetramer is in the manufacture of

(*a*) surface-active agents, benzene and propylene tetramer reacting to give dodecyl benzene, which is subsequently sulfonated to produce dodecylbenzene.
(*b*) plasticizer alcohol by Oxo process.
(*c*) petroleum additives.
(*d*) solvents used by paint industry.

329. Manufacture of styrene-butadiene rubber (SBR) is one case where butadience is consumed to a large extent. SBR is used mainly in making

(*a*) carry-bags (*b*) toys
(*c*) tyres
(*d*) eraser (for removing pencil marks)

330. Nitrile rubber is a copolymer of butadiene and

(*a*) acetonitrile (*b*) adiponitrile
(*c*) acrylonitrile (*d*) styrene

331. Maleic anhydride can be produced by gas phase oxidation of

(*a*) n-butenes (*b*) isobutene
(*c*) ortho-xylene (*d*) naphthalene

332. Of the three polymer types, which one is the most important commercially?

(*a*) thermoplastics (*b*) thermosets
(*c*) elastomers.

333. Which one of the three polymer types is not normally considered to be a plastic?

(*a*) thermoplastics (*b*) thermosets
(*c*) elastomers.

334. Which one of the three polymer types does not involve cross-linking?

(*a*) thermoplastics (*b*) thermosets
(*c*) elastomers.

335. As the degree of crystallinity in a given polymer increases, the polymer becomes denser and stiffer, and its melting temperature decreases:

(*a*) true (*b*) false.

336. Which of the following is the chemical formula for the repeating unit in polyethylene?

(*a*) CH_2 (*b*) C_2H_4
(*c*) C_3H_6 (*d*) C_5H_8,

337. Degree of polymerization is which one of the following?

(*a*) average number of mers in the molecule chain
(*b*) proportion of the monomer that has been polymerized
(*c*) sum of the molecule weights of the mers in the molecule
(*d*) none of the above.

Answers

322. (*b*)	323. (*a*)	324. (*a*)	325. (*a*)	326. (*b*)	327. (c,d)	328. (*b*)	
329. (*c*)	330. (*c*)	331. (*a*)	332. (*a*)	333. (*c*)	334. (*a*)	335. (*b*)	336. (*b*)
337. (*a*)							

338. A branched molecular structure is stronger in the solid state and more viscous in the molten state than a linear structure for the same polymer:

(*a*) true (*b*) false.

339. A copolymer is a mixture consisting of macromolecules of two different homopolymers:

(*a*) true (*b*) false.

340. As temperature of a polymer increases, its density

(*a*) increases (*b*) decreases
(*c*) remains fairly constant.

341. Which answers complete the following sentence correctly (more than one): As the temperature of an amorphous thermoplastic polymer is gradually reduced, the glass transition temperature T_g is indicated when

(*a*) the polymer transforms to a crystalline structure
(*b*) the coefficient of thermal expansion increases markedly
(*c*) the slope of specific volume versus temperature changes markedly
(*d*) the polymer becomes stiff, strong, and elastic
(*e*) the polymer solidifies from the molten state.

342. Which of the following plastics has the highest market share?

(*a*) phenolics (*b*) polyethylene
(*c*) polypropylene (*d*) polystyrene,

343. Which of the following polymers are normally not thermoplastic?

(*a*) acrylics (*b*) cellulose acetate
(*c*) nylon (*d*) polychloroprene
(*e*) polyethylene

344. Polystyrene (without plasticizers) is amorphous, transparent, and brittle:

(*a*) true (*b*) false.

345. The fiber ***rayon*** used in textiles is based on which of the following polymers:

(*a*) cellulose (*b*) nylon
(*c*) polyester (*d*) polyethylene
(*e*) polypropylene.

346. The basic difference between low density polyethylene and high density polyethylene is that the latter has a much higher degree of crystallinity:

(*a*) true (*b*) false.

347. Among the thermosetting polymers, the most widely used commercially is which of the following:

(*a*) epoxies (*b*) phenolics
(*c*) silicones (*d*) urethanes.

348. Polyurethanes can be which of the following:

(*a*) thermoplastic (*b*) thermosetting
(*c*) elastomeric.

349. The chemical formula for polyisoprene in natural rubber is which of the following:

(*a*) CH_2 (*b*) C_2H_4
(*c*) C_3H_6 (*d*) C_5H_8.

350. The leading commercial synthetic rubber is which of the following:

(*a*) butyl rubber (*b*) isoprene rubber
(*c*) polybutadiene
(*d*) styrene-butadiene rubber

351. The most important rubber product is:

(*a*) footwear (*b*) conveyor belts
(*c*) pneumatic tires (*d*) tennis balls.

352. The chemical name of the ingredient recovered from the latex of the rubber tree is which one of the following?

(*a*) polybutadiene (*b*) polyisobutylene
(*c*) polyisoprene (*d*) polystyrene.

353. Of the following rubber additives, which one would rank as the single most important?

(*a*) antioxidants (*b*) carbon black
(*c*) clays and other hydrous aluminum silicates
(*d*) plasticizers and softening oils
(*e*) reclaimed rubber.

354. Which one of the following molding processes is the most important in the production of products made of conventional rubber?

(*a*) compression molding
(*b*) injection molding
(*c*) thermoforming
(*d*) transfer molding.

Answers

338. (*a*) 339. (*b*) 340. (*b*) 341. (*c*, *d*) 342. (*b*) 343. (*d*) 344. (*a*) 345. (*a*)
346. (*a*) 347. (*b*) 348. (*a*, *b*, *c*) 349. (*d*) 350. (*d*) 351. (*c*) 352. (*c*) 353. (*b*)
354. (*a*)

355. Which of the following ingredients do not contribute to the vulcanizing process (more than one)?

(*a*) calcium carbonate (*b*) carbon black
(*c*) stearic acid (*d*) sulfur
(*e*) zinc oxide.

356. How many minutes are required to cure (vulcanize) a modern passenger car tire?

(*a*) 5 (*b*) 15
(*c*) 25 (*d*) 45.

357. When is the tread pattern imprinted onto the circumference of the tire?

(*a*) during preforming
(*b*) while building the carcass
(*c*) during molding
(*d*) during curing.

358. Which of the following are not normally used in the processing of thermoplastic elastomers (more than one)?

(*a*) blow molding
(*b*) compression molding
(*c*) extrusion
(*d*) vulcanization.

359. Screw threads are not normally molded into rubber parts:

(*a*) true (*b*) false.

360. The shear viscosity of a polymer melt is affected by which of the following?

(*a*) degree of polymerization
(*b*) polymer type (*c*) rate of flow
(*d*) temperature.

361. Which three of the following are sections of a conventional extruder barrel for thermoplastics?

(*a*) compression section
(*b*) die section (*c*) feed section
(*d*) heating section (*e*) metering section
(*f*) shaping section.

362. Which of the following processes is not associated with the production of plastic sheet and film (more than one)?

(*a*) blown-film extrusion process
(*b*) calendering (*c*) chill-roll extrusion
(*d*) doctor blade method

363. ***Spinning in*** the production of synthetic fibers refers to which of the following:

(*a*) extrusion of polymer melt through small die openings
(*b*) drawing the strands to elongate and thin them
(*c*) both of the above
(*d*) none of the above.

364. The principal components of an injection molding machine are which two of the following?

(*a*) clamping unit (*b*) hopper
(*c*) injection unit (*d*) mold

365. The ***parting line*** in injection molding is which one of the following:

(*a*) the lines formed where polymer melt meets after flowing around a core in the mold
(*b*) the narrow gate sections where the parts are separated from the runner
(*c*) where the clamping unit is joined to the injection unit in the molding machine
(*d*) where the two mold halves come together

366. The function of the ejection system is to (one best answer):

(*a*) move polymer melt into the mold cavity
(*b*) open the mold halves after the cavity is filled
(*c*) remove the molded parts from the runner system after molding
(*d*) separate the part from the cavity after molding
(*e*) none of the above.

367. A three-plate mold offers which of the following advantages when compared to a two-plate mold?

(*a*) automatic separation of parts from runners
(*b*) gating is usually at the base of the part to reduce weld lines
(*c*) sprue does not solidify
(*d*) stronger molded parts
(*e*) none of the above.

368. Which of the following defects or problems is associated with injection molding?

(*a*) bambooing (*b*) die swell
(*c*) drag flow (*d*) flash

Answers

355. (*a*, *b*)	356. (*b*)	357. (*c*)	358. (*b*, *d*)	359. (*a*)	360. (*a*, *b*, *c*, *d*)	361. (*a*, *c*, *e*)
362. (*d*)	363. (*c*)	364. (*a*, *c*)	365. (*d*)	366. (*d*)	367. (*a*, *b*)	368. (*d*, *f*, *g*)

(*e*) melt fracture (*f*) short shots
(*g*) sink marks.

369. In rotational molding, centrifugal force is used to force the polymer melt against the surfaces of the mold cavity where solidification occurs:
(*a*) true (*b*) false.

370. Use of a parison is associated with which one of the following plastic shaping processes?
(*a*) bi-injection molding
(*b*) blow molding
(*c*) compression molding
(*d*) pressure thermoforming.

371. A thermoforming mold with a convex form is called which one of the following (may be more than one)?
(*a*) a die (*b*) a negative mold
(*c*) a positive mold (*d*) a three-plate mold.

372. The term ***encapsulation*** refers to which one of the following plastics shaping processes?
(*a*) casting
(*b*) compression molding
(*c*) extrusion of hollow forms
(*d*) injection molding in which a metal insert is encased in the molded part
(*e*) vacuum thermoforming using a positive mold.

373. Which of the following terms applies to the processing of foam plastics (more than one)?
(*a*) chemical blowing agents
(*b*) open cell structure
(*c*) powder injection molding
(*d*) sandwich molding
(*e*) structural foam molding

374. The two most common polymer foams are which of the following?
(*a*) polyacetal (*b*) polyethylene,
(*c*) polystyrene (*d*) polyurethane
(*e*) polyvinylchloride.

375. In which of the following property categories do plastic parts compare favorably with metals (more than one)?
(*a*) impact resistance
(*b*) resistance to ultraviolet radiation
(*c*) stiffness (*d*) strength,
(*e*) strength-to-weight ratio
(*f*) temperature resistance.

376. Which of the following processes are generally limited to thermoplastic polymers (more than one)?
(*a*) blow molding
(*b*) compression molding
(*c*) reaction injection molding
(*d*) thermoforming (*e*) transfer molding
(*f*) wire coating.

377. Which of the following processes would be applicable to produce hulls for small boats (more than one)?
(*a*) blow molding
(*b*) compression molding
(*c*) injection molding (*d*) rotational molding
(*e*) vacuum thermoforming.

378. Which one of the following is the most common polymer type in fiber-reinforced polymer composites?
(*a*) elastomers (*b*) thermoplastics
(*c*) thermosets.

379. Most rubber products are properly classified into which of the following categories (more than one)?
(*a*) elastomer reinforced with carbon black
(*b*) fiber-reinforced composite
(*c*) particle-reinforced composite
(*d*) polymer matrix composite
(*e*) pure elastomer
(*f*) pure polymer.

380. Hand lay-up is classified in which of the following general categories of PMC shaping processes (more than one)?
(*a*) closed mold process
(*b*) compression molding
(*c*) contact molding (*d*) filament winding
(*e*) open mold process.

381. A positive mold with a smooth surface will produce a good finish on which surface of the laminated product in the hand lay-up method?
(*a*) inside surface (*b*) outside surface.

Answers

369. (*b*) 370. (*b*) 371. (*c*) 372. (*a*) 373. (*a, b, d, e*) 374. (*c, d*) 375. (*a, e*)
376. (*a, d*) 377. (*a, d, e*) 378. (*c*) 379. (*a, c, d*) 380. (*c, e*) 381. (*a*)

382. SMC molding is a form of which one of the following?
(*a*) compression molding
(*b*) contact molding
(*c*) injection molding
(*d*) open mold processing
(*e*) pultrusion
(*f*) transfer molding.

383. Filament winding involves the use of which one of the following fiber reinforcements?
(*a*) continuous filaments
(*b*) fabrics (*c*) mats
(*d*) prepregs (*e*) short fibers
(*f*) woven rovings.

384. In filament winding, when the continuous filament is wound around the cylindrical mandrel at a helix angle close to 90°, it is called which of the following (one best answer)?
(*a*) bi-axial winding (*b*) helical winding
(*c*) hoop winding
(*d*) perpendicular winding
(*e*) polar winding (*f*) radial winding.

385. Pultrusion is most similar to which one of the following plastic shaping processes?
(*a*) blow-molding (*b*) extrusion
(*c*) injection molding (*d*) thermoforming.

386. Water jet cutting is one of several ways of cutting or trimming uncured or cured FRPs; in the case of cured FRPs, the process is noted for its reduction of dust and noise:
(*a*) true (*b*) false.

387. Anisotropic means which one of the following:
(*a*) composite materials with composition consisting of more than two materials
(*b*) properties are the same in every direction
(*c*) properties vary depending on the direction in which they are measured
(*d*) strength and other properties as a function of curing temperature.

388. The reinforcing phase is the matrix within which the secondary phase is imbedded:
(*a*) true (*b*) false.

389. Which one of the following reinforcing geometries offers the greatest potential for strength and stiffness improvement in the resulting composite material?
(*a*) fibers (*b*) flakes
(*c*) particles (*d*) infiltrated phase.

390. Wood is which one of the following composite types?
(*a*) CMC (*b*) MMC
(*c*) PMC.

391. Which of the following materials is not used as fibers in fiber-reinforced plastics:
(*a*) aluminum oxide (*b*) boron
(*c*) carbon/graphite (*d*) epoxy
(*e*) Kevlar 49 (*f*) S-glass
(*g*) unsaturated polyester.

392. Which of the following metals are most commonly used as the matrix material in fiber-reinforced MMCs (name three)?
(*a*) aluminum (*b*) copper
(*c*) iron (*d*) magnesium
(*e*) titanium (*f*) zinc.

393. Which of the following metals is used as the matrix metal in nearly all WC cemented carbides?
(*a*) aluminum (*b*) chromium
(*c*) cobalt (*d*) lead
(*e*) nickel (*f*) tungsten
(*g*) tungsten carbide.

394. Ceramic matrix composites are designed to overcome which of the following weaknesses of ceramics (more than one)?
(*a*) compressive strength
(*b*) hardness (*c*) hot hardness
(*d*) modulus of elasticity
(*e*) tensile strength (*f*) toughness.

395. Which of the following polymer types are most commonly used in polymer matrix composites?
(*a*) elastomers (*b*) thermoplastics
(*c*) thermosets.

396. Which one of the following is the most common reinforcing material in FRPs?
(*a*) Al_2O_3 (*b*) boron
(*c*) carbon (*d*) SiO_2.

Answers

382. (*a*)	383. (*a*)	384. (*c*)	385. (*b*)	386. (*a*)	387. (*c*)	388. (*b*)	389. (*a*)
390. (*c*)	391. (*d*)	392. (*a, d, e*)	393. (*c*)	394. (*e, f*)	395. (*c*)	396. (*d*)	

Glossary

A

ABS: A polymer produced by the copolymerization of acrylonitrile, butadiene, and styrene.

Accelerator: A catalyst for the vulcanization of rubber.

Acetal: A polymer produced by the polymerization of formaldehyde.

Acetate rayon: Cellulose diacetate.

Alkyd: A polyester produced by the condensation of a dihydric alcohol, such as ethylene glycol, and a dicarboxylic acid, such as phthalic acid, in the presence of controlled amounts of an unsaturated monofunctional organic acid, such as oleic acid.

Acrilan: Polyacrylonitrile-based fibers.

Acrylamide: A polymer formed from monomers based on acrylic acid amides or methacrylic acid amides, e.g., polyacrylamide.

Acrylic: A polymer formed from monomers based on acrylic acid and its esters or methacrylic acid and its esters, e.g., Poly (acrylic acid) (Carbopol®, Carbomer), poly (methyl methacrylate-co-methacrylic acid

Acrylic fibers: Fibers based on polyacrylonitrile.

Acrylonitrile: A common monomer used in free-radical polymerization.

Addition polymerization: Also known as chain-growth polymerization. The mechanism in which large numbers of usually identical small molecules are joined together to rapidly form a single large molecule. This involves the addition of reactive centre (anion, cation, or unpaired electron) to a multiple bond to form a new bond and a new reactive centre - which reacts with another multiple bond . The finished chain then hangs around without reacting while more of the starting material reacts to form new polymer chains.

Adiprene: Polyurethane elastomer.

Alkyds: Originally used to describe oil-modified polyesters, but now used for many polyester plastics and coatings.

Alternating copolymer: a copolymer consisting of macromolecules comprising two species of monomeric units in alternating sequence.

Alternating copolymerization: a copolymerization in which an alternating copolymer is formed.

Amorphous: Having no ordered arrangement. Polymers are amorphous when their chains are tangled up in any old way. Polymers are *not* amorphous when their chains are lined up in ordered crystals.

Amorphous polymer: Amorphous means irregular, having no discernible order or shape. In the context of solids, the molecules are randomly arranged, as in glass, rather than periodically arranged as in a crystalline material. Amorphous polymer has a glass like structure with tangled chain and no long range order.

Amylose: a linear starch polymer.

Anionic polymerization: A polymerization initiated by an anion.

Antioxidant: A compound that retards the degradation of a polymer

Antiplasticization: The hardening and stiffening effect observed when small amounts of a plasticizer are added to a polymer such as PVC.

Antistat: An additive that reduces static charges on polymers.

Araldite: Trade name for an epoxy resin.

Atactic: A polymer in which there is a random arrangement of pendant groups on each side of the chain, as in atactic Polypropylene

Atactic polymer: a substance composed of atactic macromolecules.

B

Bakelite: A polymer produced by the condensation polymerization of phenol and formaldehyde.

Balata: A naturally occurring plastic (trans-1,4-polyisoprene).

Biocide: An additive that retards attack by microorganisms.

Biopolymer: A naturally occurring polymer, such as protein.

Bingham plastic: A plastic that does not flow until the external stress exceeds a critical threshold value.

Block copolymer: A polymer comprising more than one type of monomer repeat unit, in which each type of monomer is found in a homopolymerized "block" within the polymer chain, e.g., poly(ethylene oxide-b-propylene oxide)

Block polymer: a substance composed of block macromolecules.

Branched polymer: A polymer having extensions of the polymer chain attached to the polymer backbone, such as LDPE. Polymers having pendant groups, such as the methyl groups in polypropylene, are not considered to be branched polymers.

Bound rubber: Rubber adsorbed on carbon black that is insoluble in benzene.

Bulk polymerization: The polymerization of monomer without added solvents or water.

Buna: Copolymer of butadiene and styrene or acrylonitrile.

Buna-N: An elastomeric copolymer of butadiene and acrylonitrile.

Buna-S: An elastomeric copolymer of butadiene and styrene.

Butyl rubber: A copolymer of isobutylene and isoprene.

C

Capping: Reacting the end groups to produce a stable polymer.

Ceiling temperature: A characteristic temperature above which polymerization does not take place and polymers decompose.

Cellophane: Regenerated cellulose sheet.

Cellulose: A naturally occurring carbohydrate polymer consisting of repeating glucose units.

Cellulose acetate: Product formed from the acetylation of cellulose; properties vary with degree of substitution.

Cellulose nitrate: Also known as nitrocellulose and gun cotton; made from the reaction of cellulose and concentrated nitric acid.

Chain growth (addition) polymerization – a polymerization occurring by addition of successive monomer units, one at a time, to a reactive growing polymer chain, e.g., free-radical polymerization, ring-opening polymerization, ionic polymerization.

Chain polymerization: a chain reaction in which the growth of a polymer chain proceeds exclusively by reaction(s) between monomer(s) and reactive site(s) on the polymer chain such that the reactive site(s) are regenerated on the same polymer chain by the end of each growth step. A chain polymerization consists of initiation and propagation reactions, and may also include termination and chain transfer reactions. The adjective "chain" in "chain polymerization" denotes "chain reaction". In a chain polymerization, the average degree of polymerization remains constant with monomer conversion (e.g., in steady-state, radical polymerizations) or may increase with monomer conversion (e.g., in the formation of living polymers).

Chemical blowing agents: Volatile liquids and gases that expand and/or volatilize during processing of a polymeric material, creating pockets leading to lighter weight materials.

Coextruded film: One produced by the simultaneous extrusion of two or more polymers.

Cold drawing: The stretching of a fiber or fibers to obtain products with high tensile strength.

Compression molding press: A press that uses external pressure to force a heat-softened molding compound into a die to produce a molded article.

Condensation reaction: A reaction in which two molecules react to produce a third molecule and a byproduct such as water.

Comonomer : A monomer that is polymerized along with one or more other monomers to make a copolymer. All the different comonomers used in a copolymerization are incorporated into each chain.

Composites: Materials that contain strong fibers embedded in a continuous phase called a matrix or resin.

Condensation polymerization : Also known as Step-Growth Polymerization. A way of making polymers in which every polymer chain grows continuously through the course of the reaction, remaining quite small until almost all the monomer has reacted.

Continuous phase: The resin in a composite.

Co-oligomer: An oligomer derived from more than one species of monomer.

Co-oligomerization: Oligomerization in which a co-oligomer is formed.

Copolymer: A polymer derived from more than one species of monomer. Note: copolymers that are obtained by copolymerization of two monomer species are sometimes termed bipolymers, those obtained from three monomers terpolymers, those obtained from four monomers quaterpolymers, etc.

Copolymerization: Polymerization in which a copolymer is formed.

Coupling agents: Products such as silanes or organic titanates that improve the interfacial bond between filler and resin.

Critical molecular weight: The threshold value of molecular weight required for chain entanglement.

Cross linking: A process in which bonds are formed joining adjacent molecules. At low density, these bonds add to the elasticity of the polymer and at high density eventually produce rigidity in the polymer.

Crystalline polymer: A polymer with ordered structure that has been allowed to disentangle and form crystals such as HDPE. Thus, isotactic polypropylene, cellulose, and stretched rubber are crystalline polymers.

Cuprammonia rayon: Rayon produced from a solution of cellulose in a copper ammonia hydroxide solution.

Cure: The process of changing property of polymer into a more stable and usable condition. This is accomplished by the use of heat, radiation or reaction with chemical addition.

Curing: Crosslinking to produce a network polymer.

Curing agent: An additive that causes crosslinking.

Cyclic polymer (ring polymer): A polymer that is a continuous chain of chemically linked monomer repeat units that are joined at the ends to form a ring, e.g., cyclodextrin (strictly a cyclic oligomer).

Cyclized rubber: Isomerized rubber containing cyclohexane rings.

D

Dacron: Trade name for a polyester fiber.

Dead polymer: A polymer in which chain growth has been terminated.

Degree of polymerization: The number of repeating monomeric units in a polymer chain.

Depolymerization: The chemical reaction which results in a polymer chain being broken up into monomer units. For most polymers made by addition polymerization, this is done by heating the polymer above its ceiling temperature in the absence of oxygen. Some polymers, like styrene and vinyl chloride, will be difficult to depolymerise because the bonds between the side-groups (the phenyl ring and chlorine in these examples) are weaker than the C-C bonds between ex-monomers, and the polymer will degrade into different species than the starting materials. Other polymers, such as those made by step growth polymerization may be more easily depolymerised by undoing the condensation or elimination reaction that caused the monomer units to join.

Dielectric constant: Ratio of the capacitance of a polymer to that of air or a vacuum.

Dielectric strength: Maximum applied voltage that a polymer can withstand.

Dispersion polymerization: A polymerization carried out initially in a one-phase system with monomer, polymer and initiator dissolved in a common solvent, in which the polymer is insoluble. The polymer precipitates during the course of the reaction to yield a dispersion of polymer particles in the liquid.

Drier: A catalyst that aids the reaction of polymers with oxygen.

Drying: The crosslinking of an unsaturated polymer chain.

Dynel: Trade name for a copolymer of vinyl chloride and acrylonitrile.

E

Ebonite: Hard rubber; highly crosslinked natural rubber (NR).

Elastomer: A rubber or elastomer is any material that can be stretched many times its original length without breaking, *and* will snap back to its original size when it is released.

Elongation: How long a sample is stretched when it is pulled. Elongation is usually expressed as the length after stretching divided by the original length.

Emulsion polymerization: A polymerization performed in an oil- in-water emulsion (monomer in the oil droplets, surfactant and initiator in the aqueous continuous phase) using a water-soluble initiator where initiation occurs in the aqueous continuous phase. The polymerization occurs mainly in the monomer swollen surfactant micelles in the continuous phase, fueled by diffusion of monomer from the droplets to the micelles, and not in the monomer droplets themselves. The result is a polymer latex; a dispersion of polymer particles usually less than 1 μ in diameter, with a narrow size distribution.

Engineering plastics: Those plastics with physical properties good enough for use as structural materials.

Epoxy resin: A polymer produced by the condensation of epichlorohydrin and a dihydric alcohol or by the epoxidation of an unsaturated compound.

Eutactic: An isotactic or syndiotactic polymer.

Extrusion: A fabrication process in which a heat-softened polymer is forced continually by a screw through a die.

F

Fibrillation: Process for producing fiber by heating and pulling twisted film strips.

Filament: The extrudate when a polymer melt is forced through a hole in a spinneret.

Filament winding: A process in which resin-impregnated continuous filaments are wound on a mandrel and the composite is cured.

Filler: Usually a relatively inert material used as the discontinuous phase of a resinous composite.

Fill Pattern: It is a visual history of how a mould fills under a specific set of moulding conditions.

Fill Time: Fill time is the time in seconds that it takes to just fill the mould with polymer melt during the filling phase (up to switchover point) of the moulding cycle.

Flame retardant: An additive that increases the flame resistance of a polymer.

Flow Balancing: Flow balancing is the process of choosing a mould design strategy, which promotes even filling of mould. First, a gating strategy is selected which will promote ease of filling. Second, a runner layout is developed to feed material to the gates. Finally, runner dimensions are identified which will cause all of the flow paths to fill at the same time. Even wall thickness can be manipulated by flow leader or flow deflectors.

Flow path: A flow path describes the route that is traveled by a polymer melt front as melt fills a section of a mould.

Free radical: An atom or molecule which has at least one electron which is not paired with another electron.

Functionality: The number of active functional groups in a molecule.

G

Gating strategy: It is the approach we use to choose the number, dimension and location of gates in a mould.

Gel: A crosslinked polymer which has absorbed a large amount of solvent. Crosslinked polymers usually swell a good deal when they absorb solvents.

Gel point: The point at which crosslinking begins.

Glass transition temperature - the temperature at which a polymer changes from hard and brittle to soft and pliable.

Glyptal: A polyester produced by the condensation of glycerol and phthalic anhydride.

Graft copolymer: A branched polymer in which the branches have a different chemistry than the main chain, e.g., poly (L-lysine)-g-poly (ethylene oxide).

Gutta-percha: While natural rubber from the rubber tree *Hevea brasiliensis* is cis-poly(isoprene), the Malaysian sapodilla tree *Palaquium oblongifolia* produces a latex that is trans-poly(isoprene). This material is called gutta-percha and is more brittle and harder than natural rubber.

H

HDPE: High-density polyethylene.

Heat stabilizers: Additives that retard the decomposition of polymers at elevated temperatures .

High-density polyethylene: Polyethylene that contains a relatively high degree of crystallinity with a low degree of branching, typically made using solid state catalysts.

Hold on pressure: Follow up pressure: It is pressure on polymer melt after the switch over point in the moulding process.

Homopolymer: A polymer made up of only one repeating unit, in contrast to a copolymer, which is made up of more than one repeat unit.

Homopolymerization: polymerization in which a homopolymer is formed.

Hydrocellulose: Cellulose degraded (depolymerized) by hydrolysis.

I

Impact modifiers: Materials that improve the resistance of materials to stress.

Internal plasticization: The flexibilization resulting from the introduction of bulky groups in a polymer by copolymerization.

Initiation: The first step in a chain growth polymerization in which an active species reacts with the first monomer of the polymer chain.

Initiator: molecule that can react with a monomer molecule to start a chain growth polymerization. Usually a free-radical generator (e.g., peroxide, peroxydicarbonate, azo-compound), or in some cases a carbon centered cation

Injection rate: It is the flow rate of polymer melt (cc/sec) coming out of nozzle. Melt comes out in the form of jet. Then it spreads inside the mould in shape of the space between the core and cavity.

Injection speed/velocity profile: Before the switch over point the set pressure remains constant but set injection speed is varied with stroke position. After switch over point injection speed should remain at lower value and pressure is changed with time till the mould is just filled without over packing. The variations in the set speed throughout the injection stroke is called Injection speed profile.

Injection pressure profile: Before the switch over point the set pressure remains constant but set injection speed is varied with stroke position. After switch over point injection speed should remain at lower value and pressure is changed with time till the mould is just filled without over packing. This set pressure for filling stage and follow up / hold on pressure change with respect to time is called Injection pressure profile.

Inorganic polymers: Those containing elements other than typically carbon, nitrogen, and oxygen in their backbone or pendant groups.

Isomerization: Term often applied to cyclization reactions of polymers.

Isotactic macromolecule: a macromolecule comprising only one species of configurational base unit (having chiral or prochiral atoms in the main chain) in a single arrangement with respect to its adjacent constitutional units. Note: in an isotactic macromolecule, the configurational repeating unit is identical with the configurational base unit.

Isotactic polymer: A substance composed of isotactic macromolecules.

L

Lacquers: Polymer solutions to which pigments have been added.

Latex: A latex is a dispersion of water-insoluble polymer in water. The dispersion is usually of particles (not single molecules) that are around 100 nm (10^{-4} mm) in size. The particles are kept suspended in the water by thermal convection (which keeps them from settling out) and surfactants (which keep them from sticking together to form bigger and bigger lumps). Another word for latex is *polymer colloid*.

Linear low-density polyethylene: A copolymer of ethylene and about 10% of an olefin; it has densities and properties between LDPE and HDPE.

Linear polymer: A polymer like HDPE that consists of a linear chain without chain-extending branches.

Linear copolymer: A copolymer, the macromolecules of which are linear chains.

Living polymerization: A polymerization reaction in which there is no termination, and the polymer chains continue to grow as long as there are monomer molecules to add to the growing chain.

Low-density polyethylene (LDPE): Formerly called high-pressure polyethylene, a branched polymer produced by the free radical-initiated polymerization of ethylene at high pressure.

M

Macromolecule: A molecule of high relative molecular mass, the structure of which essentially comprises the multiple repetition of a number of constitutional units.

Macromonomer: A monomer which is a polymer of an oligomer.

Melamine-formaldehyde resin: A resin produced by the condensation of melamine and formaldehyde

Melt expansion: During the injection stroke polymer melt is compressed by about 5 to 15%. After melt enters the mould it gets the chance to relax and expand.

Melt flow index: It is the weight of polymer melt in grams extruded in 10 minutes through a standard nozzle or die under standard load condition at a certain temperature.

Melt front: As the mould is filling, the melt at the leading edge of flow is called the melt front or stream.

Melt front velocity: It is not same as injection piston (screw velocity) velocity. As melt enters mould it spreads in all direction depending on the resistance to flow in each direction. Melt front velocity is determined by the area of flow, which does not remain constant with time.

Melt spinning: Process of obtaining fibers by forcing molten polymer through holes in a spinneret and cooling the filaments produced.

Mercerized cotton: Cotton fiber that has been immersed in caustic solution, usually under tension, and washed with water to remove the excess caustic.

Melting temperature (T_m) – a temperature at which crystalline or semi-crystalline polymers melt to form an amorphous phase. T_m is higher than T_g.

Mer: The repeating unit in a polymer chain.

Metering stroke: It is stroke of screw that determines the quantity of polymer melt to be injected to the mould.

Microorganism inhibitors: Compounds that impart the ability to resist habitation and destruction by microorganisms.

Mold release agent: A lubricant that prevents the polymer from sticking to the mold cavity.

Monodisperse: A system consisting of molecules of one molecular weight only.

Monomer: A small molecule which may react chemically to link together with other molecules of the same type to form a large molecule called a polymer.

Monomeric unit/mer: The largest constitutional unit contributed by a single monomer molecule to the structure of macromolecule or oligomer molecule.

Mould cooling: Mould cooling describes the process by which the melt temperature is reduced to the point where part can be removed from the mould.

Mould packing/follow up pressure/Hold on phase: Mould packing is the process of delivering an additional amount of melt to the mould, to compensate for the shrinkage after filling.

Mould temperature: It refers to temperature of mould surface in contact with melt. It varies from point to point on surface if cooling design does not provide uniform heat extraction. It also fluctuates if heat is not balanced.

Moulding cycle: The moulding cycle is the series of steps that result in the machine producing a part. The cycle is usually described by breaking down into four separate phases. The amount of time that it takes to complete one cycle is called cycle time.

N

Natural rubber: Polyisoprene obtained from rubber plants.

Novolac: A thermoplastic phenol-formaldehyde resin prepared by the condensation of phenol and a small amount of formaldehyde on the acid side. Novolacs may be advanced to thermoset by heating with formaldehyde, which is usually obtained by the thermal decomposition of hexamethylene tetramine.

Neoprene: Polychloroprene.

Network polymer: Infusible crosslinked polymer.

No flow temperature: No flow temperature is the temperature at which the viscosity of the polymer melt is so high that it effectively cannot be made to flow.

Nonwoven textiles: Sheet produced by binding fibers with a heated thermoplastic.

Novolac: Polymers prepared by the condensation of phenol and formaldehyde under acidic conditions

Nylon-4,6: Engineering material formed from 1,4-butanediamine and adipic acid.

Nylon-66 (or nylon-6,6): A polyamide produced by the condensation of adipic acid and hexamethylenediamine.

Nylon-610 (PA): A polyamide synthesized from a 6-carbon diamine and a 10-carbon dicarboxylic acid.

Nylon salt: A salt of a diamine and a dicarboxylic acid used as the precursor of nylon.

O

Oligomer: A very low molecular weight polymer in which the number of repeating units (*n*) equals 2 to 10.

Oligomeric polyesters: Low molecular weight polyesters.

Oligomerization: The process of converting a monomer or a mixture of monomers into an oligomer

Orientation: Orientation is the change in shape that polymer molecules can undergo when they are made to flow.

P

Parlon: Trade name for chlorinated rubber.

Parting line: Contour line on the part separating core and cavity.

Periodic copolymer: A copolymer consisting of macromolecules comprising more than two species of monomeric units in regular sequence.

Periodic copolymerization: A copolymerization in which a periodic copolymer is formed.

PET: Short for poly(ethylene terephthalate), a condensation polymer that is commonly used in soft-drink bottles. It can be prepared by the reaction between ethylene glycol and terephthalic acid to give polymer and water:

Phenoxy resin: A polymer with hydroxyl pendant groups resembling an epoxy resin without epoxy groups.

Plasticizer: A small molecule that is added to polymer to lower its glass transition temperature.

Plastisol: A suspension of finely divided polymer in a liquid plasticizer. The latter penetrates and plasticizes the polymer when heated.

Plastomers: It is the name given to copolymers of ethylene that have a little crystallinity, but are largely amorphous. They are also called very low-density polyethylene (VLDPE). They are more elastic than LLDPE but less stiff. They are used as a sealing layer in film applications and controlled permeation packaging for vegetables and fruits.

Polyacetylene: polymer whose conductivity increases when doped.

Polyaddition: a polymerization in which the growth of a polymer chain proceeds by addition reactions between molecules of all degrees of polymerization, not accompanied by the formation of low-molar-mass by-product(s).

Polyamide: A polymer formed by reaction of poly (carboxylic acids) with polyamines, whose monomer repeat units are linked together by amide groups, e.g., poly (hexamethylene adipamide) (Nylon 6, 6).

Poly (butylene terephthalate) (PBT): Engineering polyester that quickly crystalizes ; formed from 1,4-butanediol and terephthalic acid.

Polycarbonate: A polymer whose monomer repeat units are linked together by carbonate groups, most commonly by reaction of bisphenol-A with phosgene, e.g., Lexan™.

Polycondensation: A polymerization in which the growth of a polymer chain proceeds by condensation reactions between molecules of all degrees of polymerization accompanied by the formation of low-molar- mass by-products(s). In a polycondensation where the total amounts of the monomers are present from the beginning of the polymerization, the average degree of polymerization increases with conversation of reactive groups.

Polydisperse: A polymer consisting of molecules of many different molecular weights, such as commercial HDPE.

Polyester: A polymer formed by reaction of poly (carboxylic acids) with polyols, whose monomer re- peat units are linked together by ester groups, e.g., polyethylene terephthalate (PET)

Polyester plastic: A name used to describe polymers produced by the polymerization of a solution of an unsaturated polyester in styrene.

Polyether: A polymer whose monomer repeat units are linked together by ether linkages, e.g., poly (ethylene glycol) PEG).

Polymer: A substance composed of macromolecules.

Polymerization: The process of converting a monomer or a mixture of monomers into a polymer.

Poly (ethylene terephthalate) (PET): A linear polyester used for fibers and for blow-molding soft drink bottles.

Polyolefin: A polymer of olefins, molecules that have an alkene (double bond) functionality. These polymers include polystyrene and poly (vinyl chloride). They are normally formed by free radical polymerization (a form of addition polymerization).

Polyoxymethylene (POM): Engineering material formed from the polymerization of formaldehyde.

Polyurethane: A polymer formed by reaction of polyisocyanates with polyols, whose monomer repeat units are linked together by urethane (carbamate) groups,

Prepolymer: A low molecular weight material (oligomer) capable of further polymerization produced by step-reaction polymerization.

Propagation: The chemical reaction of monomer units to a growing chain to incorporate them.

Pultrusion: A process in which bundles of resin-impregnated filaments are passed through an orifice and cured.

R

Radical copolymerization: A copolymerization which is a radical polymerization.

Random copolymer: A copolymer in which there is no definite order for the sequence of the different mers or building blocks ($MM_1M_2M_1M_1M_2M_1M_2M_2M$)

Random copolymerization: A copolymerization in which an alternating copolymer is formed.

Rayon: Cellulose regenerated by acidification of a cellulose xanthate (viscose) solution.

Regenerated cellulose: Cellulose obtained by precipitation from solution.

Regular polymer: A substance composed of regular macromolecules.

Reinforced plastic: A composite whose additional strength is dependent on a fibrous additive.

Resole: A condensation product produced by the reaction of phenol and formaldehyde under alkaline conditions.

Ring-opening polymerization: A polymerization in which cyclic monomer is converted into a polymer which does not contain rings. The monomer rings are opened up and stretched out in the polymer chain.

Rotational molding: Polymer added to a warm, rotating mold; centrifugal force distributes the polymer evenly.

Roving: A bundle of untwisted strands (a bundle of filaments).

Ryton: Trade name for poly (phenylene sulfide).

S

Saran: Trade name for polymers of vinylidene chloride.

SBR: A rubbery copolymer of styrene and butadiene.

Shear thinning: Shear thinning is the description for the physical effects of orientation and affect the flow behavior of the polymer. Shear thinning causes the melts viscosity to drop when it is made to flow within a certain shear rate range.

Side-chain crystallization: A stiffening effect noted when long, regularly spaced bulky pendant groups are present on a polymer chain.

Silk: Natural protein fiber.

Solution polymerization: A polymerization carried out in a one- phase system with monomer, polymer and initiator dissolved in a common solvent, in which the final polymer product is also soluble.

Solvent molding: Immersion of a mold into a polymer solution resulting in the mold being coated with a polymeric film.

Spandex: An elastic fiber consisting of a block copolymer of a polyurethane and a polyester.

Spinneret: Small holes through which a solution or molten polymer is extruded in order to form filaments for fibers.

Star polymer: A polymer in which three or more linear polymer chains radiate from a central, multifunctional moiety to which they are all chemically attached, either by coupling or by being grown from that moiety in the polymerization step.

Step growth (condensation) polymerization: A polymerization that proceeds by linking together multi-functional monomers to first dimers (two monomers) then trimers (three monomers), etc., which themselves link together to form high molecular weight polymer towards the end of the polymerization, e.g., polyesteri- fication, polyamidation.

Step-reaction polymerization: Polymerization in which polyfunctional reactants react to produce larger units in a continuous stepwise manner.

Styrofoam: Foamed polystyrene.

Suspension polymerization: A process in which liquid monomers are polymerized as liquid droplets suspended in water.

Syndiotactic Polymer: A polymer in which the pendant groups are arranged alternately on each side of the polymer backbone, as in syndiotactic Polypropylene

T

Tacticity: The orderliness of the succession of configurational repeating units in the main chain of a regular macromolecule (or oligomer or block).

Telechelic polymers: Polymers containing two reactive end groups.

Tenacity: A term for the tensile strength of fibers.

Termination: In a chain growth polymerization, the reaction which causes the growing chain to stop growing. Termination reactions are reactions in which none of the products may react to make a polymer grow.

Terpolymer: A copolymer comprising three different monomer repeat units, e.g., poly (acrylonitrile-co-butadiene-co-styrene).

Thermal stability: It is polymers thermal characteristic. It is the time interval for which polymer remains stable at certain temperature. Beyond that time if the polymer is exposed to longer duration for a given temperature, it degrades.

Terylene: Poly (ethylene terephthalate).

Thermoforming: The shaping of a hot thermoplastic sheet.

Thiokol: A polyethylene sulfide elastomer.

Thermoplastic: A material that can be molded and shaped when it's heated.

Thermoset: A hard and stiff crosslinked material. Thermosets are different from *thermoplastics*, which become moldable when heated. Thermosets are crosslinked, so they don't. Also, they are different from crosslinked *elastomers*. Thermosets are stiff and don't stretch the way elastomers do..

Tornesit: Trade name for chlorinated rubber.

TPE: Thermoplastic elastomer

Trommsdorff effect: The decrease in termination rate in viscous media that results in higher molecular weight polymers.

Toughness: A measure of the ability of a sample to absorb mechanical energy without breaking, usually defined as the area underneath a stress-strain curve.

Transfer molding: A process in which a preheated briquette or preform is forced through an orifice into a heated mold cavity.

Tyrin: Trade name for chlorinated polyethylene

U

Ultra-accelerator: A catalyst that cures rubber at low temperatures.

Ultrahigh molecular weight polyethylene (UHMWPE): High-density PE with over 100,000 ethylene units; the size allows the chains to intertangle, increasing its strength.

Ultra linear polyethylene: Polyethylene produced with soluble stereo regulating catalysts; has a narrow molecular weight distribution, high crystallinity, low branching.

Ultraviolet stabilizer: An additive that retards degradation caused by UV radiation.

Unsaturated polyester: A term used to describe alkyds with unsaturated chains, particularly those produced by the condensation of maleic anhydride with ethylene glycol.

Urea-formaldehyde resin: A resin produced by the condensation of urea and formaldehyde.

V

Vinoflex: Trade name for copolymers of vinyl chloride and vinyl isobutyl ether.

Vinyl: A polymer made by polymerizing monomers that contain the vinyl group, e.g., poly (vinyl pyrolidinone) (PVP), poly (vinyl alcohol) (PVA), polytetrafluoroethylene (PTFE, Teflon).

Vinylite: Trade name for copolymers of vinyl chloride and vinyl acetate.

Vinyon: Trade name for copolymers of vinyl chloride and acrylonitrile.

Viscose: A solution of cellulose xanthate produced by the reaction of alkali cellulose and carbon disulfide.

Viscose process: Regeneration of cellulose fibers by precipitation of the sodium salt of cellulose xanthate in acid.

Viscosity response: The viscosity response is a way of describing how the viscosity of a particular polymer responds to the changes in temperature and shear rate.

Viton: Polyfluorocarbon elastomer.

Vulcanization: The process in which natural rubber is crosslinked by heating with sulfur to strengthen it.

W

Warpage: Uneven bending, twisting etc. on account of differential cooling, differential shrinkage, or non-uniform freezing of melt in the mould can cause these conditions in the moulded part.

Wasted loops: The formation of cyclic compounds instead of polymers.

Weld Lines: Weld lines are locations in the moulded part where two met fronts meet.

Z

Zwitterionic polymerization: Copolymerization between nucleophilic and electrophilic comonomers.

References

1. Dryden C. E., *Outlines of Chemical Technology*, East-West Press.
2. R. NORRIS SHREVE, *The Chemical Process Industries*, McGraw-Hill Book Company, Inc.
3. Kirk–Othmer, *Encyclopedia of Chemical Technology*, John Wiley & Sons Inc.
4. Ullmann's *Encyclopedia of Industrial Chemistry*.
5. James, A. K., *Riegel's Handbook of Industrial Chemistry*, 9th edition.
6. Groggins P.H., *Unit Processes in Organic Synthesis*, 5th Edition, McGraw-Hill Book Company, New Delhi.
7. "Chemtech" Volume I-IV, *Chemical Engineering Education Development Centre*, I.I.T., Madras.
8. Shukla S. D and Pandey G. N., *A Textbook of Chemical Technology*. Vol. 1 & 2 (Inorganic/Organic). Vikas publishing House PVT Ltd. New Delhi.
9. Gowariker, V. R., Viswanathan, N. V. and Shreedhar, *J. Polymer Science*, New Age International, New Delhi, **2005**
10. Chanda M., *Introduction to Polymer Science and Chemistry*, CRC Press, Taylor and Francis Group, FL, USA, **2006**
11. Relevant Internet Resources